Mechanisation and Automation in Dairy Technology

Sheffield Food Technology

Series Editors: P.R. Ashurst and B.A. Law

A series which presents the current state of the art of chosen sectors of the food and beverage industry. Written at professional and reference level, it is directed at food scientists and technologists, ingredients suppliers, packaging technologists, quality assurance personnel, analytical chemists and microbiologists. Each volume in the series provides an accessible source of information on the science and technology of a particular area.

Titles in the Series:

Chemistry and Technology of Soft Drinks and Fruit Juices
Edited by P.R. Ashurst

Natural Toxicants in Food
Edited by D.H. Watson

Technology of Bottled Water
Edited by D.A.G. Senior and P.R. Ashurst

Environmental Contaminants in Food
Edited by C.F. Moffat and K.J. Whittle

Handbook of Beverage Packaging
Edited by G.A. Giles

Technology of Cheesemaking
Edited by B.A. Law

Mechanisation and Automation in Dairy Technology
Edited by Adnan Y. Tamine and B.A. Law

Mechanisation and Automation in Dairy Technology

Edited by

ADNAN Y. TAMIME
Food Systems Division
SAC Auchincruive
Ayr, UK

and

BARRY A. LAW
Research & Development Consultant
Reading, UK

CRC Press

First published 2001
Copyright © 2001 Sheffield Academic Press

Published by
Sheffield Academic Press Ltd
Mansion House, 19 Kingfield Road
Sheffield S11 9AS, England

ISBN 1-84127-110-1

Published in the U.S.A. and Canada (only) by
CRC Press LLC
2000 Corporate Blvd., N.W.
Boca Raton, FL 33431, U.S.A.
Orders from the U.S.A. and Canada (only) to CRC Press LLC

U.S.A. and Canada only:
ISBN 0-8493-0509-8

All rights reserved. No part of this publication may be reproduced, stored in a retrieval system or transmitted in any form or by any means, electronic, mechanical, photocopying or otherwise, without the prior permission of the copyright owner.

This book contains information obtained from authentic and highly regarded sources. Reprinted material is quoted with permission, and sources are indicated. Reasonable efforts have been made to publish reliable data and information, but the author and the publisher cannot assume responsibility for the validity of all materials or for the consequences of their use.

Trademark Notice: Product or corporate names may be trademarks or registered trademarks, and are used only for identification and explanation, without intent to infringe.

Printed on acid-free paper in Great Britain by
Bookcraft Ltd, Midsomer Norton, Bath

British Library Cataloguing-in-Publication Data:
A catalogue record for this book is available from the British Library

Library of Congress Cataloging-in-Publication Data:
Mechanisation and automation in dairy technology / edited by Adnan Y. Tamime, Barry A. Law.
     p. cm -- (Sheffield food technology)
  Includes bibliographical references (p.).
  ISBN 0-8493-0509-8 (alk. paper)
    1. Dairy processing. I. Tamime, A. Y. II. Law, Barry A. III. Series.
SF250.5.M43 2000
637--dc21

00-029730

Preface

Like any major and highly developed sector of food manufacturing, the dairy foods industry is now highly mechanised, from raw material production through to processing and product manufacture. This book has been written and compiled to show the extent and advantages of mechanisation across the range of dairy technology, and to explain the principles and rationale of introducing automated systems. In this way, we hope to demonstrate by example how combinations of state-of-the-art machinery and dedicated control systems bring to the dairy industry increased efficiency, higher quality and consistency, and opportunities for product innovation through its technological base.

Although automation and mechanisation are blended by dairy technologists to create unitary plant and process operations, we have chosen to open the volume with an introduction to the principles on which automated systems themselves are based. The scene is thus set for an appreciation of how these principles have been adopted and adapted to most aspects of milk and dairy technology. Subsequent chapters then take the reader progressively through milk production technology on the farm, the distribution of milk to the processing plant, and the diverse dairy food technologies used to make the enormous range of products we enjoy as consumers.

We begin the dairy-specific coverage with a description of milking systems and storage, distribution and tracing technology. This leads naturally to a separate chapter describing the options available for producing and packaging liquid milk products in automated processing lines. Next we consider the special requirements for plant to produce and handle concentrated and dried products, and also those dairy foods containing a high fat content. Both are areas that have benefited greatly from process automation based on new insights into the fundamental behaviour of milk components. Concentration and drying, in particular, can now be fine-tuned in automated plant to make products and ingredients to very exacting specifications of functionality.

Fermented dairy product technology is already highly advanced in terms of the availability of consistent and reliable processing aids and ingredients, such as coagulants and microbial cultures. The reliability of the 'biology' of milk fermentation has gone hand-in-hand with the development of fermentation plant and product modification machinery that can be programmed to run the unit operations of cheese and yoghurt manufacture. This is such a fascinating and diverse area of dairy product technology that we have devoted no less than five separate chapters to it, written by experts in the science and technology of fermented milks, hard, semi-hard and soft cheeses, and the very specialised pasta filata-type cheeses.

Membrane processes have become an integral part of modern dairy technology, and they lend themselves easily to automation. Add to this the extensive applied research base that is driving innovative membrane use in the processing of dairy materials, and the reader will appreciate why we have dedicated a chapter to the topic. The volume concludes with an overview of the application of mechanisation and automation to the 'non-product' technologies that support dairy technology through operations such as cleaning, de-fouling and waste management.

This book is for dairy technologists seeking greater understanding of existing plant under their control, together with ideas on further improvements to its operation. Through such understanding, we hope that new and more extensive mechanisation and automation options will emerge for the reader. Through our choice of contributors to the volume, we have been able to include not only the nuts and bolts of the systems under review, but also some of the scientific principles and research data which underpin them. We hope, therefore, that we have produced a volume that will also enlighten and inspire advanced students and experienced researchers in dairy technology.

Finally we wish to acknowledge the authors whose expertise, precious time and patience have come together to make this book possible. If it has any shortcomings, we—and not the authors—must bear the responsibility. We thank them wholeheartedly for their unstinting support and expert contributions.

Barry A. Law
Adnan Y. Tamime

Contributors

Mr Ir G. van den Berg	Irenelaan 57, 6713 MT Ede, The Netherlands,
Mrs Ir H.C. van der Horst	NIZO food research, PO Box 20, 6710 BA Ede, The Netherlands
Dr P. de Jong	NIZO food research, PO Box 20, 6710 BA Ede, The Netherlands
Dr A. Kelly	Department of Food Science, Food Technology and Nutrition, University College Cork, Cork, Ireland
Mr W. Kirkland	APV UK, PO Box 4, Gatwick Road, Crawley, West Sussex RH10 2JB, UK
Dr E. Latrille	INRA, Centre de Biotechnologies Agro-Industrielles, F-78850 Thriverval-Grignon, France
Professor B. Law	3 Sun Gardens, Burghfield Common, Reading RG7 3JB, UK
Professor D.D. Muir	Hannah Research Institute, Ayr KA6 5HL, UK
Mr H. Pointurier	14 rue le Sueur, 75116 Paris, France
Dr K.K. Rajah	74 Chadwick Road, Westcliff-on-Sea, Essex SS0 8LD, UK
Dr H.M.P. Ranjith	Milk Marque, Reaseheath, Nantwich CW5 6TA, UK
Mr L. Robertson	New Zealand Dairy Research Institute, Private Bag 11029, Palmerston North, New Zealand
Dr R.K. Robinson	University of Reading, Department of Food Science and Technology, Whiteknights, PO Box 226, Reading RG6 6AP, UK
Dr O. Salvadori del Prato	School of Dairy and Cheese Technology Specialization, via Besana 8, 26900 Lodi, Italy

	Address for correspondence: via Sporting Mirasole 20, 20090 Opera, Milan, Italy
Dr A.Y. Tamime	SAC Auchincruive, Food Systems Division, Ayr KA6 5HW, UK
	Address for correspondence: 24 Queens Terrace, Ayr KA7 1DX, UK
Ir. R.E.M. Verdurmen	NIZO food research, PO Box 20, 6710 BA Ede, The Netherlands

Contents

1 Principles of automation in the dairy industry   1
W. KIRKLAND

1.1	Introduction and historical development	1
1.2	Automation and control of dairy processes	7
	1.2.1 Process equipment	8
	1.2.2 Sensors and actuators	8
	1.2.3 Electrical cabling, fieldbus technology and smart devices	11
	1.2.4 Programmable logic controllers	11
	1.2.5 Soft programmable logic controllers and embedded controllers	15
	1.2.6 Supervisory control and data acquisition	17
	1.2.7 Network communications and systems integration	17
	1.2.8 Manufacturing execution systems	20
	1.2.9 Enterprise resource planning	25
1.3	Designing an automated process line	26
	1.3.1 User requirements specification	26
	1.3.2 Functional design specification	27
	1.3.3 Design implementation: project management	27
1.4	The future	28
	Further reading	29

2 Primary milk production   30
A. L. KELLY

2.1	Introduction	30
	2.1.1 Global milk production trends	30
	2.1.2 Farm production trends	31
2.2	Husbandry management and milk quality	32
	2.2.1 Introduction	32
	2.2.2 Lactation cycle and milk quality	32
	2.2.3 Effect of diet on milk composition	34
	2.2.4 Influence of genetic factors and breed on milk quality	35
	2.2.5 Mastitis, somatic cell counts and milk quality	36
2.3	Milking and feeding systems	38
	2.3.1 Milking machines and effects on milk quality	38
	2.3.2 Automated concentrate feeding systems	43
2.4	Bulk storage, collection and transportation	43
	2.4.1 Milk cooling and storage	43
	2.4.2 Milk collection and handling in developing countries	45
2.5	Quality payment schemes and quality optimization	45
	2.5.1 Mastitis control strategies	45
	2.5.2 Other animal welfare issues	47
	2.5.3 Milk payment and acceptance schemes	47

	Acknowledgements		48
	References		48

3 Liquid milk — 53
D. D. MUIR and A. Y. TAMIME

3.1	Milk composition	53
	3.1.1 Proteins in milk	53
	3.1.2 Lactose and minerals	54
	3.1.3 Milk fat	55
3.2	Heat-treated milk products	56
	3.2.1 Chemical effects	57
	3.2.2 Destruction of microorganisms and enzymes	57
	3.2.3 Effects on other milk constituents	58
3.3	From farm to factory	59
	3.3.1 Milk collection	59
	3.3.2 Milk distribution	59
	3.3.3 Delivery to the factory	59
	3.3.4 Extension of the shelf-life of raw milk	59
	3.3.5 At the factory	60
3.4	Milk handling in dairies	61
	3.4.1 Reception of milk	64
	3.4.2 Milk processing	65
	3.4.3 Pasteurisation systems	65
	3.4.4 Extended-shelf-life milk	73
	3.4.5 High-temperature pasteurisation	77
	3.4.6 In-container sterilisation	77
	3.4.7 Ultra high temperature (UHT)	80
3.5	Recombination technology	86
3.6	Packaging lines and storage	89
3.7	Statistical process control	91
	Acknowledgement	93
	References	93

4 Concentrated and dried dairy products — 95
P. DE JONG and R. E. M. VERDURMEN

4.1	Introduction	95
	4.1.1 Evaporation	95
	4.1.2 Drying	97
4.2	Product and process technology	99
	4.2.1 Evaporated and dried products	99
	4.2.2 Process design and operation	101
4.3	Quality control	111
	4.3.1 Control of process conditions	112
	4.3.2 Control of product properties	115
	References	117

5 High fat content dairy products
H. M. P. RANJITH and K. K. RAJAH
119

5.1	Introduction	119
	5.1.1 Properties of milk fat	120
	5.1.2 Melting and crystallisation	122
5.2	High fat content emulsions: oil-in-water type	122
	5.2.1 Centrifugal separation	123
	5.2.2 Control of fat content in creams	124
	5.2.3 Cleaning of milk separators	127
	5.2.4 Description of creams	127
	5.2.5 Processing of cream	128
	5.2.6 Factors affecting cream quality	132
5.3	Processing recommendations for high fat content products	135
	5.3.1 Properties required of high fat emulsions for table spreads	136
	5.3.2 Butter manufacture	140
	5.3.3 Anhydrous milk fat	143
	5.3.4 Ghee	147
	5.3.5 Butterschmalz	147
5.4	Fractionation of milk fat	148
	References	150

6 Yoghurt and other fermented milks
A. Y. TAMIME, R. K. ROBINSON and
E. LATRILLE
152

6.1	Background	152
6.2	Classification of fermented milks	153
	6.2.1 Mesophilic microfloras	153
	6.2.2 Thermophilic and/or therapeutic microfloras	153
	6.2.3 Microfloras including yeasts and lactic acid bacteria	154
	6.2.4 Microfloras including moulds and lactic acid bacteria	155
6.3	Manufacture of fermented milks	156
	6.3.1 Raw materials	156
	6.3.2 Fortification of the milk	158
	6.3.3 Heat treatment of the milk	159
	6.3.4 Microbiology of the processes	160
	6.3.5 Fermentation	161
	6.3.6 Final processing	162
	6.3.7 Retail products	162
6.4	Options for automation and mechanisation	163
	6.4.1 Introduction	163
	6.4.2 Reception of milk	165
	6.4.3 Processing plants	165
	6.4.4 Quick chilling, cold storage and retrieval of products	177
	6.4.5 Product recovery	181
	6.4.6 Automation in handling systems for finished product	185
6.5	Recent developments in some fermented-milk products	186
	6.5.1 Long-life yoghurt	186
	6.5.2 Strained or concentrated yoghurt	188

		6.5.3	Dried fermented milks	192
		6.5.4	Frozen yoghurt	193
		6.5.5	Drinking yoghurt	193
	6.6	Process control systems		193
		6.6.1	Introduction	193
		6.6.2	Controlled variables	194
		6.6.3	New reliable sensors for fermentation monitoring	194
		6.6.4	Advanced monitoring: prediction of the final process time	200
		6.6.5	Statistical process control and future trends	201
	Acknowledgement			201
	References			201

7 Cheddar cheese production — 204
B. A. LAW

7.1	Introduction		204
7.2	Cheesemaking as process engineering		205
7.3	Coagulation of milk and curd formation		206
	7.3.1	Vat design	208
	7.3.2	Cutting and stirring	209
	7.3.3	Theoretical aids to the optimisation of the cutting and scalding stage	212
7.4	Curd draining, cheddaring, milling and salting		215
7.5	Production of pressed cheese blocks ready for maturation		220
7.6	Storage and maturation of cheese		222
Acknowledgements			223
References			223

8 Semi-hard cheeses — 225
G. VAN DEN BERG

8.1	Introduction		225
	8.1.1	Cheese varieties involved	225
	8.1.2	General technology	225
	8.1.3	General historical background	226
8.2	Basic technology		227
8.3	Milk handling and processing		228
	8.3.1	Milk fat standardisation	228
	8.3.2	Control of sporeformers by bactofugation and microfiltration	228
	8.3.3	Pasteurisation	231
8.4	Cheese vats and curd production		231
	8.4.1	Horizontal vats	232
	8.4.2	Vertical vats	233
	8.4.3	Preparation of the curd	234
	8.4.4	Instrumentation to control and automate curd cutting time	235
8.5	Curd drainage and moulding		236
	8.5.1	Buffer tanks	236
	8.5.2	Casomatic® systems	237
	8.5.3	Pre-pressing vats	240
8.6	Pressing		241
	8.6.1	Cheese pressing	241
	8.6.2	Mould handling	242

	8.7	Brining	243
		8.7.1 Brine composition	243
		8.7.2 Hygiene measures	243
		8.7.3 Brining systems	244
		8.7.4 Dry salting	244
	8.8	Treatment during natural ripening	245
		8.8.1 Cheese handling systems	245
		8.8.2 Conditioning of the ripening room	247
		References	248
9	**Soft fresh cheese and soft ripened cheese**		**250**
	H. POINTURIER and B. A. LAW		
	9.1	Introduction	250
	9.2	Characteristics of ripened and fresh soft cheeses	250
		9.2.1 Soft ripened cheeses (*les fromages a pâte molle*)	250
		9.2.2 Fresh cheese (fromage frais)	252
	9.3	The key phases in the process plant for soft cheese manufacture	253
		9.3.1 Soft ripened cheeses	253
		9.3.2 Soft fresh cheeses	256
		9.3.3 Cottage cheese	258
	9.4	Mechanisation and automation solutions	259
		9.4.1 Soft ripened cheeses	259
		9.4.2 Soft fresh cheeses	263
		References	265
10	**Pasta Filata cheeses**		**266**
	O. SALVADORI DEL PRATO		
	10.1 General introduction and basic classification		266
	10.2 Technology of Pasta Filata cheeses		269
		10.2.1 Mozzarella and soft Pasta Filata cheeses	269
		10.2.2 Provolone and hard Pasta Filata cheeses	276
	10.3 Mechanisation and control of Pasta Filata cheese production		279
		10.3.1 Coagulators or cheese vats	279
		10.3.2 Filatrici and moulding machines	281
		10.3.3 Hardening and brining	283
		10.3.4 Packaging	285
		10.3.5 Miscellaneous systems	286
	10.4 Quality control of Pasta Filata cheese processing		288
		10.4.1 Rheological properties	288
		10.4.2 Microstructure	291
		10.4.3 Hazard analysis critical control points	292
		References	293
11	**Membrane processing**		**296**
	H.C. VAN DER HORST		
	11.1 Principles of membrane processes		296
	11.2 Process control and automation of membrane processes		300

11.3 Membrane applications for milk	303
11.3.1 Milk concentration by reverse osmosis	303
11.3.2 Demineralisation by nanofiltration	304
11.3.3 Milk protein standardisation by ultrafiltration	304
11.3.4 Milk protein concentration by ultrafiltration and microfiltration	305
11.3.5 Removal of bacteria, spores and somatic cells from raw milk by mirofiltration	305
11.4 Applications to cheese	308
11.4.1 Soft and hard cheese varieties	308
11.5 Applications for whey	310
11.5.1 Concentration of whey by reverse osmosis	310
11.5.2 Demineralisation of whey by nanofiltration	311
11.5.3 Whey protein concentrate production by ultrafiltration	313
11.5.4 Whey protein fractionation	314
11.6 Miscellaneous processes	316
11.6.1 Clarification of brine	316
11.6.2 Recycling of cleaning solutions	316
References	316

12 Nonproduct operations, services and waste handling — 318
L. ROBERTSON

12.1 Nonproduct operation and maintenance	318
12.1.1 Plant commissioning	318
12.1.2 Start-up and shut-down	319
12.1.3 Maintenance, including predictive or planned maintenance	320
12.1.4 Cleaning-in-place operation, control and automation	321
12.2 Supply and control of services	324
12.2.1 Water quality	324
12.2.2 Electricity	324
12.2.3 Steam	326
12.2.4 Hot water	327
12.2.5 Chilled water	327
12.2.6 Compressed air	327
12.2.7 Dryer air	328
12.2.8 Cogeneration	328
12.2.9 Waste heat recovery and re-use	331
12.3 Waste handling	331
12.3.1 Legal issues	331
12.3.2 Waste minimisation	332
12.3.3 Waste characterisation	333
12.3.4 Waste product and by-product treatment	334
12.3.5 Nutrient and biological oxygen demand reduction	335
Acknowledgement	338
Further reading	338
References	339

Index — **340**

1 Principles of automation in the dairy industry
W. Kirkland

1.1 Introduction and historical development

Automation is a key technology in the dairy industry. It would be impossible to meet the required levels of hygiene, food quality and plant utilisation, and to achieve the product costs and distribution targets demanded by retailers without a highly automated plant and business infrastructure. There is a drive towards full integration of the automation technologies that support business and processing needs, allowing the elimination of tasks which add little value and speeding up decision-making processes at all levels. The fundamental elements of automation technology, information and communication, and the power, performance and accessibility of personal computer (PC) technology are the basis of the rapid changes that are taking place throughout dairy process plant.

In a little over thirty years, since the early 1960s, the dairy industry has evolved from largely manual plant operations to fully integrated control and information systems at the heart of the modern plant. Manual plant operation involved dismantling the process plant after production to clean pipework and equipment by hosing and soaking in baths of cleaning chemicals. Cleaning-in-place (CIP) was the first major step towards fully automated systems. Early control systems were based on electromagnetic relays, hardwired logic and motor-driven devices for timing or sequential functions (Table 1.1). Figure 1.1 shows a typical control panel for such a dairy. The operator is holding the programming cards used to set up the timing sequences for the CIP programmes. Each card had 16 tracks, each track controlling a valve or pump on the CIP station. The card was fed through the reading device at a constant speed, enabling a timed sequence of operations to be programmed onto the card. This was achieved by sticking tape onto the tracks at the times at which the valves or pumps were required to operate.

With product and CIP chemicals flowing through adjacent pipework, significant developments in valve technology were required to avoid product contamination and injury to operators. Figure 1.2 shows a hygienic valve designed for dairy plants. The valve head contains a solenoid valve, which switches the air supply, and two feedback sensors, which enable the control system to determine whether the valve is fully shut, fully open or faulty. The use of this type of valve increased the complexity of the control system and its design.

Table 1.1 Historical aspects that influenced automation in the latter part of the twentieth century

Decade	Technology used	Business and market environment	Operator characteristics
1960s	Electromagnetic relay systems; hardwired logic	There were many small manufacturing sites serving a local market	Operators were very specialised: only one specific area of the process was handled
	Optical, drum or cam motor driven devices for sequence logic	Most dairy products were delivered to the doorstep	Operators had a strong process 'know-how' and years of craft training
	Pneumatic instrumentation for continuous control of process variables	Dairy products were sold at the corner shop or by small local traders	Strong in-house support was provided for fault or problem diagnosis
	Switches, push-buttons and lamps for the operator interface to the plant	Manufacturers dictated price and distribution	Most operations were manual and physically demanding
	Hardwired mimic displays for plant status and operation		Plant had to be dismantled to clean pipework; there was very little cleaning-in-place
	Very few international standards for system design		
1970s	Minicomputers, with ferrite core memory; hard drives were fast but of very low capacity	Large manufacturing sites were built for national distribution of products	Technology skills become more important than process skills
	Software developed on paper tape for early systems; later, magnetic tape, floppy disks and hard disks were used	Plants were geared up for long production runs	The process 'know-how' was stored in the computer program
	Printers with keyboards and monochrome text displays for operators	A small variety of products was produced	A single operator ran all parts of the plant for each shift
	Mimic displays driven from the plant for status information	Most areas of the plant were automated, including cleaning-in-place	Most users required manual back-up for the system
	Electronic instrumentation for continuous control and measurement		

1980s	Microprocessors and programmable logic controllers (PLCs)	Supermarkets developed and exerted buying pressure and introduced own brands	Operators provided front-line problem diagnosis
	Colour graphics for plant status; reduced number of instruments	Manufacturing plants needed to be more versatile, handling a wider range of products	Technology skills became more important than process skills
	Direct digital control for continuous processes	Lowest-cost production was essential; plants were highly automated	A process control system was used to measure production performance
	Distributed architectures for PLCs and input–output cabinets	Multi-skilling was introduced for maintenance staff	
	Relay ladder a standard for most PLC software development		
1990s	Windows™ technology	Greater international competition for dairy products	Plant supervisors and operators were required to drive plant to maximum throughput
	Ethernet network communications	Just-in-time delivery demanded accurate forecasting	Control systems measure plant utilisation
	Supervisory control and data acquisition	Price squeeze by dominant supermarkets put pressure on costs and losses	Complex production schedules had to be managed
	Production relational databases for recording operations	Consolidation and mergers and consequent factory closures required remaining plants to be reconfigured to handle many products	Product recovery systems and recipe information were managed by the operator
	Integraton of control and business systems	Reductions in in-house engineering capability occurred	Operators were required to fix simple faults

Figure 1.1 Dairy control panel with relays and card programming logic.

Figure 1.3 shows examples of relay circuits and relay logic modules used within the panel. This type of system was relatively difficult to change once installed as it required rewiring of panels and of the mimic diagrams used to inform the operator of plant and equipment status.

The success of these automated plants, the emergence of minicomputers and the desire to consolidate production into larger manufacturing units brought about rapid growth of computer-based automation in the dairy industry. Table 1.1 summarises the key drivers during the 1970s. The main benefits of these plants were product consistency, and safety and ease of operation. In the past, many operators had been skilled in dairy technology; the new computer-based systems had this process know-how stored in the software which controlled operations. The computer-based control system allowed plant operations to be changed by reprogramming the software; this was then pretested on simulators and installed with minimal disruption to plant operation. Figure 1.4 shows the control room of a computer-controlled dairy, which processed around 2 million litres of milk per day and catered for around 10 000 input and output signals. The process routing and CIP was fully automated and was managed by a single operator.

The 1980s saw the introduction of the microcomputer and the emergence of the industrial programmable logic controller (PLC). The PLC used the relay ladder logic design methodology as the basis of its user applications programming design. This approach stemmed from the need to be able to support installation, a role that was usually the function of the site maintenance department. As PLC became more pervasive and critical to business operations within dairies, there came a drive to merge the operations, production and maintenance roles through use of multiskilled technicians. This happened slowly and, coupled with the lack of well-defined standards for implementing automated control systems, frustrated progress; nevertheless, this era saw the introduction of distributed systems and networked architectures for control systems (see Table 1.1).

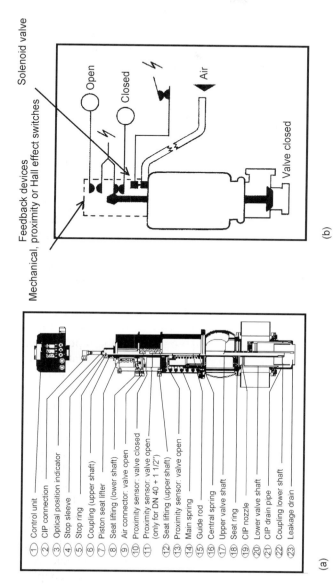

Figure 1.2 Hygienic valve design for dairy applications: (a) section through a double-seat safety break valve; (b) electrical schematic of the valve control unit. CIP, cleaning-in-place.

6   MECHANISATION AND AUTOMATION IN DAIRY TECHNOLOGY

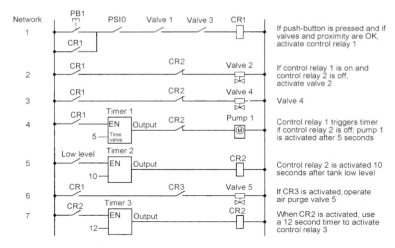

Figure 1.3 A relay ladder logic sequence diagram for controlling the emptying of a milk silo. CR, control relay; EN, enable; PB, push-button; PS, pressure switch.

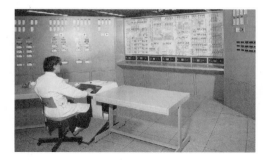

Figure 1.4 Control room for a large computer-controlled dairy.

Windows™ technology and use of the PC as a key component in the architecture of an automated plant accelerated the integration of dairy process automation and business and office systems throughout the 1990s. Tighter links between planning, scheduling and plant operation and the need to establish audit trails for CIP and batch production have led to the use of relational databases and the introduction of international standards for the design and execution of automation systems. Open systems standards for network communications and fieldbus technology have driven this standardisation, and this has enabled dairy manufacturers to combine systems based on different hardware and software into an integrated control and information management system for their plant. Table 1.1

describes the technology, business and operational impacts that have prevailed over the past 10 years in the dairy industry. PLC technology is beginning to blur with PC technology, and IEC (International Electro-technical Commission) 1131-3 programming languages are becoming the norm for control system design.

1.2 Automation and control of dairy processes

Figure 1.5 defines the levels of automation which combine to form a totally integrated control system for dairy plant. Depending on the system requirements and objectives of the end user, the architecture for a system may cut off at any of these levels. Figure 1.6 sets out the options for the degree of automation applied and notes its impact on the purchase price of the system and on operating costs, looking in detail at the levels identified in Figure 1.3.

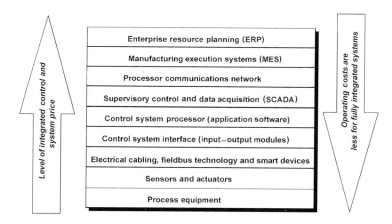

Figure 1.5 Automation technology.

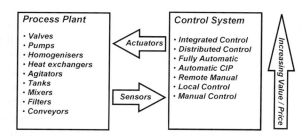

Figure 1.6 Different levels of automation used in the dairy industry. CIP, cleaning-in-place.

1.2.1 Process equipment

The most common items of equipment found in a modern dairy are identified in Figure 1.6. For fully automated operation, these items must be designed to meet hygiene standards, which in the majority of cases require CIP. A group of users and manufacturers, supported by various research organisations, have formed the European Hygienic Equipment Design Group (EHEDG) to establish standards and measurements for compliance in the design of process equipment. Equipment must include sensors and actuators that effectively bring these devices to life by connecting them to the automation system.

Fieldbus technology developments mentioned below have brought about the introduction of smart valves and pumps as well as intelligent controllers for homogenisers and milk fat standardisation. Such controllers are embedded in the process equipment and provide a high-level interface to the control system through communications technology. A typical milk fat standardisation system will include flow measurement and control and will use embedded algorithms to compensate for temperature and seasonal variations in milk supply. It will connect directly into any existing system by using fieldbus technology. A smart valve with an embedded controller that measures valve performance—for example, number of operations, travel time from open to close and close to open, power supply voltages and valve position—is illustrated in Figure 1.7. The valve also uses fieldbus technology to communicate with the automation system.

1.2.2 Sensors and actuators

A summary of the most commonly found sensors and actuators in the dairy industry is given in Table 1.2. These provide the fundamental interface for all plant with the automation system, and over the past 10 years there has been

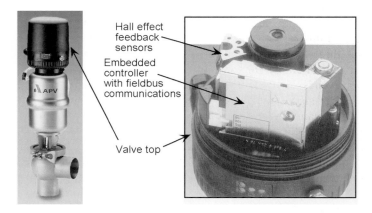

Figure 1.7 A smart hygienic valve with embedded controller.

Table 1.2 Sensors and actuator technology for dairy process equipment

Device	Basis of technology
Sensor:	
Temperature	Platinum resistance
Level	Conductivity, vibration, pressure, capacitance
Pressure	Resistance, capacitance or piezoresistance
Flow (rate or volume)	Electromagnetism
Weight	Load cells (strain gauges)
Mass/density	Vibrating tubes
Conductivity	Electrodes (current flow or induced current)
Turbidity	Infrared or visible light
Dissolved oxygen	Membrane diffusion
Actuator:	
Pneumatic on–off valves	Solenoids valves
Pneumatic modulating valves	Current-to-pressure converters
Motors	Relays or contactors
Speed	Frequency converters

a drive towards the use of low-voltage signals (24 V direct current) for digital or on–off devices, and 4–20 mA for analogue signals which represent process variables such as temperature, level or pressure. Most of these measurement devices are available with fieldbus-compliant interfaces.

The utilisation of feedback from sensors to determine the setting of a control element, such as a feed pump or a steam valve, is the basis of accurate continuous control of process variables such as flow, temperature, pressure and level in the dairy process. For example, a typical temperature control loop using a local controller connected to the temperature sensor and steam valve is illustrated in Figure 1.8. However, the control output to the actuator is determined by comparing the set point with a measured variable and by generating an error signal that is processed by a control method or algorithm. The most commonly used control algorithms for dairy processes are on–off control, time proportioning control and three-term or proportional–integral–derivative (PID) control.

Typical control curves for these different methods are shown in Figure 1.9. The responsiveness of the control is governed by the tuning parameters that act on the error signal, and these parameters, together with the set point, are either entered manually at the local controller or downloaded from a master station. PLCs and other computer-based control systems implement these control algorithms in software which simplifies the integration of these loops into start-up and shut-down operations and which is significantly cheaper than local stand-alone controllers. Self-tuning and adaptive control algorithms have been developed for simplifying system set-up and process-performance improvement, and advanced model-based controllers are used for optimal control of critical processes, such as evaporation and drying.

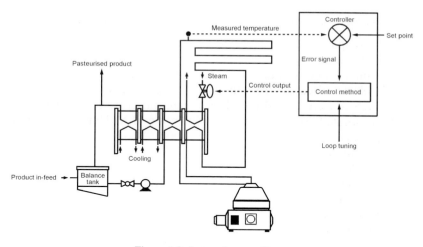

Figure 1.8 Automatic control loop.

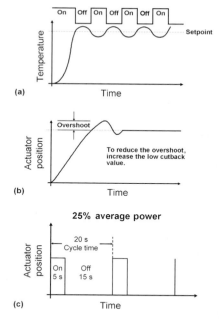

Figure 1.9 Responses for different control methods. (a) On–off control: the control output is turned on if the measured variable is below the setpoint and is turned off when it is above the setpoint; this is termed 'discrete' or 'digital' control. (b) proportional–integral–derivative (PID), or three-term control: the control output is a weighted value of the error signal (proportional), the integral of the error signal and its derivative; the weighting is known as loop tuning and some controllers have self-tuning algorithms. (c) Time-proportional control: the control output is turned on for a period proportional to the size of the error signal; for the remainder of the time period it is turned off; this is an improvement on the simple on–off control method.

1.2.3 Electrical cabling, fieldbus technology and smart devices

Inputs and outputs from sensors and actuators are translated into electrical signals for the control system. In the past 10 years, these electrical systems have changed from having a centralised to a distributed configuration. In the former approach, electrical cables from the plant devices are fed via local junction boxes, which convert the individual feeds into multicore cables, and then on to centrally located control panels containing the PLC. However, distributed systems position the control system in local panels and use a communication cable to provide the information to the PLC. In the past 5 years, the control industry has developed standards for the communications cables and protocols, which have come to be known as fieldbus systems.

Fieldbus systems fall into the category of 'open systems' technology and allow users to integrate devices from different vendors into the control system via a common communications cable. Systems based on fieldbus technology have the potential to save up to 50% of installation and design costs and have the ability to connect directly to smart devices. Figure 1.10 compares the architecture of a centralised system with one based on fieldbus technology. The fieldbus system includes smart process valves which have the ability to measure valve travel time (and include an alarm if this goes out of specification), count the number of valve operations (indicating when preventative maintenance is due) and monitor electrical supply against specification. Multipoint sensors, which measure more than one process variable at a single sensing point and smart sensors also use fieldbus technology. In the dairy process industry, Profibus™, DeviceNet™, ASI™ and CANbus™ are the most common fieldbus technologies used. These standards were originally developed by major manufacturing companies, such as Siemens, Rockwell and Bosch, and are now supported by independent user-funded organisations.

1.2.4 Programmable logic controllers

The PLC forms the basis of most control systems in the dairy process industries. An illustration of the physical construction of a typical PLC and a schematic diagram of the architecture of the electrical modules are shown in Figure 1.11. The support structure holds a backplane that enables different electrical modules to be connected together. A power supply module supplies the different voltages required for the system, and there is usually a circuit on the module which monitors the incoming voltage supply and which provides a power failure and restart signal to power down or initiate the central processing unit (CPU).

The CPU holds the stored programs and data required to control the process; this is usually called the application program. The CPU operation is organised by a program often called the operating system. It is not accessible to the user, is often stored in read only memory and is referred to as 'firmware'. The basic functions of the operating system are shown in Figure 1.12.

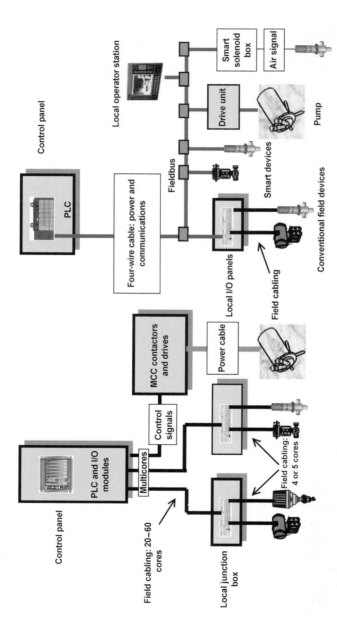

Figure 1.10 Comparison of centralised and fieldbus technology. I/O, input–output; MCC, motor control centre; PLC, programmable logic controller.

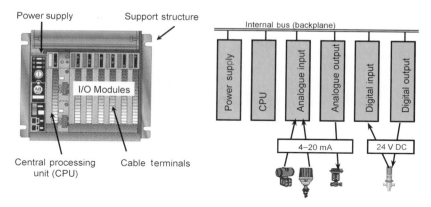

Figure 1.11 Architecture of a programmable logic controller. I/O, input–output.

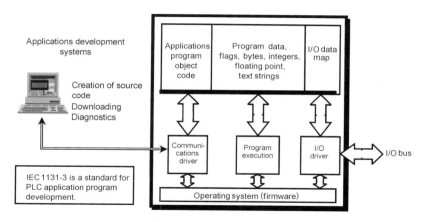

Figure 1.12 Central processing unit software organisation in a programmable logic controller (PLC). I/O, input–output.

The input–output (I/O) driver manages the communications with the backplane and reads and writes information to the I/O modules connected into the backplane, building up a data map of the plant equipment status that is used by the application program. The majority of I/O modules in a dairy system are either analogue or digital (Figure 1.9). Some modules count pulses, typically used for volumetric flow measurement, whereas others may collect data from smart devices that use message packets to pass data in a serial stream of binary information, using standards for electrical connection such as RS232 and RS485.

The program execution module runs the application programs. The development of the programs makes use of a programming language to describe the functional requirements of the control system. International standards groups have specified programming languages for process control and these are documented in the IEC 1131-3 standard. Figure 1.13 shows examples of the three languages that form the IEC 1131-3 standard and Figure 1.14 shows an example of a programming language developed by APV, UK for secure operation of dairy processes. Most PLCs require a program development system, usually a PC and a suite of software to create and compile these programs; the compiled program is then downloaded into the CPU via a communications driver, which is also

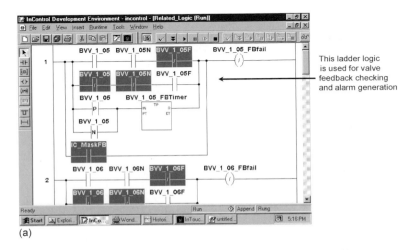

(a)

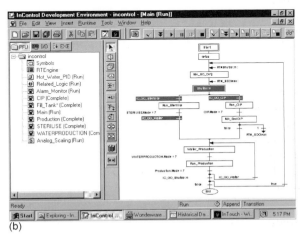

(b)

Figure 1.13 The IEC 1131-3 standard: (a) relay ladder logic; (b) sequence flow chart; (c) structured text program.

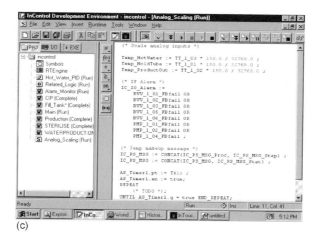

Figure 1.13 (*Continued*).

used to provide program and I/O status to the program development system for de-bugging or to an operator station for running the process.

PLCs have other modules that fit into the backplane for network and fieldbus communications. These modules will have their own on-board processor and are often referred to as smart or intelligent modules. PLCs that are built up from a CPU and smart modules are referred to as multiprocessor systems and provide very high performance in terms of speed of monitoring I/O and systems communication. Process control systems are classified as real-time systems because they must respond to events within acceptable and definable time periods. Typical time periods are less than 50 ms for total program execution of relatively large systems of several hundred I/O points. Some systems, typically machine control systems, require much faster action than this, and it is possible to structure PLCs to provide very fast scan rates and program execution for critical devices, sometimes less than 1 ms.

To meet the price and performance criteria demanded by the dairy industry, manufacturers produce a range of products with varying capabilities in terms of numbers of I/O, size of program, network capability, etc. There are no hardware standards, as such, for PLCs and so it is not possible to interchange modules from different suppliers; however, the use of fieldbus communications allows systems to have I/O configurations using multiple vendors for field devices. This is a major step forward in the drive for open systems architecture.

1.2.5 Soft programmable logic controllers and embedded controllers

The IEC 1131-3 standard, together with advances in the security and robustness of the PC, have led to the use of soft PLC. Figure 1.15 shows a typical

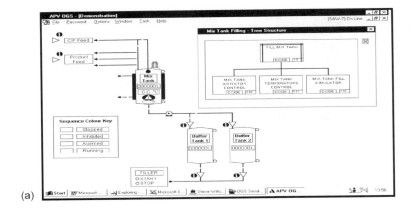

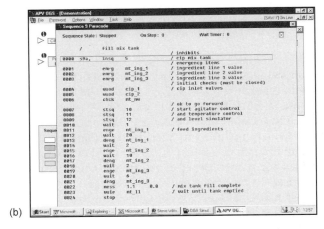

Figure 1.14 Sequence structured process control software language: (a) sequence tree structure; (b) paracode sequence.

configuration of a control system using this technology combined with fieldbus for the connection of I/O devices. Soft PLC allows the functions of program control, graphical interface and program development to be merged into a single hardware device. Also shown in Figure 1.15 is a derivative of the soft PLC called the embedded controller, which makes use of PC technology housed in an industrial enclosure for mounting on DIN terminal rails within a control panel. Soft PLCs provide an open systems environment and allow users to integrate specialist software (e.g. optimisation or model-based control) into the PLC and provide communications capabilities for ease of integration of the control into the dairy factory infrastructure.

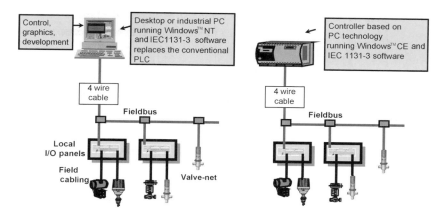

Figure 1.15 The IEC 1131-3 standard and the soft programmable logic controller (PLC): open system. I/O, input–output.

1.2.6 Supervisory control and data acquisition (SCADA)

Supervisory control and data acquisition (SCADA) is the name given to the systems that 'sit over the top' of a PLC and provide the operator and other users access to the system. SCADA systems work in real-time and provide graphical status displays. These displays are organised in a such a way that the operator can 'drill down' from a total plant overview to a specific plant device. Instrument faceplates that mimic conventional devices and include trend displays of critical process variables are embedded in the displays for easy access. Alarm and event logs are used to provide a complete history of operations and replace the conventional hard-copy printer; in addition, status displays that provide information or reports of processes or process equipment and plant operation are used to aid fault diagnosis or to support operator actions. These functions are shown graphically in Figure 1.16.

There is often confusion concerning the difference between SCADA and distributed control systems (DCSs). DCSs are more commonly found in the petrochemical, oil and gas industries, where there is a predominance of analogue control loops. DCSs combine the functions of the PLC and SCADA systems, although DCSs often use PLCs as subsystems for discrete or digital control functions.

1.2.7 Network communications and systems integration

The hardware architecture of a typical dairy plant SCADA system is shown in Figure 1.17. SCADA systems are supplied with I/O drivers that allow them to connect with different types of PLC and are consequently used to integrate systems within the dairy into a cohesive control unit. Networks, typically based

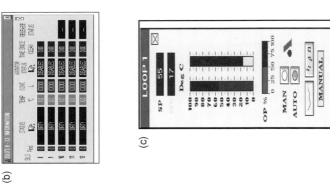
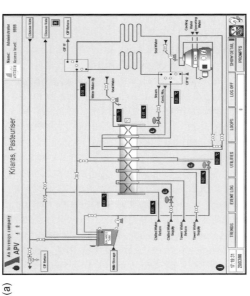

Figure 1.16 Supervisory control and data acquisition (SCADA) functions: (a) master screen; (b) production report; (c) instrument.

PRINCIPLES OF AUTOMATION IN THE DAIRY INDUSTRY   19

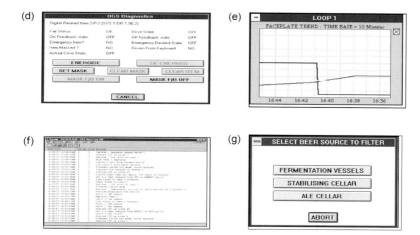

Figure 1.16 (*Continued*) (d) plant diagnostics; (e) trend display; (f) alarm log; (g) plant operation.

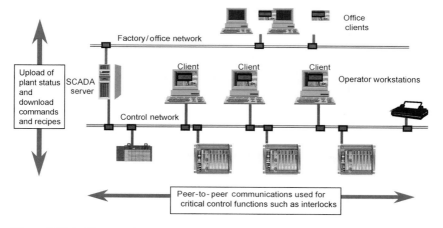

Figure 1.17 Architecture of a typical supervisory control and data acquisition (SCADA) system.

on Ethernet standards, are used to communicate between PLCs, and are called peer-to-peer communications. They may also be used to upload plant status, and to download commands and recipes between the SCADA system and the PLC.

The Ethernet is a nondeterministic system, meaning that it is not possible to state the speed of response to individual network transactions. For some time, this was seen by some experts as an issue; however, the speed of modern Ethernet networks is so fast (i.e. 100 Mbits s^{-1}) that even the biggest and most complex dairies are not at risk.

Where several operators or users are required, the SCADA system is organised into a client–server configuration, with the server often acting as a bridge between the control network and the office or business network. These networks are separated to avoid conflicts with downtime and maintenance requirements that often occur with office networks, as well as with the high loading that often arises in the PLC networks, which must run 24 hours a day, every day. This type of network architecture allows complete visibility of the control system to users throughout the factory, and with modern communications and Internet technology access can be unlimited. This puts great emphasis on the design of the security and access systems to avoid abuse.

1.2.8 Manufacturing execution systems

Figure 1.18 shows how the manufacturing execution level provides a bridge between the real-time control supplied by PLCs and SCADA and the business enterprise systems carrying out order processing, purchasing, stock control and production planning. The principal role of MESs is to integrate the planning functions with the real-time data from the control system to improve and speed up the decision-making of dairy operators and supervisors.

Figure 1.18 also shows the most common applications provided by the MES level, many of which fall into the category of historical reporting or process audits; other applications involve the integration of laboratory and quality control functions to provide tighter product and ingredient management. Driven by specifications developed by major retailers, a significant requirement for dairy companies has been the management of CIP and the ability to provide

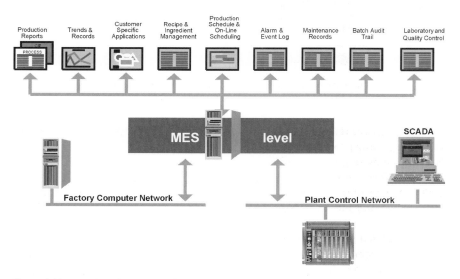

Figure 1.18 Manufacturing execution systems (MESs). CIP, cleaning-in-place; SCADA, supervisory control and data acquisition.

PRINCIPLES OF AUTOMATION IN THE DAIRY INDUSTRY 21

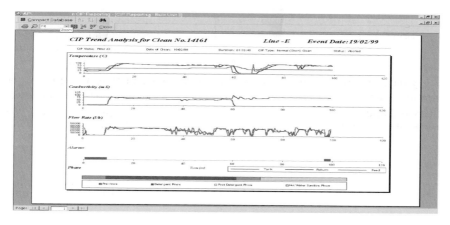

Figure 1.19 Cleaning-in-place (CIP) reporting.

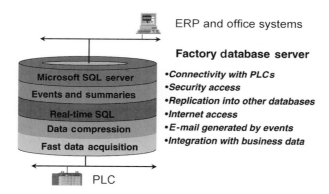

Figure 1.20 Real-time data collection for process reporting. ERP, enterprise resource planning; PLC, programmable logic controller; SQL, structured query language.

a comprehensive record of all operations and key parameters. A typical example of CIP reporting provided by APV to its dairy customers is shown in Figure 1.19. Such systems require a relational database and reporting tools to capture and display information. Figure 1.20 shows how the data flow into the database as well as some other features, such as data compression, replication and links with e-mail and Internet access.

1.2.8.1 Batch process control

There have been significant developments made by standards groups working in Europe and North America in the development of model forms for defining and specifying the construction and operation of recipe-based batch processes. Software packages for developing the models for both functional specification and process control are having a major impact on control system design within the dairy industry. The North American standard for batch process control systems was developed by the Instrument Society of America (ISA) and is known as the ISA S88 standard; the European standard is known as IEC 6512-1.

The model-building process involves first breaking the process plant down into process units and defining the allowable connections between these units; this defines the physical model of the plant. The processes that are carried out within the process units, such as, heating, cooling, agitation or mixing, are called phases, and the combination, order and control parameters for these phases varies for each recipe. The recipe for each unique product comprises a procedure for describing which process units are used to make the product, the procedure for the allocation of these units, the phases required and the unique formulae or parameters for the product (see Figure 1.21). To run the plant, the operator has to enter a schedule of the different production batches for the shift, day or week. Each batch will specify a product recipe (CIP programs can be considered to be product recipes) and a production batch. The batch control software will then run the plant, either selecting the equipment required or prompting the operator to make a selection and then initiating the process phases that are normally within the PLC. As each phase is processed, the batch control software compiles a complete history of the batch operation.

From a system design standpoint, the use of batch process control models can reduce design effort by 40–60%, and the use of graphical development tools enables the models to be constructed and maintained by the users rather than by software experts. The models can be validated by using simulation tools, without the need for the PLC or SCADA systems, enabling rapid prototyping of new products or changes in operational procedures, which significantly reduces the time to bring new products to the market. In the majority of cases, the new products or recipes are developed without any impact on the software within the PLC.

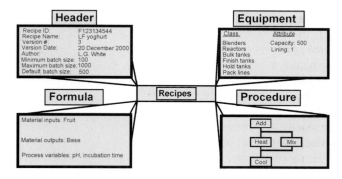

Figure 1.21 Batch control recipe.

PRINCIPLES OF AUTOMATION IN THE DAIRY INDUSTRY

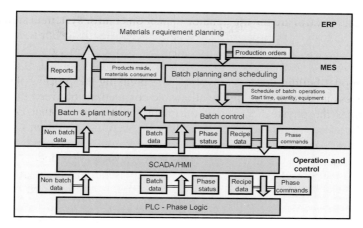

Figure 1.22 Information flow in an integrated system. ERP, enterprise resource planning; HMI, human machine interface; MES, manufacturing execution system; PLC, programmable logic controller; SCADA, supervisory control and data acquisition.

Batch process control software packages incorporate monitoring of all events and changes made to a batch, which automatically builds up batch records in a history database. The database is used to provide complete material genealogy, batch reporting, and an interface with ERP software. The model-building process creates comprehensive documentation of the system, which imposes a very tight and clear structure on the PLC software. Yoghurt, cheese, powders, ice-cream and most other dairy processes benefit from the use of batch process control principles.

This model-building approach has been extended to planning and scheduling operations, and ISA standards committees are working on a model called SP95 that addresses the integration of real-time controls with ERP systems. Figure 1.22 illustrates the flow of information between the different levels of ERP, MES and real-time systems to achieve integration of control and management information.

1.2.8.2 Batch planning and scheduling

In most dairy plants, the operator is expected to determine the precise utilisation of equipment and process plant. This is usually based on custom and practice and is driven by the finished product requirements and filling patterns for each packaging line. The results are unscheduled production overruns, lost production capacity as a result of poor planning of CIP, high losses as a result of small production batches, product not available at the right time, and failure to satisfy order requirements.

Model-based solutions are now in use for synchronising production requirements with plant operation. The models take the schedule of orders from the

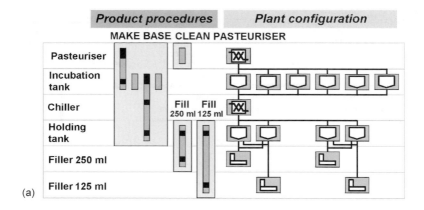

Figure 1.23 Model-based batch planning and scheduling: (a) plant model; (b) production orders. CIP, cleaning-in-place.

materials requirement planning (MRP) systems and transform them into a utilisation schedule for each major item of equipment. The schedule of batches respects the constraints of services, equipment and manpower within the model and produces an achievable schedule identifying where production requirements cannot be met. The planner or operator has the option to review the schedule and change priorities to achieve more acceptable results.

The model-building process involves building a physical model of the plant and its interconnections and defining the procedures for making each unique product; a scheduling algorithm then translates the schedule of orders into the detailed operating plan. A model of this type is invaluable when building a new process plant, where it is used to determine the best mix of process and CIP equipment and line capacities to meet production demands. The model can also be used to predict losses that will result from plant operating schedules.

Recent developments have resulted in these models replacing or supporting the operator in the running of the plant. This is called on-line scheduling and uses the detailed plan produced by the scheduling algorithm either to prompt the

PRINCIPLES OF AUTOMATION IN THE DAIRY INDUSTRY 25

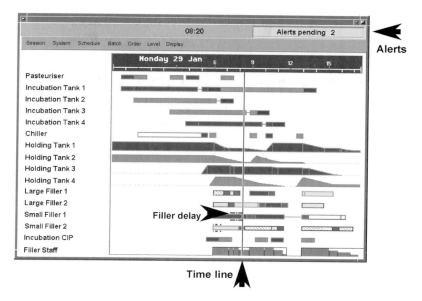

Figure 1.24 Production schedule with time line for on-line operation. CIP, cleaning-in-place.

operator or to send commands to the control system directly. The system also monitors the plant status; if operations do not match the plan, the system will inform the operator and reschedule plant utilisation by identifying future events that will now fall beyond their due date. Figure 1.23 shows the graphical model-building process used to generate the schedule, and Figure 1.24 illustrates how this is used for on-line operations.

1.2.9 Enterprise resource planning

ERP sits at the top of the integrated structure described in Figure 1.5. Its role is to manage the demands of business in terms of:

- financial accounting
- sales and distribution of products or services
- materials management
- production planning
- quality management
- plan maintenance
- human resources
- asset management

The systems that support these services are usually integrated into the dairy infrastructure, providing word-processing, analysis tools or marketing support, enabling users to switch seamlessly between the different information systems. In new plants, this integration may also include the control and decision support systems described above, and many of the software packages for SCADA, batch process control or batch planning and scheduling are supplied with modules that allow exchange of data with ERP systems. In these totally integrated environments, the dairy process floor is effectively linked to the consumer via the facilities that are available for product purchases through the Internet. As the power of these communications capabilities are realised there will be enormous change in the configuration, management and operation in the dairy of the future.

1.3 Designing an automated process line

The steps involved in designing and building an automated process line are outlined below. The complex requirements of modern systems described above demand strict adherence to these steps. In each case, there is not a single product or solution supplier that will satisfy all the needs of a fully automated dairy and therefore, as well as defining requirements, the design process must recognise the need to integrate different systems. This must be achieved with minimal risk to the implementation and at the lowest possible cost to ensure an effective payback on the investment.

1.3.1 User requirements specification

The user requirements specification (URS) defines the overall objectives of the automated process line and should cover the following:

- definition of terminology to avoid ambiguity as the full specification is developed
- the overall objectives of the automated production and process line
- key performance indicators (KPIs) for the automated line—this could be specified in terms of availability, downtime, losses, throughput or system response to commands
- degree of automation required as a minimum to satisfy the objectives
- the role and responsibility of the line operators and other users
- manual actions
- performance reports
- the degree of integration with other automated lines or the dairy business infrastructure
- degree of compliance with international standards for design
- company design standards and requirements
- preferred suppliers of equipment or software
- documentation and training requirements
- maintenance and support requirements for the automated process line

In some cases, the end user of the plant may not feel sufficiently confident to produce the URS and will seek the help of specialist suppliers or consultants. This is often achieved by simplifying the URS and generating a request for information (RFI), which is then used to generate a more comprehensive URS. A URS is usually developed by a team of users representing the different aspects of process, operation, maintenance, quality control, logistics and product management.

1.3.2 Functional design specification

The next step in the design process is to develop a functional design specification (FDS). This step is driven by the URS document, which interprets the process and operational requirements and translates them into a form that can be converted into a detailed design. For hardware, the URS document will be in the form of electrical and mechanical drawings, showing how equipment fits into control panels and how the electrical connections are made. For software, the situation is much more complicated and involves defining how data are stored in the PLCs, SCADA or MES, the organisation and structure of the software that uses these data, and how this software communicates with other software programs. The FDS is therefore critical to a successful design; however, other than batch process control models and some specialised software tools for system design there has been very little activity in this area to standardise the approach to generating an FDS. The steps involved are to:

- define the technology, hardware and software to meet the URS
- define the hardware and software architecture for the automation system
- define the design tools and methodology
- define the documentation for detailing hardware and software design specification
- define the test procedures for validating the design
- complete the design specification documentation

1.3.3 Design implementation: project management

The URS and FDS are critical activities within an overall project. Every project involves a complex interplay between the members of the various teams, whether they are implementers or users, customers or suppliers. The elements of a successful project include the following aspects:

- a quality plan which sets out the project organisation, roles and responsibilities of team members, communication procedures, document control procedures, change management and risk assessment
- a project plan that sets out the milestones for each activity
- a design review to monitor progress and analyse hazards

- factory acceptance tests (FATs) to validate the functionality of the automation on a simulated plant
- site acceptance tests (SATs) to validate the performance and functionality of the automation on the process line
- handover to operators and support staff, including training and as-built documentation

1.3.3.1 Simulation and factory acceptance tests
FATs are a critical part of the design process for automatic process lines. The majority of the functionality of an automated line is determined by software, whether this is configured in a SCADA, PLC, MES or smart device. Rigorous testing is therefore essential and in many cases within the dairy industry this cannot be achieved without the use of plant simulators, because the majority of the mechanical items are assembled on the factory site itself. Progress in the development of simulation tools has been particularly disappointing over the past 10 years, although this is beginning to change with the help of batch process control software and the use of SCADA systems with I/O simulation capability.

1.3.3.2 Design reviews, risk assessment and reduction of failure costs
To improve project implementation and to guarantee a low level of failure, management systems and tools such as 'hazard analysis critical control point' (HACCP) and 'good automation manufacturing practice' (GAMP) have been developed for use within the dairy and other process industries.

HACCP systems are directed at food safety and form part of the European Union Food Hygiene Directive. It is a systematic and documented quality management tool designed to produce safe products, with the emphasis on pre-emptive control as opposed to end-product testing. HACCP procedures should be developed in conjunction with the functional design specification and FATs of the automation system to ensure food safety criteria are met.

GAMP was developed for the pharmaceutical industry and provides guidelines for both the supplier and the operator of an automation system. It covers design procedures, documentation, testing and validation requirements to ensure safe operation and product quality. Strict adherence to GAMP adds significantly to project costs, and for dairy industry applications many users and suppliers operate GAMP principles but do not produce all the documentation to back up the validation and change control.

1.4 The future

Dairy plants are becoming more complex, handling many products, with a huge mix of production requirements. Dairies will increasingly be offering flexibility and speed of response to customer needs at the lowest operating costs. Inevitably, they will be highly automated and fully integrated with their suppliers' and customers' business systems, playing a major role in e-commerce and business-to-business developments via the Internet. Remote diagnostics and predictive maintenance prompted by smart devices will ensure that plants run efficiently, with minimum downtime, and advanced technology and design tools will ensure high reliability of the hardware and software of the automation systems. Plant operation will be more tightly planned and scheduled, and the role of the operator will become more passive as plants move to 'lights-out' production.

Further reading

Internet sites

APV Internet site: http://www.apv.com
Invensys Internet site: http://www.invensys.com

Internet pages

Activa. GAMP3: good automation manufacturing practice for validation of automation systems in pharmaceutical manufacture, http://www.activa.co.uk/gamp/gamp.hmt
European Hygienic Equipment Design Group (EHEDG). Guidelines and test methods for the Safe and Hygienic Processing of Foods Guidelines from the European Hygienic Equipment Design Group (EHEDG), http://sofa.dartnet.co. uk/www-campden/www/publ/pubfiles/ehedg.htm
Institute of Food Science and Technology (UK) (1998) Food and drink good manufacturing practice—a guide to its responsible management, http://www.ifst.org/guides.htm
ISA (1995) ISA 88 batch control, http://isa.org/standards/index/bynumber
ISA (2000) ISA 95 enterprise/control system integration, http://isa.org/ standards/index/bynumber

Other publications

Kirkland, B. (1995) *APV Automation Marketing Bulletin*, APV UK.
Lewis, R.W. (1998) Programming industrial control system, using IEC 1131-3. *Control Engineering Series*, **50**.
Institute of Brewing/ABTA (1995) HACCP: the problem or the solution?
EHEDG (European Hygienic Equipment Design Group) (1995) *Trends in Food Science and Technology*, Elsevier Science Ltd.

2 Primary milk production
A.L. Kelly

2.1 Introduction

2.1.1 Global milk production trends

In recent years, milk production has changed dramatically as a result of many developments in farming, markets and milk processing technologies. Global milk production has remained virtually static in developed countries while continuing to grow in the developing world. There have been reductions in the barriers to free trading and a consequent emergence of big retailing firms, trading globally and using information and communications technologies to transcend national boundaries. In parallel to this, there has been an increasing segmentation of consumer markets for milk-based foods. Consumer values have also changed, with increasing priorities on food safety and quality, the promotion of health and an insistence on assurances that food is produced in an environmentally friendly manner and with due care for animal welfare.

Worldwide milk production in 1998 was estimated at 557 million tonnes. Of this amount, 496.0 million tonnes was cows' milk and the rest was largely from buffalo (59.0 million tonnes), goats (11.9 million tonnes) and sheep (8.6 million tonnes). The major producers of cows' milk were the European Union (EU) (120.7 million tonnes) and North America (88.6 million tonnes) (Table 2.1). Between 1994 and 1998, this production increased in South America, Asia and Oceania, increased moderately in North America, was static in the EU because of the quota regime, and decreased in the Confederation of Independent States. World trade in dairy products is, in general, dominated by the EU and Oceania.

For at least the past 10 years, the continuation of global economic progress and the related political stability of nations has been directly related to the lowering of international trade barriers, particularly in agricultural commodities (FAO, 1996). The trend towards the dismantling of trade barriers became the basis for agreement between many of the largest of the world's trading blocks in preparation for more liberalised international trading opportunities. These trends were reflected in the General Agreement on Tariffs and Trade (GATT, 1995–2001) negotiations and progress, in Europe, towards the establishment of a single market.

The GATT agreement was of immense importance to world trade in dairy foods and was complemented by agreements to ensure the safety of the food

Table 2.1 Cows' milk production (million tonnes) by world regions, 1994–98 (IDF, 1998)

Region	1994	1997	1998[a]
World	465.7	471.2	496.0
Africa	4.8	4.7	4.8
North America	85.0	87.5	88.6
South America	28.4	33.5	35.2
Asia	44.7	49.7	51.0
European Union	120.0	121.0	120.7
Central and Eastern European Countries	33.0	33.9	35.0
Commonwealth of Independent States	66.4	52.6	50.0
Other Western Europe	5.9	5.8	5.8
Oceania	16.8	20.1	20.7

[a] Estimate.

supply and to prevent the use of national or regional technical requirements as unjustified barriers to trade (Boutrif, 1998). Subsequently, the objective of consumer protection was provided in the Codex Standards and Recommendations, prepared by the Food and Agriculture Organization (of the United Nations) and the World Health Organization. The 1992 reform of the EU Common Agricultural Policy (CAP) was part of the overall changes that were necessary alongside the GATT agreements. Essentially, the CAP reform shifted EU farm policy from price supports to income supports. The advent of increased international trading facilitated the emergence of multinational firms, trading globally, with a consequent consolidation of firms in the world dairy industry.

2.1.2 Farm production trends

Milk production technologies are now linked to the concept of a sustainable agriculture that is rich in technology and information, as distinct from the intensive use of energy and market-purchased inputs. There are, additionally, concerns for animal welfare and for the preservation of genetic diversity. There has been a lowering of priorities for milk production output and a higher priority for food safety, the environment, the landscape and nature. EU Directive 92/46/EEC in June 1992 laid down health regulations for the production of milk and dairy products, including standards for farm buildings and equipment, and set maximum acceptable levels for total bacterial counts (TBCs) and somatic cell counts (SCCs) of milk.

The requirement for assurances as to the mode by which food is produced greatly limited the potential progress that became possible when biotechnological techniques, such as the use of recombinant bovine somatotropin (bST) in milk production, emerged in the mid-1990s and has acted as a major influence on how food may be produced.

These changing markets and requirements have all been reflected in recent developments in milk production technology, particularly in the areas of control of milk composition, and quality, automation and control of milk production, (discussed in this chapter).

2.2 Husbandry management and milk quality

2.2.1 Introduction

Milk consists of a mixture of components that are synthesized in the udder of the cow and components derived from digestion of the food consumed by the cow that enter milk from the blood. Typical average composition of milk is: fat 4.0 g 100 g^{-1}; protein 3.3 g 100 g^{-1}; lactose 4.8 g 100 g^{-1}; salts or minerals 0.7 g 100 g^{-1}; and total solids 12 g 100 g^{-1}. Proteins in milk are divided into hydrophobic caseins (α_{s1}, α_{s2}, β and κ), which are found in calcium-rich micelles, and globular whey proteins (including β-lactoglobulin, α-lactalbumin, bovine serum albumin and immunoglobulins). There are also a number of enzymes in milk, including lipoprotein lipase, which liberates free fatty acids (FFAs) from milk fat, and the heat-stable alkaline serine proteinase, plasmin.

A number of factors influence the composition of milk, specifically, the stage of lactation (time from calving), diet, genetic factors, breed of cow and mastitis. The quality of milk is similarly influenced by a range of factors, such as gross composition of the milk, SCC and enzymatic hydrolysis of proteins and lipids. Throughout the rest of this section, the factors that influence milk composition and quality are discussed.

2.2.2 Lactation cycle and milk quality

The average bovine milk production cycle, or lactation, lasts for 9–10 months, over which time the composition and processing characteristics of milk can change considerably. Milk yield is typically highest in early lactation and subsequently decreases to a certain level at the end of lactation, when the cow is 'dried off'.

In certain countries, such as Ireland and New Zealand, the majority of calving traditionally occurs in spring, for economical maximum milk production from summer grass. This, however, leads to a grossly lactational pattern for national milk production, with an extreme peak-to-trough ratio between summer and winter milk yields and seasonal fluctuations in manufacturing milk quality. This causes problems for producers, as milk from different times of the year can have different manufacturing properties and it may not be possible to make some products at certain times of the year, which limits product diversification and reduces the ability of companies to react to market forces (Auldist et al., 1998). It is acknowledged that spreading calving throughout the year lessens seasonal variation in the gross composition of milk supplied to factories and leads to more consistent product yield and quality; thus any variation in manufacturing milk quality must be due to seasonal, as opposed to lactational, factors.

However, some compositional parameters, such as ratios of protein:fat and casein:whey protein, as well as fatty acid composition, are affected more by seasonal than by lactational variation (Auldist et al., 1998). Seasonal variation in the composition of milk may be due to changing availability and quality of pasture throughout the year, physiological changes associated with the stage of lactation and changing incidence of mastitis (Auldist et al., 1998). In most countries, milk production continues throughout the year but calving peaks before the most favourable season for milk production, with a resultant moderate effect of lactation stage on seasonal variation.

During any individual cow's lactation, the first milk produced, colostrum, is very different from normal milk, with low pH, reduced lactose content, high levels of immunoglobulins and certain minerals (copper and iron) and vitamins, and altered levels of fat and casein.

Milk produced towards the end of the milk production cycle, or late lactation milk, has increased total protein levels, but altered levels of individual proteins, with increased levels of whey proteins, particularly immunoglobulins (Auldist et al., 1996a). Casein micelles in late lactation milk are smaller, richer in minerals and more highly hydrated than those in mid-lactation milk (Donnelly et al., 1984). Changes in composition and processability of milk in late lactation are summarized in Table 2.2. Late lactation milk exhibits an elevated level of proteolysis, mainly as a result of increased plasmin activity, which reduces levels of intact casein, and may affect processability. A further factor which may influence the quality of very late lactation milk is decreasing frequency of milking, which may exaggerate the changes in milk quality encountered at this time (Auldist and Hubble, 1998; Kelly et al., 1998).

In late lactation, fat globules are small and have a weaker protective membrane, which may lead to high FFA content (Phelan et al., 1982). Late lactation milk also exhibits reduced lactose levels, altered salts levels, elevated pH and increased SCC, and is generally associated with poor dairy product quality (Kefford et al., 1995; Auldist et al., 1996a). For instance, late lactation milk exhibits poor coagulation properties and tends to produce a Cheddar cheese with a high moisture content (Kefford et al., 1995). Strategies for improving the suitability of late lactation milk for cheese manufacture include addition of calcium chloride, adjustment of pH, varying rennet levels and blending with mid-lactation milk (Lucey and Fox, 1992).

Recent studies of the quality of late lactation milk suggest that diet is a highly significant factor influencing the quality of milk at this time, with improved nutrition significantly improving the quality of such milk (Kefford et al., 1995; Lacy-Hulbert et al., 1995).

Table 2.2 Changes in milk composition and processing characteristics associated with late lactation and mastitic milk

Parameter	Effect of late lactation	Effect of mastitis
Milk composition:		
Milk yield	Decreases	Decreases
Total protein level	Increases	Unclear
Casein level	Unclear	Decreases
Casein number	Decreases	Decreases
Whey protein level	Increases	Increases
Fat	Unclear	Unclear
Lactose	Decreases	Decreases
Calcium	Increases	Unclear
Sodium	Increases	Increases
Potassium	Decreases	Decreases
Chloride	Increases	Increases
Dairy product quality:		
Ethanol stability	Decreases	Unknown
Heat stability	Decreases	Decreases
Rennet coagulation time	Increases	Increases
Rennet gel formation and strength	Decreases	Decreases
Cheese yield	Increases	Decreases
Cheese moisture content	Increases	Increases
Stability of UHT milk	Decreases	Decreases
Shelf-life of butter	Decreases	Decreases
Viscosity of yoghurt	Decreases	Decreases
Stability of cream liqueurs	Decreases	Unknown

Note: UHT, ultra heat-treated.

2.2.3 Effect of diet on milk composition

The diet of the lactating cow has a significant effect on the yield and composition of milk. Increased stocking densities reduce milk fat and protein levels and negatively affect many processing characteristics; concentrate supplementation, at all stages of lactation, increases total protein, casein and whey protein levels without significant effects on processing quality (Macheboeuf *et al.*, 1993; O'Brien *et al.*, 1996, 1997, 1999). The method of harvesting grass silage and the method and level of concentrate supplementation also influence milk yield (Petit *et al.*, 1993).

The composition of milk may be manipulated by feeding strategies and, in late lactation, this may offer an important tool for improvement of milk and dairy product quality. A balanced approach to manipulation of milk composition is, however, necessary, as it is not practical to increase concentrations of individual components to the detriment of total milk yield, unless for specific functional advantage. Overall, the relationship between feed constituents and milk composition is complex.

Milk fat levels, although more innately variable than other milk constituents, are far easier to control than milk protein levels, and milk fat was traditionally one of the components targeted for control because of the nature of milk payment schemes. The amount of roughage, the carbohydrate composition of concentrates, meal frequency and lipid supplementation all affect milk fat levels (Sutton, 1989). The fatty acid composition of milk fat may also be manipulated by feeding. For example, levels of nutritionally important compounds, such as conjugated linoleic acid (CLA) and oleic acid, may be increased by feeding the cows with soybeans and rapeseed, respectively. However, it has proved difficult to alter levels of long-chain omega-3 fatty acids in milk successfully through feeding.

Recent trends in food consumption, for example towards reduced fat intake, have driven changes in milk payment schemes, which now, in many areas, value protein above fat. For cows on a grass-silage-based diet, the quality of the silage influences milk composition, with increases in dry-matter digestibility clearly increasing milk protein concentration and milk yield. Low-protein diets reduce milk protein levels, while high-protein diets may increase milk nonprotein nitrogen levels. Supplementation of diet with maize silage, pressed pulp, fodder beet or brewers' grains all increase milk protein concentration. Grass feeding in early lactation and incorporation of molasses and fish meal in the diet have been linked with increasing total protein yield (Murphy and O'Brien, 1997). There is also interest in using protected amino acids (such as methionine and lysine) as dietary supplements to increase milk protein levels.

Milk from organic farms, where use of synthetic chemicals is prohibited, has higher vitamin levels and lower unsaturated fatty acids levels than conventionally produced milk and, for some breeds of cattle, may exhibit improved renneting properties (Lund, 1991), but it may also have higher nitrate levels (Guinot-Thomas *et al.*, 1991).

2.2.4 *Influence of genetic factors and breed on milk quality*

Certain important yield and compositional traits are influenced by genetic factors, and it is recognized that understanding this relationship may create strategies for controlling milk yield and composition. The degree to which specific traits are passed through calving is reflected in heritability scores. Yields of milk, fat, protein and total solids have low heritability scores, whereas scores for percentages of milk constituents are higher. Moderate to high heritability scores indicate that improvement of those traits would be expected through proper selection and breeding strategies. However, selection for one trait may lead to changes in other traits genetically correlated with the selected trait, either positively or negatively. It appears that milk yield is the best trait on which to base selection, as increased milk yield increases the yield of other milk solids, although their percentage levels may decrease (Harding, 1995).

Another factor that influences milk composition and processing characteristics is the genetic variant of individual proteins expressed in the cows milk (Macheboeuf et al., 1993). For example, β-lactoglobulin and κ-casein may both exist in three variants (AA, AB and BB), which differ in total protein content, casein concentration and cheesemaking characteristics, and thus milk from cows with specific variants could be selected for optimal cheese manufacturing properties. However, further research is needed to evaluate the benefit of breeding for cows of this genotype.

The breed of cow has a major influence on milk yield and a lesser influence on milk composition. Heavier breeds tend to produce more milk, with Holsteins, or Friesians, having highest mean lactation milk yields, followed by Brown Swiss, Ayrshire, Guernsey and Jersey (Harding, 1995). Milk from Jersey and Guernsey cows tends to have higher, but also more variable, fat levels than other breeds, and Jersey cows' milk tends to have high protein levels.

With regard to the possible impact of biotechnology on milk production, it has been suggested that transgenic techniques may allow manipulation of the composition of milk, for example through reducing levels of allergenic constituents, such as lactose or β-lactoglobulin, increasing levels of β-casein to improve coagulation properties or increasing nutritional or 'neutraceutical' properties of milk (Houdebine, 1998). Studies of the effects of growth hormones, such as bST, on milk composition have shown that, although milk yield may be increased by bST administration, the effects on composition are minor. Milk from bST-treated animals has been deemed safe for human consumption by medical and health authorities in several countries (Bauman et al., 1994).

2.2.5 Mastitis, somatic cell counts and milk quality

Mastitis is an inflammatory immune reaction of the mammary tissue in response to the introduction and multiplication of pathogenic microorganisms, which is characterized by an influx of white blood cells (somatic cells) into the milk, and is accompanied by an increase in endogenous protease activity and altered composition and quality of milk. Common mastitis pathogens include the contagious pathogens *Staphylococcus aureus* (the most common mastitis pathogen), *Streptococcus agalactiae* and *Streptococcus dysgalactiae* and the environmental pathogens *Escherichia coli* and *Streptococcus uberis*. Some strains of *E. coli* may cause toxaemia, and such infections result in death of the infected cow. *Mycoplasma bovis* can also cause a very intractable form of mastitis. Cases of mastitis may be classified based on causative microorganism, duration and persistence (chronic or acute), cow behavior and appearance, milk appearance (clinical or subclinical) or nature and origin (contagious or environmental). *S. aureus* is found in the udder and teat lesions and spreads readily from cow to cow, whereas *E. coli* mastitis is related to housing conditions, hygiene and machine milking; *S. dysgalactiae* mastitis is related to nutrition and milking technique, and *S. uberis* mastitis is linked with housing, nutrition and machine milking (Barkema *et al.*, 1999). In a small number of cases of mastitis, no causative agent can be cultured and this is referred to as nonspecific mastitis.

Milk from a cow with clinical mastitis has high SCCs, a watery or off-colour appearance, may contain flakes or clots and is easily recognized, facilitating exclusion from the bulk tank and treatment of the infected animal. The principal characteristic change in milk during mastitis is a significant increase in SCC, owing to an influx of polymorphonuclear leucocytes (PMNs), which account for ca. 10% of cells in normal milk, but more than 90% of cells in mastitic milk. The other principal somatic cell types are macrophages (the predominant cell in normal milk) and lymphocytes. There is currently interest in the potential use of differential, as opposed to total, milk SCC for diagnostic or veterinary purposes and perhaps as an indicator of milk quality.

Sub-clinical mastitis is the less obvious but most common type of mastitis and, with increasing cases of sub-clinical mastitis in a herd, the bulk-tank SCC increases dramatically (Auldist and Hubble, 1998). It is generally accepted that an SCC greater than 500 000 cells ml^{-1} milk is indicative of mastitis (Senyk *et al.*, 1985), and EU Directive 92/46/EEC set maximum acceptable geometric mean SCC of 400 000 cells ml^{-1} for liquid milk and 500 000 cells ml^{-1} for manufacturing milk. Although mastitis is the principal cause of elevated milk SCC, factors such as the cow's age, the stage of lactation, stress, season, udder injury and milking procedure all influence SCC (Brolund, 1985).

Mastitis is associated with significant decreases in milk yield and changes in milk composition, and elevated milk SCC is associated with poor quality of many dairy products (Munro *et al.*, 1984; Auldist and Hubble, 1998). During mastitis there is an increase in the permeability of the mammary epithelium, allowing an influx of serum components into milk and leakage of components such as lactose out of milk. Different mastitis pathogens may differ in nature of infection and patterns of changes in milk composition (Auldist and Hubble, 1998). Changes in composition and processing characteristics of milk during mastitis, resulting both from changes in the composition of the secreted milk and from post-secretory enzymatic activity, are summarized in Table 2.2.

Mastitic milk has elevated plasmin activity (Auldist and Hubble, 1998), and somatic cells themselves are associated with a range of proteolytic enzymes, including the acid proteinase cathepsin D (Verdi and Barbano, 1991). Elevated proteinase activity in the mastitic udder decreases milk casein level (Senyk *et al.*, 1985), which influences the processing properties of the milk. The bacterium, *S. uberis*, may manipulate the milk plasmin system to produce growth peptides (Leigh, 1993).

Mastitic milk has elevated whey protein levels, impaired rennet coagulation properties and reduced levels of β-casein (Politis and Ng-Kwai-Hang, 1988). High SCC in milk has been linked to early gelation of ultra high temperature

(UHT) treated milk (Auldist *et al.*, 1996b) and reduced yield and quality of Cheddar cheese (Auldist *et al.*, 1996a).

The changes in milk quality during mastitis are similar to those in late lactation milk and there is an additive effect of stage of lactation and SCC, with elevated SCC in late lactation being associated with further deterioration in milk quality. It has been suggested that many of the problems associated with late lactation milk could be overcome by containing mastitis at this time (Auldist *et al.*, 1996a, b). Strategies for control of mastitis will be discussed in Section 2.5.

2.3 Milking and feeding systems

2.3.1 Milking machines and effects on milk quality

2.3.1.1 Milk production and secretion

Immediately following milking, secretion of milk by alveolar epithelial cells proceeds at a constant rate until the alveoli expand and exert pressure on the milk producing cells, whereupon synthesis stops. Thus, the interval between milkings influences the yield and composition of milk (Sapru *et al.*, 1997). For example, the interval between twice-daily milkings may influence milk fat levels (O'Brien *et al.*, 1998b). Short time intervals between milkings may increase the susceptibility of milk to lipolysis.

When milking intervals are longer than 15 hours, rates of secretion change considerably, and milk composition changes, with decreases in concentrations of lactose, potassium salts and total non-fat solids, and increases in percentages of fat, whey proteins, sodium and chloride (Harding, 1995). Once-daily milking reduces milk yield owing to a loss in integrity of tight junctions, small extracellular structures that form a tight seal around secretory cells and control traffic of blood components into milk and vice versa (Stelwagen *et al.*, 1994). Application of bST during once-daily milking can increase milk yield significantly (Stelwagen *et al.*, 1994).

Many stimuli applied to the udder, such as the touch of the milker's hands, wash towels or teat cups, result in nervous impulses to the brain, causing release of the hormone oxytocin into the blood, stimulating milk ejection by increasing intramammary pressure and contraction of the alveoli. Proper pre-milking stimuli and timing of attachment of teat cups is important to take advantage of oxytocin release and milk let-down for optimal milking. A conditional reflex becomes established, and the normal milking routine should then be followed because if the schedule becomes irregular this reflex may be lost.

2.3.1.2 Milking machine design and operation

Milking systems may be broadly divided into bucket systems (not widely used any more), pipeline systems (milk flows directly from the cow into the bulk tank) and recorder systems (where an intermediate milk recording jar between cow and bulk tank allows measurement of yield and sampling of milk for testing). The most common milking parlor designs are herringbone or side-by-side systems (where cows enter in turn into a two-sided parlor) or rotary systems (where cows are variously entering the milking unit, being milked or leaving at any one time). Herringbone parlors, which are very common, are flexible and easy and cheap to install and have high throughputs; the more complex rotary parlors are common for large herds, have high throughputs and are very efficient.

The basic constituents of a milking machine are a vacuum pump, a vacuum vessel that also serves as a milk collection vessel, teat cups connected by hoses to the vacuum chamber, and a pulsator that alternately applies a vacuum (50 kPa, suction phase) and atmospheric pressure (massage phase, allowing a rest and avoiding pain), at a frequency of 40–60 cycles per minute, to the teat cups. The teat cup has an inner rubber liner, and the four teat cups, attached to a manifold called the milk claw, are held on the udder by suction. The process of milking may be preceded by washing or sanitization and drying of the udder to minimize bacterial contamination, although this may be avoided if hygiene and management are good.

The rate of milk ejection during milking depends on teat canal diameter and tightness of sphincter muscles (Harding, 1995). The pulsation rate and vacuum level within the milking machine also affect milking performance, with high pulsation rates at low vacuum levels increasing milking rate more than at high vacuum levels. Slow pulsation rates may provide insufficient teat-end stimulation and lead to pain. Fluctuations in vacuum should be avoided to allow maximum milk flow and prevent swelling of the teat end. The ratio of milking phase to rest phase (pulsation ratio) affects milk flow, as longer ratios increase milking rate but increase milking time. Increasing the vacuum pressure and pulsation ratios reduces milking time (O'Callaghan, 1998).

There is some dispute over optimum design and operation of milking machines and over the relative importance of various components and operating parameters within the system as a whole. It is acknowledged that there is a complex relationship between pulsation characteristics, vacuum conditions at the teat end and liner behavior (O'Callaghan, 1998).

Milking systems must be cleaned after the milking operation is complete. A typical cleaning protocol may involve fully draining the system of milk, pre-rinsing lines, vessels, the vacuum system and exterior surfaces with warm water (40–50°C), cleaning with an alkaline or chlorinated alkaline detergent solution and rinsing with clean water or an acid solution (preferable if in a hard-water area). Typical cleaning-in-place (CIP) cycle times are of the order of 8–10 min, and cleaning should be followed, just before next use of the equipment, by sanitizing (with chlorine-, iodine- or ammonia-containing compounds or very hot water) to destroy residual bacteria.

2.3.1.3 *Changes in milk composition during milking*

The composition and quality of milk varies during milking. The first milk (foremilk) may be high in bacteria and may be separately removed, inspected and discarded (called 'forestripping' or 'foremilking') before washing of the udder. However, this practice is becoming less common, particularly as a result of the rise in the use of in-line systems for detection of abnormal milk.

Fat globules, being of lower density than milk serum, tend to accumulate in the highest regions of the quarter, and so the percentage fat in milk increases significantly during milking. If milking is incomplete, the percentage fat for that milking will be reduced, and that for the next milking will rise accordingly. A certain amount of milk [$10–25$ ml 100 ml^{-1} or $10–25\%$ of total volume], called residual milk, always remains in the udder, even under ideal milking conditions, and may interfere with secretion during the intermilking period. Increased milk removal may be achieved by increasing downward and forward tension on the milking machine, allowing better drainage. Residual milk levels are increased by stress factors, such as excessive vacuum, stray electrical voltages and excessive physical handling of cows.

FFAs are released by enzymatic lipolysis, which is enhanced by physical rupture of the milk fat globule membrane and may result in unacceptable off-flavours in dairy products. To minimize FFA levels in milk, a milking system should have flow-controlled pumping, low air admission and float valves in the recorder jars (O'Brien *et al.*, 1998a).

2.3.1.4 *Automated milking systems*

A number of automated or robotic milking systems have been designed to reduce labor input and increase efficiency and throughput. These range from systems that merely automate the act of attachment of cups to teats (Frost, 1990), to unmanned, fully automatic milking systems (AMSs), which include a milking stall, a teat-locating system, a teat-cup attachment system and milking equipment (Ipema, 1997). Selection of cows to be milked may take place either in special selection areas or within the milking stall itself, although the latter option reduces the milking efficiency of the AMS, as cows will enter if no system exists to prevent unwarranted visits (Stefanowska *et al.*, 1999a, b).

Modern selection and segregation units based on cow weight have vacuum-powered entry and exit gates controlled by photoelectric beams to allow access for one cow at a time, and ensuring smooth operation and accurate weight registration. Segregation systems are an intrinsic part of automated systems and have several different functions, including selecting cows for attention (for health reasons), separating cows into groups for feeding or grazing, and ensuring that cows which have been milked do not return to the milking unit. Cow identification may be by means of ear tags, transponders or pedometers, with pedometers having the added advantage that step counting may be a very useful means of monitoring cow welfare and detecting heat.

AMSs must be able to locate a cow's teats reliably, and problems may arise as a result of morphological differences in udders and movement of the cow during the attachment operation (Frost, 1990). In an AMS, milking time is not fixed, which offers greatly improved efficiency and flexibility to the farmer. Also, the AMS may be controlled such that cows will access it only if a set interval has elapsed since the previous milking (Prescott *et al.*, 1998). AMSs are flexible enough to manage variation in milking frequency, feedstuff allocation and cow traffic, with data being collected and immediately used for on-line decisions (Devir *et al.*, 1996). It is accepted that cow behavioral characteristics influence the success of AMS units, with cows gradually, at variable rates, becoming accustomed to the use of such systems (Winter and Hillerton, 1995). Music may be used to stimulate cows to readiness for milking in an AMS (Uetake *et al.*, 1997).

AMSs are usually linked to automated concentrate feeding systems (see Section 2.3.2), and the total automation concept requires the use of integrated dairy control and management systems (DCMSs). DCMSs control the daily milking and feeding regime, ultimately facilitate the maximum production of milk for the minimum input of resources, such as feed and labor, and allow automated decision-making to control short-term milk production management (Devir *et al.*, 1993). Automated management regimes use three principal functions by which the farmer may control milk production: milking frequency, individual concentrate allocation and cow traffic control (Devir *et al.*, 1997). A schematic diagram of a DCMS is shown in Figure 2.1.

Milking in an AMS where cows present themselves for milking, with minimal human intervention, is a continuous process which does not therefore facilitate standard cleaning protocols, designed for batchwise cleaning after each milking session. AMSs may have automated cleaning protocols that activate after a set number of cows have been milked, or after a certain time of standing idle. The teat may be sprayed automatically with disinfectant after cluster removal. Between cows, back-flush cleaning systems, if in place, use compressed air to send a jet of water through the milking chambers, removing residues and avoiding the possibility of cross-contamination. Some cleaning systems may use boiling water instead of detergent.

2.3.1.5 Milking machines and mastitis

Milking machines may contribute to mastitis through transmission of pathogens, both from the environment to the teat and from cow to cow, and by damaging the teat and lowering the physical resistance to bacterial invasion. High bulk milk SCCs have been linked to poor use of post-milking teat dips and irregular maintenance of the milking machine. Automatic cluster removal is associated

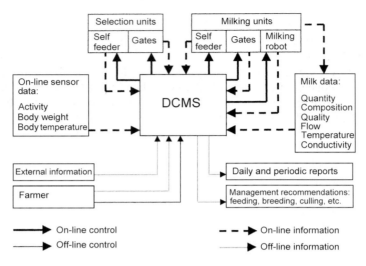

Figure 2.1 Schematic diagram of a dairy control and management system (DCMS) (Devir et al., 1993)

with low SCCs in bulk milk, perhaps because vacuum shut-off before removal reduces vacuum fluctuations and backflow of milk (Hutton et al., 1990; Fenlon et al., 1995). Pulsation failure and impact phenomena, due to abrupt losses of vacuum, have been linked to infection incidence (Spencer, 1989).

The morphology of the udder and teats may affect milking performance, in particular in terms of liner slippage and requirement for manual adjustment, with wide teats, long teats and tilted udders being associated with increased liner slips, which may cause mastitis (Rogers and Spencer, 1991). Models have been developed for diagnosis of mastitis problems due to malfunctioning milking machines or incorrect milking techniques (Hogeeven et al., 1995). Low SCCs in the bulk tank have been linked to milking the highest producing cows first and clinical cows last, using automatic cluster removal, keeping bedding moisture contents low and disinfecting teat ends prior to antibiotic treatment (Hutton et al., 1990).

Measurement of milk electrical conductivity during milking may detect changes in milk salt levels associated with developing mastitis, allowing early identification of cases requiring immediate treatment. Combined with antibiotic treatment, this may lead to rapid recovery (Hillerton and Semmens, 1999). Other factors, including milk yield, composition and temperature, could be monitored during milking and integrated into a milking control system, enhancing overall control of milk production (Figure 2.1; see also Devir et al., 1993). Milk yield may be measured in the line from the teat cluster by near-infrared technology, which has no moving parts to restrict milk flow and may automatically control cluster removal when cessation of milk flow is detected.

2.3.2 *Automated concentrate feeding systems*

A modern approach to individual concentrate allotment uses an integrated computerized identification and feeding system that feeds cows according to their energy requirements. Such systems, which provide the right type and quantity of concentrate to individual cows, when linked to automated milking, greatly improve efficiency of management and milk production (Livshin et al., 1995). Automated concentrate feeding is usually a part of an AMS, with the reward of feeding acting as a motivation to visit the AMS. The feeder may be located at the exit of the AMS so that cows must pass through the milking system to reach the feeder (Ipema, 1997). Automated feeding systems may allot an amount of feed proportional to the time since the previous feeding (variable-time feeding), or use fixed feeding intervals for all cows in the herd (fixed-time routines). Automated feeders are frequently gravity-fed, using spiral auger conveyors, and may handle and flexibly mix several different concentrates simultaneously. The software controlling automatic concentrate feeding may also adjust automatically for factors such as milk production and stage of lactation, and may dispense feed at the eating speed of the individual cow.

The behavioral responses of cows to such systems, and the pattern of visits to the feeding station, are complex and depend on biological factors, psychological factors and the layout of feeding units, although cows can adapt well to such systems (Livshin *et al.*, 1995).

2.4 Bulk storage, collection and transportation

2.4.1 *Milk cooling and storage*

Milk leaving the cow has a temperature of around 37°C and is almost free of bacteria but, to avoid growth of contaminating bacteria, it should be cooled as quickly as possible. EU standard 96/46/EEC recommends cooling milk to 8°C or less in cases of daily collection and to 6°C or less in cases where collection is not daily. Cooling of milk may be achieved in direct expansion systems (in which the milk comes into direct contact with an evaporator), ice bank systems (in which iced water is circulated through the bulk tank jacket), immersion coolers or plate heat exchangers. Rapid chilling in continuous systems results in lower bacterial growth and oxidation of fat than does cooling to similar temperatures in batch processes. Cooling to 2.5°C extends shelf-life more than does cooling to 4°C (GuulSimonsen *et al.*, 1996).

Physical rupture of milk fat globule membranes, leading to rancidity, must be avoided in milk handling. Milk pipelines may cause foaming and damage to globules if there is excessive air intake, too many or too high risers, or excessively long pipelines. Pipeline construction must also avoid 'dead spaces', which are impossible to clean effectively and where build-up of milk solids can lead to bacterial growth. Washing routines for pipelines should provide sufficient volume and turbulence to ensure adequate cleaning of the top of the pipeline.

The frequency of milk collection may depend on the on-farm bulk storage facilities, creamery requirements and economics of transportation. Milk produced under good hygienic conditions is generally stored for collection every 1–3 days. The traditional system of milk storage in churns has been largely replaced by storage in refrigerated bulk tanks. Farm bulk tanks should be made of polished stainless steel for long-term durability and ease of cleaning, with a double-wall construction and a suitable CIP system. Cleaning regimes usually involve sequentially washing the drained tank with alkaline, acidic and sanitizing solutions. Displacement cleaning systems, where milk residues are removed from the tank with a small amount of water at the start of the cleaning cycle, are preferred to dilution systems, where residues are mixed with clean recirculating rinsing water. Bulk tanks must be fitted with suitable agitation for mixing and heat transfer, good temperature monitoring and have ports for sampling of raw milk.

Bulk tank milk temperature may be controlled such that, before each milking, stored milk is further cooled to reduce cooling time for the next milking, although this must be carefully controlled to avoid freezing in the tank, which reduces the cooling efficiency of the tank and may lead to off-flavours.

Growth of psychrotrophic bacteria during refrigerated storage of raw milk should be minimized. Such bacteria, which include *Pseudomonas*, *Bacillus*, *Micrococcus* and *Acinetobacter* species, produce heat-stable lipases and proteases which survive pasteurization, although the bacteria themselves do not and may cause quality defects in milk and dairy products, such as high FFA content (Shah, 1994). Psychrotroph growth can be avoided by good farm hygiene, by adding low levels of lactic acid bacteria (for milk to be used for cheesemaking), by addition of preservatives, such as hydrogen peroxide, and by activation of the milk lactoperoxidase system, as described in Section 2.4.2 (Shah, 1994). An alternative approach involves carbonation of the chilled bulk milk (Espie and Madden, 1997). High TBCs in bulk tank milk have been linked to confinement housing, poor hygiene, exposure to environmental pathogens and use of post-milking dips (Goldberg et al., 1992).

Concentration of milk on the farm by ultrafiltration or reverse osmosis systems to reduce transportation costs and provide new options for dairy processors has been examined (Zall and Chen, 1984) but has not been widely adopted because of the high capital cost of equipment and strict hygiene requirements (Harding, 1995).

High FFA content in milk may result from excessive agitation of relatively small volumes of milk in bulk tanks and from plate coolers running under capacity. Small milk volumes in bulk tanks, as may be encountered in late lactation, may also result in apparent increased added water levels in milk as a result of less dilution of residual water in lines and tanks, which may lead to penalties.

2.4.2 Milk collection and handling in developing countries

An International Dairy Federation report in 1985 found that in many developing countries dairying was generally controlled by governmental rather than private organizations. The milk tended, as in more developed counties, to be collected over wide distances and processed at centralized locations, although the standard of transportation was frequently undesirable (IDF, 1985). In warm countries, a key difficulty is suboptimal maintenance and operation of chilling centers, which effectively determine the quality of the final milk for consumption and processing. Also, many milk testing methods tend not to be suited to use in small rural dairies (IDF, 1985).

A critical issue in warm climates is the holding time before processing, with morning milk, delivered rapidly to the processing center, being typically more acceptable than evening milk, stored at warm temperatures overnight. Activation of the native milk lactoperoxidase system (by addition of sodium thiocyanate and sodium carbonate peroxyhydrate), combined with evaporative cooling of milk containers covered with layers of fabric saturated in water, although not as efficient as refrigerated control, may have possible application for preservation of milk in rural areas (Ridley and Shalo, 1990). The effectiveness of preservative systems such as lactoperoxidase depends, however, on the initial hygiene quality of the milk, which is influenced by the practices and environment for milking. It is recognized that education and training for sanitary milk production are prerequisites for dairy production in developing countries.

2.5 Quality payment schemes and quality optimization

2.5.1 Mastitis control strategies

The primary goals of mastitis treatment are achievement of maximum clinical and bacteriological cure with minimum treatment and milk discard time and a rapid return to normal production of acceptable quality milk. However, mastitis is notoriously difficult to eliminate, and prevention is a critical part of mastitis control programmes. Hygienic farm practices, such as hygienic milking (including use of germicidal post-milking teat dips), are important in controlling contagious forms of mastitis, and provision of clean, dry, cool and comfortable housing are important in controlling environmental forms of mastitis.

Use of antibiotics, applied either through intramammary infusion (the most common method) or intramuscular injection, is the most common mastitis treatment for lactating cows with an SCC greater than 200 000 cells ml^{-1}. A recent study of 9000 cases of mastitis found that the best antibiotic cure rates were associated with infections due to streptococci (particularly *Streptococcus agalactiae*) and coagulase-negative staphylococci. Also, that amoxycillin was the antibiotic most associated with high cure rates (followed by erythromycin and cloxacillin) and that the overall cure rate for antibiotic-treated mastitis was 75%, compared with a spontaneous recovery rate of 65% in untreated cows (Wilson et al., 1999). Application of antibiotics at very early stages of infection, detected by changes in milk conductivity during milking, is particularly effective at eliminating pre-clinical S. uberis infections, allowing reduced use of antibiotics (Hillerton and Semmens, 1999).

Antibiotics may also be applied at the start of the dry period, which has the advantage that the antibiotic is not removed from the gland by milking, there are no negative effects on milk production and the concentration in the gland remains high for a long time. This is a particularly effective means of eliminating S. aureus, which is very hard to eliminate by antibiotic usage during lactation (Nickerson et al., 1995).

It is imperative to avoid antibiotics entering milk supplied by the farm, both for public health reasons and because of the negative impact of antibiotic residues on the manufacture of fermented dairy products. Segregation strategies may involve labeling or tagging of cows undergoing treatment, treating only in the dry period or withholding milk for a certain period following treatment. The cost of antibiotics and the loss of milk production due to withdrawal of milk from cows being treated often result in a poor cost–benefit ratio for antibiotic therapy.

Intramuscular injection of oxytocin, increasing efficiency of stripping milk from the udder, has been suggested as a treatment for mastitis (Hillerton and Semmens, 1999). There is also increasing interest in the use of specific microbicidal proteins that have no associated toxicity problems and are inactivated by digestive enzymes in the intestine, thus presenting less of a hazard than do antibiotics. Examples include the peptide, nisin, already a recognized food preservative, which has demonstrated potential as a nonirritant germicidal sanitizer for teats (Sears et al., 1992), and the recombinant bactericidal enzyme, lysostaphin, which may offer an effective alternative to antibiotics for the treatment of mastitis (Oldham and Daley, 1991). Also, a natural food-grade inhibitor derived from lactic acid bacteria, lacticin 3147, reduces the incidence of S. dysgalactiae mastitis during the dry period (Ryan et al., 1999).

Immunomodulatory substances, such as interleukins and interferon, which stimulate the natural defense systems of the cow, may provide a future basis of therapy (Quiroga et al., 1993). A further promising new area for combating mastitis is the use of vaccines, particularly for bacteria such as S. aureus that do not respond well to antibiotic treatment (Nickerson, 1994). Vaccines against clinical coliform mastitis are an extremely effective means of controlling environmental mastitis (Yancey, 1998). Feeding selenium and vitamin E, or treating with these nutrients during the dry period, may lower the SCC and increase resistance to mastitis (Hemingway, 1999). Finally, the use of products such as homeopathic remedies, herbs and ointments, which have no milk withdrawal time, but questionable effectiveness, remains controversial (Egan, 1998).

2.5.2 Other animal welfare issues

Cow welfare is an issue of increasing concern, primarily in terms of health but also in terms of provision of proper production conditions, including bedding space, housing, conditions of road and yard surfaces, environmental conditions, and control of movement and activity. Stress factors, which vary from cow to cow, possibly linked to oxytocin-mediated physiological mechanisms, may influence the bovine immune system and milk production and quality (Algers, 1998).

The major health problems associated with dairy herds are clinical mastitis, lameness, reproductive problems [such as postparturient metritis (inflammation of the uterus)], retained placentas, milk fever, bowel stasis, diarrhoea, displaced abomasa and respiratory problems. Other health issues of concern include bovine spongiform encephalopathy (BSE), grass tetany (staggers), leptospirosis, ketosis (imbalance in carbohydrate and fat metabolism), parasitic diseases, such as liver fluke and hoose, and lice. Infectious diseases of continuing concern include brucellosis (caused by the bacterium *Brucella abortis*), tuberculosis (caused by *Mycobacterium bovis*) and salmonellosis. Diseases of increasing concern include bovine viral diarrhoea (BVD), viral pneumonia and bovine leucosis virus. Culling of cows is most often a result of reproductive problems, mastitis, lameness or injury, disease, chronic mastitis, infertility or aggressiveness.

A relatively new concern for milk producers, processors and consumers alike is *Mycobacterium paratuberculosis*, which causes Johne's disease in cattle and has been tentatively linked to the similar human syndrome, Crohn's disease. It has been suggested that milk from infected animals can contain the live bacterium and act as a carrier for infection and that holding times on commercial pasteurization processes should be extended to ensure inactivation (Grant *et al.*, 1998). However, extreme practical difficulties in culturing the microorganism have rendered systematic studies difficult, and it is unclear whether or not inactive killed cells may also cause the disease.

2.5.3 Milk payment and acceptance schemes

The bases by which payment to farmers for milk is calculated vary between countries and individual dairy processors. Common criteria include weight of milk, weight of total solids, levels of individual constituents or hygiene parameters. Schemes which allot payment on a milk composition basis typically use protein or fat levels. Compositional payment has the advantage of fair remuneration for the standard of milk produced, while acting as an incentive for the production of milk with higher levels of beneficial solids (Lemétayer and Aireau, 1998). Milk fat is the oldest and most widespread compositional payment criterion, but crude protein level (or sometimes true protein content) is increasingly being used as a payment basis. Average international frequency of milk testing for payment purposes is reported to be 2 to 4 tests per month (de Wet, 1998).

Hygiene criteria which may be applied by processors include TBC, presence of spores, lipolysis, presence of antibiotics, pesticides and aflatoxin and presence of specific undesirable microorganisms, such as *Salmonella* or *Listeria* spp. (de Wet, 1998). In the case of many hygiene requirements, when specific tests are positive, milk is rejected outright or financial penalties are imposed. Payment on the basis of milk quality impresses on primary producers the importance of quality control, indirectly leading to minimization of risks to public health and improving the processing quality of the milk. Another very common acceptance–rejection criterion is SCC, for which general standards are set by directives such as EU directive 92/46/EEC but for which individual processors may have even more stringent requirements.

Payment for milk may vary in frequency, method, principle of determination, producer prices, purchase contracts and premiums, which may be paid for particular milk quality or volume. Special incentives may be offered in certain countries for production of milk at certain times of the year, to address national seasonal patterns in milk volume.

Acknowledgements

The author wishes to thank Dr John Walsh (Teasgasc; retired) and Mr David Waldren (Department of Food Science and Technology, University College Cork) for their help in the preparation of this chapter, and Dr Michael Keane (Department of Food Economics, University College Cork), Dr Bernadette O'Brian (Dairy Production Research Centre, Moorepark, Fermoy, County Cork) and Mr Diarmuid Doorge (MRCVS, Acting Chief Veterinary Officer, Cork Corporation) for their critical reading and helpful comments.

References

Algers, B. (1998) *What is animal stress and how does it affect milk production?* Proceedings, Future Milk Farming, 25th International Dairy Congress, Aarhus, Denmark, September 1998, pp. 27-31.

Auldist, M.J. and Hubble, I.B. (1998) Effects of mastitis on raw milk and dairy products. *Aust. J. Dairy Tech.*, **53** 23-36.

Auldist, M.J., Coats, S., Sutherland, B.J., Mayes, J.J., McDowell, G.H. and Rogers, G.L. (1996a) Effects of somatic cell count and stage of lactation on raw milk composition and the yield and quality of Cheddar cheese. *J. Dairy Res.*, **63** 269-80.

Auldist, M.J., Coats, S., Sutherland, B.J., Mayes, J.J., McDowell, G.H. and Rogers, G.L. (1996b) Effects of somatic cell count and stage of lactation on the quality and storage life of ultra high temperature milk. *J. Dairy Res.*, **63** 377-86.

Auldist, M.J., Walsh, B.J. and Thomson, N.A. (1998) Seasonal and lactational influences on bovine milk composition in New Zealand. *J. Dairy Res.*, **65** 401-11.

Barkema, H.W., Schukken, Y.H., Lam, T.J.G.M., Beiboer, M.L., Benedictus, G. and Brand, A. (1999) Management practices associated with the incidence rate of clinical mastitis. *J. Dairy Sci.*, **82** 1643-54.

Bauman, D.E., McBride, B.W., Burton, J.L. and Sejrsen, K. (1994) Somatotropin (bST): International Dairy Federation technical report. *Int. Dairy Fed. Bull.*, **293** 2-7.
Boutrif, E. (1998) *Food safety and risk issues in a changing international context*. Proceedings, Quality and Risk Management, 25th International Dairy Congress, Aarhus, Denmark, September 1998, pp. 199-211.
Brolund, L. (1985) Cell counts in bovine milk: causes of variation and applicability for diagnosis of subclinical mastitis. *Acta Vet. Scand.*, **80** 1-123.
Devir, S., Renkena, J.A., Huirne, R.B.M. and Ipema, A.H. (1993) A new dairy control and management system in the automatic milking farm: basic concepts and components. *J. Dairy Sci.*, **76** 3607-16.
Devir, S., Hogeveen, H., Hogewerf, P.H., Ipema, A.H., Ketelaar-De Lauwere, C.C., Rossing, W., Smits, A.C. and Stefanowska, J. (1996) Design and implementation of a system for automatic milking and feeding. *Can. Agric. Eng.*, **38** 107-13.
Devir, S., Maltz, E. and Metzm J.H.M. (1997) Strategic management planning and implementation at the milking robot dairy farm. *Computers and Electronics in Agriculture*, **17** 95-110.
Donnelly, W.J., Barry, J.G. and Bucheim, W. (1984) Casein micelle composition and synergetic properties of late lactation milk. *Ir. J. Food Sci. Tech.*, **8** 121-30.
Egan, J. (1998) A questionnaire survey on the uptake of homeopathic mastitis remedies in Irish dairy herds. *Irish. Vet. J.*, **51** 141-3.
Espie, W.E. and Madden, R.H. (1997) The carbonation of chilled bulk milk. *Milchwissenschaft*, **52** 249-53.
EU (European Union) (1992) Council Directive 92/46/EEC. http://europa.eu.int/eur-lex/en/lif/dat/1992/en_392L0046.html.
FAO (Food and Agricultural Association) (1996) *FAO Agriculture series, 28: The State of Food and Agriculture 1995. Agricultural Trade: Entering a New Era?* Food and Agriculture Organization, Rome.
Fenlon, D.R., Logue, D.N., Gunn, J. and Wilson, J. (1995) A study of mastitis bacteria and herd management practices to identify their relationship to high somatic cell counts in bulk tank milk. *Br. Vet. J.*, **151** 17-25.
Frost, A.R. (1990) Robotic milking: a review. *Robotica*, **8** 311-18.
Goldberg, J.J., Wildman, E.E., Pankey, J.W., Kunkel, J.R., Howard, D.B and Murphy, B.M. (1992) The influence of intensively managed rotational grazing, traditional continuous grazing and confinement housing on bulk tank milk quality and udder health. *J. Dairy Sci.*, **75** 96-104.
Grant, I.R., Ball, H.J. and Rowe, M.T. (1998) Effect of high temperature, short time (HTST) pasteurization on milk containing low numbers of *Mycobacterium paratuberculosis*. *Lett. Appl. Microbiol.*, **26** 166-70.
Guinot-Thomas, P., Jondreville, C. and Laurent, F. (1991) Comparison of milk from farms with biological, conventional and transitional feeding. *Milchwissenschaft*, **46** 779-82.
GuulSimonsen, F., Christiansen, P.S., Edelsten, D., Kristiansen, J.R., Madsen, N.P., Nielsen, E.W. and Petersen, L. (1996) Cooling, storage and quality of raw milk. *Acta. Agric. Scandinavica A*, **46** 105-10.
Harding, F. (1995) *Milk quality*, Blackie Academic and Professional, London.
Hemingway, R.G. (1999) The influences of dietary selenium and vitamin E intakes on milk somatic cell counts and mastitis in cows. *Vet. Res. Comm.*, **23** 481-99.
Hillerton, J.E. and Semmens, J.E. (1999) Comparison of treatment of mastitis by oxytocin or antibiotics following detection according to changes in milk electrical conductivity prior to visible signs. *J. Dairy Sci.*, **82** 93-8.
Hogeeven, H., van Vliet, J.H., Noordhuizen-Stassen, E.N., de Koning, C, Tepp, D.M. and Brand, A. (1995) A knowledge-based system for diagnosis of mastitis problems at the herd level: 2. Machine milking. *J. Dairy Sci.*, **78** 1441-55.
Houdebine, L.M. (1998) *The impact of genetic engineering on milk production*. Proceedings, Future Milk Farming, 25th International Dairy Congress, Aarhus, Denmark, September 1998, pp. 130-7.

Hutton, C.T., Fox, L.K. and Hancock, D.D. (1990) Mastitis control practices: differences between herds with high and low cell counts. *J. Dairy Sci.*, **73** 1135-43.
IDF (1985) Milk collection in developing countries. *Int. Dairy Fed. Bull.*, **205** 1-15.
IDF (1998) The world dairy situation. *Int. Dairy Fed. Bull.*, **333** 1-52.
Ipema, A.H. (1997) Integration of robotic milking in dairy housing systems: review of cow traffic and milking capacity aspects. *Computers and Electronics in Agriculture*, **17** 79-94.
Kefford, B., Christian, M.P., Sutherland, B.J., Mayes, J.J. and Grainger, C. (1995) Seasonal influences on Cheddar cheese manufacture: influence of diet quality and stage of lactation. *J. Dairy Res.*, **62** 529-37.
Kelly, A.L., Reid, S., Joyce, P., Meaney, W.J. and Foley, J. (1998) Effect of decreased milking frequency of cows in late lactation on milk somatic cell count, polymorphonuclear leucocyte numbers, composition and proteolytic activity. *J. Dairy Res.*, **61** 167-77.
Lacy-Hulbert, S.J., Woolford, M.W. and Bryant, A.M. (1995) Influence of once-daily milking and restricted feeding on milk characteristics in late lactation. *Proc. NZ Soc. Anim. Prod.*, **56** 65-7.
Leigh, J.A. (1993) Activation of bovine plasminogen by *Streptococcus uberis*. *FEMS Microbiol. Lett.*, **114** 67-72.
Lemétayer, J.-M. and Aireau, B. (1998) *Requirements for the calculation of a compositional milk price*. Proceedings, Future Milk Farming, 25th International Dairy Congress, Aarhus, Denmark, September 1998, pp. 59-65.
Livshin, N., Maltz, E. and Edan, Y. (1995) Regularity of dairy cow feeding behavior with computer-controlled feeders. *J. Dairy Sci.*, **78** 296-304.
Lucey, J.A. and Fox, P.F. (1992) Rennet coagulation properties of late-lactation milk: effect of pH adjustment, addition of $CaCl_2$, variation in rennet level and blending with mid-lactation milk. *Irish J. Agric. Food Res.*, **31** 173-84.
Lund, P. (1991) Characterisation of alternatively produced milk. *Milchwissenschaft*, **46** 166-9.
Macheboeuf, D., Coulon, J.-B. and D'Hour, P. (1993) Effect of breed, protein genetic variants and feeding on cows' milk coagulation properties. *J. Dairy Res.*, **60** 43-54.
Munro, G.L., Grieve, P.A. and Kitchen, B.J. (1984) Effects of mastitis on milk yield, milk composition, processing properties and yield and quality of milk products. *Aust. J. Dairy Tech.*, **39** 7-16.
Murphy, J.J. and O'Brien, B. (1997) *Quality milk for processing—the production technology*. Proceedings, Teagasc National Dairy Conference, Fermoy, Ireland, March 20th, 1997, pp. 56-87.
Nickerson, S.C. (1994) Bovine mammary gland: structure and function; relationship to milk production and immunity to mastitis. *Agri-Practice*, **15** 8-18.
Nickerson, S.C., Owens, W.E. and Boddie, R.L. (1995) Mastitis in dairy heifers: initial studies on prevalence and control. *J. Dairy Sci.*, **78** 1607-18.
O'Brien, B., Crosse, S. and Dillon, P. (1996) Effects of offering a concentrate or silage supplement to grazing dairy cows in late lactation on animal performance and milk processability. *Ir. J. Agric. Food Res.*, **35** 113-25.
O'Brien, B., Murphy, J.J., Connolly, J.F., Mehra, R.K., Guinee, T.P. and Stakelum, G. (1997) Effect of altering the daily herbage allowance in mid lactation on the composition and processing characteristics of bovine milk. *J. Dairy Res.*, **64** 621-6.
O'Brien, B., O'Callaghan, E. and Dillon, P. (1998a) Effect of various milking machine systems and components on free fatty acid levels in milk. *J. Dairy Res.*, **65** 335-9.
O'Brien, B., O'Connell, J. and Meaney, W.J. (1998b) Short-term effects of milking interval on milk production, composition and quality. *Milchwissenschaft*, **53** 123-6.
O'Brien, B., Dillon, P., Murphy, J.J., Mehra, R.K., Guinee, T.P., Connolly, J.F., Kelly, A. and Joyce, P. (1999) Effects of stocking density and concentrate supplementation of grazing dairy cows on milk production, composition and processing characteristics. *J. Dairy Res.*, **66** 165-76.

O'Callaghan, E.J. (1998) Effects of pulsation characteristics on machine yield, milking time and cluster stability. *Ir. J. Agric. Food Res.*, **37** 201-7.

Oldham, E.R. and Daley, M.J. (1991) Lysostaphin: use of a recombinant bactericidal enzyme as a mastitis therapeutic. *J. Dairy Sci.*, **74** 4175-82.

Petit, H.V. Tremblay, G.F., Savone, P., Tremblay, D. and Wauthy, J.M. (1993) Milk yield, intake, and blood traits of lactating cows fed grass silage conserved under different harvesting methods. *J. Dairy Sci.*, **76** 1365-74.

Phelan, J.A., O'Keefe, A.M., Keogh, M.K. and Kelly, P.M. (1982) Studies of milk composition and its relationship to some processing criteria: 1. Seasonal changes in the compositon of Irish milk. *Ir. J. Food Sci. Technol.*, **6** 1-11.

Politis, I. and Ng-Kwai-Hang, K.F. (1988) Association between somatic cell count of milk and cheese-yielding capacity. *J. Dairy Sci.*, **71** 1720-7.

Prescott, N.B., Mottram, T.T. and Webster, A.J.F. (1998) Effect of food type and location on the attendance to an automatic milking system by dairy cows and the effect of feeding during milking on their behaviour and milking characteristics. *Animal Sci.*, **67** 183-93.

Quiroga, G.H., Nickerson, S.C. and Adkinson, R.W. (1993) Histologic response of the heifer mammary gland to intramammary infusion of interleukin-2 or interferon-γ. *J. Dairy Sci.*, **76** 2913-24.

Ridley, S.C. and Shalo, P.L. (1990) Farm application of lactoperoxidase treatment and evaporative cooling for the intermediate preservation of unprocessed milk in Kenya. *J. Food Prot.*, **53** 592-7.

Rogers, G.W. and Spencer, S.B. (1991) Relationships among udder and teat morphology and milking characteristics. *J. Dairy Sci.*, **74** 4189-94.

Ryan, M.P., Flynn, J., Hill, C., Ross, R.P. and Meaney, W.J. (1999) The natural food grade inhibitor, lacticin 3147, reduced the incidence of mastitis after experimental challenge with *Streptococcus dysgalactiae* in nonlactating dairy cows. *J. Dairy Sci.*, **82** 2108-14.

Sapru, A., Berbano, D.M., Yun, J.J., Klei, L.R., Oltenacu, P.A. and Bendler, D.K. (1997) Cheddar cheese: influence of milking frequency and stage of lactation on composition and yield. *J. Dairy Sci.*, **80** 437-46.

Sears, P.M., Smith, B.S., Stewart, W.K., Gonzalez, R.N., Rubino, S.D., Gusik, S.A., Kulisek, E.S., Projan, S.J. and Blackburn, P. (1992) Evaluation of a nisin-based germicidal formulation on teat skin of live cows. *J. Dairy Sci.*, **75** 3185-90.

Senyk, G., Barbano, D.M. and Shipe, W.F. (1985) Proteolysis in milk associated with increasing somatic cell counts. *J. Dairy Sci.*, **68** 2189-94.

Shah, N.P. (1994) Psychrotrophs in milk: a review. *Milchwissenschaft*, **49** 432-27.

Spencer, S.B. (1989) Recent research and developments in machine milking—a review. *J. Dairy Sci.*, **72** 1907-17.

Stefanowska, J., Ipema, A.H. and Hendricks, M.M.W.B. (1999a) The behaviour of dairy cows in an automatic milking system where selection for milking takes place in the milking stalls. *Appl. Animal Behav. Sci.*, **62** 99-114.

Stefanowska, J., Tiliopoulos, N.S., Ipema, A.H. and Hendricks, M.M.W.B. (1999b) Dairy cow interactions with an automatic milking system starting with 'walk-through' selection. *Appl. Animal Behav. Sci.*, **63** 177-93.

Stelwagen, K., Davis, S.R., Farr, V.C. and Eichler, S.J. (1994) Effect of once-daily milking and concurrent somatotropin on mammary tight junction permeability and yield of cows. *J. Dairy Sci.*, **77** 2994-3001.

Sutton, J.D. (1989) Altering milk composition by feeding. *J. Dairy Sci.*, **72** 2801-14.

Uetake, K., Hurnik, J.F. and Johnson, L. (1997) Effect of music on voluntary approach of dairy cows to an automatic milking system. *Appl. Animal Behav. Sci.*, **53** 175-82.

Verdi, R.J. and Barbano, D.M. (1991) Properties of proteases from milk somatic cells and blood leucocytes. *J. Dairy Sci.*, **74** 2077-81.

de Wet, H. (1998) *Overview of payment schemes—results of a survey conducted*. Proceedings, Future Milk Farming, 25th International Dairy Congress, Aarhus, Denmark, September 1998, pp. 52-8.

Wilson, D.J., Gonzalez, R.N., Case, K.L., Garrison, L.L. and Gröhn, Y.T. (1999) Comparison of seven antibiotic treatments with no treatment for bacteriological efficacy against bovine mastitis pathogens. *J. Dairy Sci.*, **82** 1664-70.

Winter, A. and Hillerton, J.E. (1995) Behaviour associated with feeding and milking of early lactation cows housed in an experimental automatic milking system. *Appl. Animal Behav. Sci.*, **46** 1-15.

Yancey, R.J. (1998) *Mastitis therapy—Efficacy of antibiotics and alternative treatment of different pathogens*. Proceedings, Future Milk Farming, 25th International Dairy Congress, Aarhus, Denmark, September 1998, pp. 215-24.

Zall, R.R. and Chen, J.H. (1984) Milk treatment on the farm *or* on-farm use of membrane systems. *Int. Dairy Fed. Bull.*, **182** 13-24.

3 Liquid milk

D.D. Muir and A.Y. Tamime

3.1 Milk composition

Milk of different species of mammals has evolved to provide the appropriate nutrients to promote the development of the newborn. Because the time taken for the young of different species to mature varies widely, milk composition is diverse. Table 3.1 summarises the gross composition of human milk and that of other species widely used to provide milk for human consumption. There are striking differences between human milk and that available from buffalo, cow, goat, horse, sheep, yak and zebu. For example, the protein content of human milk is significantly lower than that of other species, and the lactose level is very high. Such differences must be accommodated when infant formulae are being prepared. The differences in composition between the milk of species used for the manufacture of dairy products are also important, but they do not impinge directly on process control procedures. The systems for automation and process control are the same as those used for bovine milk, but the set points (e.g. heating temperature and homogenisation pressure) may be substantially different.

The properties of milk determine its behaviour during processing irrespective of whether the manufacturing sequence is 'simple', such as that applied to pasteurised milk, or complex, as in the case of cheesemaking. It is important to recognise that biological effects, such as change in milk composition, occur and to compensate for such changes in manufacturing processes.

3.1.1 Proteins in milk

The proteins found in milk are usually classified into two distinct groups—caseins and whey proteins. Caseins, which form over 80% of the total protein,

Table 3.1 Proximate chemical composition (g $100\,g^{-1}$) of milk of different species of mammals

Species	Total solids	Fat	Protein	Lactose
Buffalo	17.9	8.0	4.2	4.9
Cow	13.0	4.0	3.4	4.8
Goat	13.2	4.5	2.9	4.1
Horse	11.2	1.9	2.5	6.2
Human	12.6	3.8	1.0	7.0
Sheep	15.3	7.3	5.5	4.8
Yak	17.3	6.5	5.3	4.6
Zebu	13.5	4.8	3.3	4.7

are a family of related proteins. The α- and β-caseins are phosphoproteins with 5–13 serine phosphate residues per molecule, whilst the δ-caseins are fragments of β-casein. κ-Casein is distinct, being a glycoprotein, and the sugar moiety of κ-casein contains charged residues. None of the caseins possesses an organised structure—like that of globular proteins—and therefore cannot be denatured in the conventional sense. Caseins are unusual proteins because, in α_s-, β- and κ-casein, acidic amino acids (carboxyl and ester phosphate) are unevenly distributed along the polypeptide chains. As a result, the proteins have highly charged polar regions and contrasting areas of a very hydrophobic nature. Such heterogeneity confers important functional properties on the isolated proteins, for example, the highly charged groups can be cross-linked by multivalent ions such as those of calcium. Such reactivity is the basis of cheesemaking, but can also be detrimental to stability of long-life products.

The caseins in milk form aggregates with an average size of 100 nm. Many models have been advanced for the structure of these aggregates, or micelles as they are usually known, but most models comprise a core of α_s-casein and calcium phosphate within which β-casein and some κ-casein are held. The surface of the casein micelle is rich in κ-casein, which forms a 'hairy' stabilising layer.

In contrast to the caseins, the whey proteins in milk are globular proteins with defined tertiary structures; as a result, they can be denatured on heating. However, in their native form at the natural pH of milk they are insensitive to aggregation in the presence of multivalent ions. When milk is heated above 65°C, interactions between the caseins and some whey proteins occur via sulphydryl interaction. As a result, the complex may be further sensitised to calcium-induced aggregation.

3.1.2 Lactose and minerals

Lactose constitutes 4.5–4.9 g 100 g^{-1} of the solids in milk. It is a disaccharide consisting of β-D-galactose linked to α-D-glucose. Lactose is a reducing sugar and, under appropriate conditions of pH, temperature and water activity, will react with free amino acids in milk protein. These reactions are evident as browning and limit the shelf-life of some dairy products. The concentration of lactose in milk is related to milk yield and is maximal at the peak of lactation (i.e. around 10 weeks after parturition), declining thereafter as milk yield falls.

The mineral system found in milk is almost as complex as that of the proteins. Sodium, potassium and chloride are the main monovalent ions and, together with lactose, they regulate the total ionic strength of milk. Such regulation is required to maintain milk in osmotic equilibrium with blood and intracellular fluid in the mammary gland. Chloride concentrations in milk are lowest at peak lactation and increase towards the end of the lactational cycle. The concentrations of total calcium and phosphate vary to a lesser extent and are related to the total protein content of the milk. In contrast, the citrate content of milk varies widely and this variation is associated with dietary changes. Citrate in milk is closely associated with *de novo* synthesis of fatty acids. The citrate concentration of milk is important per se because the ion is a very effective buffer of calcium ions and hydrogen ions. At near neutral pH, citrate forms a strong complex with calcium and about one third of the total calcium exists in this form. Seasonal changes in the citrate concentration of milk, induced by diet, are important because the concentrations of citrate and diffusible calcium in milk are highly correlated. As a result, seasonal changes in citrate content of milk are mirrored by changes in the concentration of 'soluble' or diffusible calcium.

3.1.3 Milk fat

Milk fat globules range in diameter from 1–12 μm, with a mean size of about 4 μm. These globules consist almost exclusively of triglyceride and are stabilised by a phospholipid membrane. The composition of milk fat depends on the route from which fatty acids are derived. When preformed dietary acids are scarce, *de novo* synthesis is activated. As a result, short-chain fatty acids (C4:0–C14:0) are produced and constitute a higher than usual proportion of the total fatty acid content. Palmitic acid (C16:0) is derived from the diet and synthesised in the mammary gland. Fatty acids of longer chain length are derived from the feed. Although it might be expected that milk fat is almost completely composed of saturated fatty acids, about $30\,g\ 100\,g^{-1}$ of the fatty acid is mono-unsaturated and comprises mainly oleic acid (C18:1). The reason for this apparent contradiction lies in the route by which C18 acids are modified by the cow. Most C18 acids leave the rumen as stearic acid (C18:0). However, within the gut and the mammary gland a specific enzyme, a C18:0 desaturase, converts the stearic acid into oleic acid. This explains the origin of high levels of mono-unsaturated fatty acids found in milk. The composition of milk fat changes with diet and stage of lactation. During the winter months, there are $40\,g\ 100\,g^{-1}$ less fatty acids derived from *de novo* synthesis than in the early summer (May and June). A similar pattern is observed in the proportions of oleic acid. In contrast, the seasonal variation in poly-unsaturated fatty acids is slight (there is a peak in May and June), and the overall level of C18:2 and C18:3 is never high.

Degradation of milk fat occurs by lipolysis or oxidation. Lipolysis is the hydrolysis of triglyceride to yield free fatty acids. Lipase preferentially hydrolyses milk fat to yield butyric and caproic acids. Unfortunately, these acids are volatile and have distinctly unpleasant smells and tastes. Low levels of lipolysis thus result in rejection of product—trained assessors detect a doubling of the natural level of free fatty acid in milk ($1.5\,mEq\ 100\,g^{-1}$ fat).

The other type of fat degradation is by oxidation of the poly-unsaturated fatty acids. Autoxidation is promoted by the formation of free radicals. The initial reaction involves formation of a peroxide. A chain of reactions then ensues, terminating in the formation of aldehydes and ketones, which give rise to oxidised off-flavours. The phospholipids in milk are rich in unsaturated fatty acid and are particularly susceptible to oxidation. Oxidation of milk products is enhanced by high storage temperatures, exposure to light, contamination with heavy metals and exposure to oxygen; it is retarded the presence of natural antioxidants.

3.2 Heat-treated milk products

Heat treatments applied to milk fall into four general categories:

- thermisation
- pasteurisation
- in-container sterilisation
- ultra high temperature (UHT)

Thermisation is the most gentle heat treatment applied and is designed to extend the shelf-life of raw milk destined for subsequent manufacture into cheese, butter or powder. There is no legal definition for thermisation but, usually, heat treatment is applied between 65°C and 70°C for 15 s and the milk is promptly cooled to less than 6°C. However, thermised milk should always be subject to a second heat treatment, adequate to destroy pathogenic bacteria, before producing any dairy product for human consumption.

Pasteurisation inactivates almost all pathogens that may be present in milk. The legal definition of pasteurised milk varies but, in Scotland, two methods are allowed: holder, or batch, pasteurisation; and continuous pasteurisation. Holder, or batch, pasteurisation involves heating the milk at between 62.8 and 65.6°C for 30 min. Continuous pasteurisation may be carried out at 71.7–78.1°C for at least 15 s (this system of processing is known as high temperature for short time [HTST]). Thereafter, the milk must be promptly cooled to less than 6°C. In some countries, the heating temperature of the milk may be raised to ca. 80°C without holding, and the process is known as 'flash' pasteurisation. Neither thermisation nor pasteurisation render milk commercially sterile.

In-container sterilisation is achieved by heating the packaged product at 120°C for up to 15 min. Commercially sterile milk can be produced by this method.

The ultra high temperature (UHT) process heats the milk to no less than 135°C for at least 1 s (see Section 3.4.7). Commercial sterility of milk can be achieved by this process.

In the present context, the effects of heat treatment of milk can be broadly summarised as: (a) destruction and/or elimination of pathogens and other microorganisms, including their enzymes; (b) inactivation of many natural milk enzymes, but not plasmin; (c) changes in the physico-chemical properties of the milk constituents, which may affect the quality of the product (however, these changes can also be influenced by the homogenisation process); and (d) effect on nutritional properties of milk constituents. Nevertheless, thermal treatment of milk has been extensively studied, and excellent reviews have been published by Burton (1988), Fox (1989, 1991, 1995), Holdsworth (1992) and Walstra et al. (1999).

3.2.1 Chemical effects

Considerable potential for interaction exists between the protein and the minerals in milk. The extent of such interactions hinges on the nature of the protein phase and on the 'activity' of calcium ions. Heating of milk to control bacterial proliferation in all long and intermediate shelf-life products induces changes both in the sensitivity of the protein and in the mineral equilibrium. Above 65°C denaturation of whey proteins occurs, which then form complexes with casein; specifically, β-lactoglobulin interacts with κ-casein (complete denaturation of whey proteins takes place at sterilisation temperature). Denaturation of whey protein reduces the subsequent heat stability of raw milk, but increases the stability of product that is subsequently concentrated. The effect of heating on the stability of milk and milk products is multifactorial and not fully defined. Furthermore, concentration, freezing and acidification also have a significant impact on stability.

Also, at sterilisation temperature, calcium phosphate precipitates on the casein micelles and, as a result of chemical degradation: (a) the pH of the milk decreases; (b) the nutritive value is reduced by de-amination and de-phosphorylation; and (c) browning occurs as a result of Maillard reaction. These undesirable chemical changes can be ameliorated by application of UHT for sterilisation. In principle, the rate of chemical reaction increases by a factor of two for every 10°C increase in temperature, whereas bacteria are killed at a very much faster rate for the same temperature increase. For example, heat treatment at 140°C results in a fourfold increase in chemical effect over treatment at 120°C. Reduction in sterilisation treatment from 5 min to 1.25 min would thus achieve the same chemical effect. In contrast, the same bactericidal effect can be achieved by heat treatment for only 2–4 s. As a result, sterility can be achieved with a great saving in the extent of chemical degradation.

3.2.2 Destruction of microorganisms and enzymes

At pasteurisation temperatures, lipoprotein lipase and alkaline phosphatase are completely deactivated. The destruction of alkaline phosphatase is a useful marker for effective pasteurisation. Bacterial exo-enzymes are much more resistant to heating. Gram-negative psychrotrophs, the predominant flora in raw milk, produce extracellular lipase and protease. All Gram-negative bacteria are killed by pasteurisation, but the extracellular enzymes are resistant to heat treatment. After pasteurisation, the residual levels of protease, lipase and phospholipase are, on average, 66%, 59% and 30%, respectively. Some bacteria found in raw milk are not killed by pasteurisation—mainly spore-forming organisms and bacteria in the coryneform group. Nevertheless, pasteurisation destroys the important pathogenic flora.

Most of the spores can be deactivated by a relatively short heat treatment at 120°C, but if *Bacillus stearothermophilus* or *Bacillus thermoacidurance* are present they are very hard to eliminate. Sterilisation at or around 120°C usually involves heat treatment for at least 5 min (i.e. 300 s). As a result, 10 decimal reductions of *Clostridium botulinum* or of *Bacillus cereus* spores may be expected. In practice, spoilage by other, more heat resistant, organisms is unusual. Although the UHT sterilisation process has fulfilled its promise in terms of limiting chemical effect, it does not destroy the extracellular enzymes of psychrotrophic bacteria. The residual activity of protease, lipase and phospholipase after treatment at 140°C for 5 s is, on average, 30%, 40% and 0.57% (depending on bacterial source), respectively. Therefore, it is crucial that milk to be UHT sterilised is of the highest quality. The psychrotroph count must never be allowed to exceed a level of 10^6 colony forming units (cfu) ml^{-1}, if subsequent degradation of the sterilised product is to be avoided.

3.2.3 Effects on other milk constituents

The relative increase in temperature during the processing of milk can cause a significant decrease in some water-soluble vitamins that are heat labile (see Table 3.2).

Table 3.2 Typical values of vitamin losses (%) from milk during different heat treatments

Vitamin	Pasteurisation		Sterilisation	
	Batch	HTST	In-container	UHT
Thiamine	10	<10	20–35	10
Riboflavin	n.a.	n.a.	<10	10
Folic acid	0	0	40–50	15
Biotin	<10	<10	<10	<10
Pantothenic acid	n.a.	n.a.	<10	<10
B_6	10	0	60–90	<10
B_{12}	20	10	40–50	10

n.a., data not available; HTST, high temperature for short time; UHT, ultra high temperature.
Source: adapted from Tamime and Robinson (1999).

3.3 From farm to factory

3.3.1 Milk collection

Raw milk is collected from the farm by bulk tanker every day or every alternate day. Alternate-day collection is advantageous in terms of the cost of transport. However, the 'safe' shelf-life of raw milk is predetermined by the initial level of bacterial contamination and the storage temperature. In this circumstance, if milk is stored at the farm for an extra day, the remaining shelf-life available at the factory is proportionately reduced.

On arrival at the farm, the tanker driver inspects the bulk tank for overt problems and checks that the temperature is below 5°C. Samples of milk are taken for testing and the milk is transferred with minimal agitation from the refrigerated farm tank to the insulated road tanker. Tankers usually collect milk from several farms before proceeding to the factory.

3.3.2 Milk distribution

The demand from retail stores for dairy products fluctuates markedly throughout the week in line with consumer shopping patterns. In addition, there are substantial peaks in demand associated with public holidays. As a result, the call from factories for raw material to satisfy this demand fluctuates markedly. The situation is complicated by the fact that raw material may be purchased from the contract suppliers to a dairy company or from a central cooperative by predetermined contract or on the 'spot' market. Milk on the spot market is residual to previously negotiated contracts and may vary widely in price. The logistics of matching the demands for raw material from the factory with the supplies available from these three sources are complicated and require a dedicated resource.

3.3.3 Delivery to the factory

On reception at the factory, a bulk tanker is subject to stringent checks because the introduction of faulty product into the factory could contaminate a large quantity of milk and have far-reaching consequences (see Section 3.4.1).

3.3.4 Extension of the shelf-life of raw milk

Bacterial growth, albeit slow, limits the shelf-life of raw milk. Several options are available to extend shelf-life. For example, bacterial growth is related to storage temperature. Shelf-life can be usefully extended by deep-cooling raw milk to 2°C on reception at the factory. Another option is to apply a modest heat treatment that inactivates the majority of the spoilage bacteria. Such heat treatments are called thermisation (see Section 3.2). They do not kill all pathogenic bacteria and are not a replacement for pasteurisation. Finally, the most effective option is to combine thermisation with deep cooling. This practice can extend the safe shelf-life of raw milk by two to three days.

3.3.5 At the factory

Raw milk is converted into a wide range of dairy products at the factory. However, only a limited number of different unit operations are required to manufacture the whole product range. The processing sequence is illustrated by the sequence of unit operations used to produce heat-treated milk.

3.3.5.1 Separation of milk fat

The first step in the process is to separate milk into skim milk (essentially fat-free) and cream. These two components can be recombined to produce skimmed milk (<0.5 g $100\,g^{-1}$ fat), semi-skimmed milk (1.5–1.7 g $100\,g^{-1}$ fat), whole milk (>3.0 g $100\,g\,fat^{-1}$) or cream products, ranging in fat content from 12–55 g $100\,g^{-1}$ fat. Separation is based on the density difference between the fat droplets (light phase) and the milk serum (heavy phase). If milk is stored, the fat globules rise naturally to the surface forming a cream layer; however, this natural process is slow. The separation time may be reduced from several hours to seconds by applying a centrifugal force in a high-speed cream separator. Separators come in various forms and are self cleaning. Modern plants can automatically regulate the fat content of the cream phase (see Section 3.4).

3.3.5.2 Pretreatment to reduce content of bacterial spores

At this point, treatments may be applied to enhance the shelf-life of the finished product. As described earlier, pasteurisation kills pathogens and most spoilage bacteria. Nevertheless, spore-forming bacteria—minor contaminants of milk—survive conventional pasteurisation and ultimately grow and degrade product. Two practical solutions have been advanced. First, the spores may be removed from the skimmed milk by a membrane filtration process called microfiltration. Second, the spore concentration may be reduced by bactofugation; this process, analogous to cream separation, uses the density difference between bacterial spores and milk serum to achieve separation. Neither process completely removes spores from milk, but both significantly enhance the shelf-life of refrigerated, pasteurised milk.

3.3.5.3 Standardisation of fat content and heat treatment

After pre-treatment, the fat phase and skimmed milk are recombined to achieve the required fat content and the blended product is then heat treated.

3.3.5.4 Control of creaming

Fat separation during storage is undesirable in milk packed in cartons. In contrast, many consumers rate milk packed in glass bottles on the extent of the cream layer. These mutually exclusive demands are met in different ways. Creaming is avoided by reducing the average size of the fat globules. Natural dispersive forces (Brownian motion) counteract the propensity of the fat to cream. Reduction of particle size is achieved by using a high-pressure homogeniser. The total surface area of fat globules increases as particle size is reduced. The newly formed surface is immediately stabilised by adsorption of milk protein. This interfacial layer of protein is very effective for inhibiting coalescence of the fat globules, but the protein layer is porous and, unlike the natural membrane, does not protect the fat globule from degradation by enzymes. Therefore, homogenised milk must not be mixed with unheated milk containing active lipoprotein lipase.

When creaming is desired, equally stringent measures have to be applied. Creaming is a form of cold agglutination and is enhanced by an immunoglobulin IgM. This protein is only an effective agglutinin in its undenatured form. At pasteurisation temperature it is partially denatured. As a result, slight increases in the severity of heat treatment reduce the cream line and are to be avoided in the production of bottled milk.

3.3.5.5 Packing and cooling

After heat treatment, the way in which the milk is handled is critical. Contamination by spoilage bacteria after heat treatment must be avoided at all costs. Over the shelf-life of good quality milk—around 14 days—an initial level of contamination of 10 organisms per litre is significant. The storage temperature is equally important and bacterial growth in pasteurised milk is similar to that in raw milk. When post-heat-treatment contamination is well-controlled (e.g. by use of aseptic packaging) the main limitation on shelf-life is the level of contamination by spore-forming bacteria capable of growing at refrigeration temperature. Potential methods of reducing the spore load are discussed in Section 3.3.5.2.

3.4 Milk handling in dairies

In industrialised countries, milk is stored in bulk refrigerated tanks on the farm and is collected from farms by road tanker. The protocols for production on farms, storage and collection of milk are detailed in Chapter 2. After the milk has been accepted at the factory (see Section 3.3), the tanker load is coded and handled from a remote control room where routing of the product is overseen. The operator enters the instructions in the computer, and valves and pumps are automatically actuated. In addition, details of the operations are stored in a database. Figure 3.1 illustrates a highly integrated system, including a Gannt chart for milk reception, batch processing of milk and waste-modelling management (see also Figure 3.3).

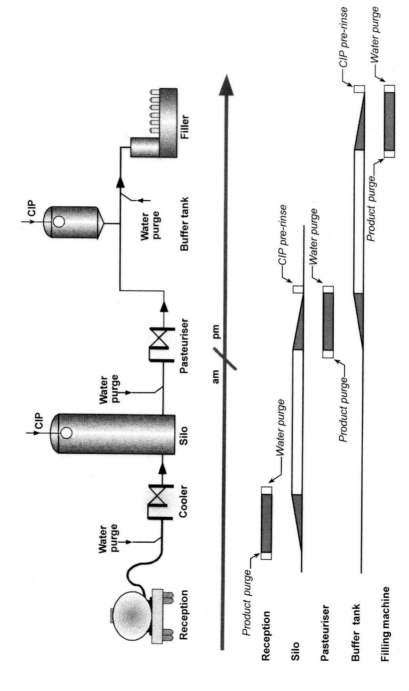

Figure 3.1 Layout of milk reception and batch processing of milk for waste-modelling work. CIP, cleaning-in-place. Source: Chester, personal communication; reproduced by courtesy of APV UK Co. Ltd, Crawley, UK.

The equipment used for processing market milk and a wide range of dairy products has much in common (as seen in this book; see also Bylund, 1995). In a highly automated plant with remote control of processes, the following basic equipment is required.

1. Large tanks or silos are used to store the in-coming milk prior to production.
2. Coarse filters remove large particles of solids that accidentally originate in milk during production on farms.
3. Mechanical centrifuges find several applications in the dairy industry, for example:
   - clarifiers have no holes in the disk stack and have a single outlet; the milk enters the separation channels at the outer edge of the disc stack, flows radially through the channels towards the axis of rotation and leaves through the outlet; the solid impurities are separated and accumulate at the periphery of the clarifier bowl from where they are discharged intermittently;
   - separators are similar to clarifiers, but the disc stacks have distribution holes and two outlets; the milk components are separated on the basis of difference in density (i.e. the lipid phase and milk serum have different densities), and the fractions may be recombined for fat standardisation or for production of products with a high content of milk fat (see Chapter 5);
   - bactofuges (e.g. with one-phase or two-phase outlets) are used to separate microorganisms, particularly spores, from milk;
   - decanter centrifuges are used in special applications to harvest precipitated caseins or crystallised lactose.
4. Membrane filtration—for example, ultrafiltration (UF)—is used to concentrate the proteins in skimmed milk selectively; microfiltration (MF) separates microorganisms from skimmed milk; for process parameters and controls of membrane filtration systems refer to Chapter 11.
5. Buffer and balance tanks: buffer tanks are used for intermediate or temporary storage of milk and balance tanks are used to provide constant inlet pressure to the pump.
6. Heat exchangers (with direct or indirect heating) are used to destroy the microorganisms present in milk and hence to extend the shelf-life of the product; the different types of heat exchangers are: (a) plate; (b) tubular (monochannel or multichannel); and (c) scraped surface.
7. Pipes, fittings and pumps are used to interconnect the elements of the processing system; pneumatically-operated valves control the product flow (see Figure 3.2).

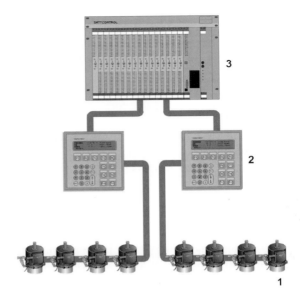

Figure 3.2 Illustration of a valve control system. 1, Valve units; 2, modem; 3, control system [programmable logic controllers (PLCs)]. Reproduced by courtesy of Tetra Pak (Processing Systems Division) A/B, Lund, Sweden.

8. Control equipment is used to regulate the flow rate, heating temperature and pressure drop in the processing plant.
9. Miscellaneous services, such as electric power, steam, water supply and compressed air, are calculated after the plant design is agreed.

3.4.1 Reception of milk

When the tanker arrives at the factory, the milk, including a sample from each individual farm, is routinely tested. These tests are categorised into two main groups. First, the intermediate quality control tests are carried out, which include measurement of temperature and pH or titratable acidity, a check for smell (tasting of milk is carried out on a sample after it has been heated in the laboratory), and measurement of freezing point depression and antibiotics residues. Second, the delayed tests are for: (a) proximate chemical composition; (b) total viable counts; and (c) somatic cell counts. Some of these tests are performed by means of automated analytical techniques in order to minimise any delays at reception.

There are two possible routes for milk handling at reception. First, the tanker outlet valve is connected to an air eliminator. The milk passes through this, then, in sequence, through a pump, filter, flow meter, cooler, a further pump, then into a silo. Second, for extended storage of raw milk, the milk is handled as above, but, after the metering stage, the milk is thermised in a plate heat exchanger (PHE), cooled and finally stored in the silo.

3.4.2 Milk processing

A simplified milk pasteurisation processing plant for milk products not requiring milk fat standardisation or extended shelf-life is shown in Figure 3.1.

Associated with the automation of the system, process information is recorded in a database which has a myriad of uses (Chester, personal communication). Some examples of the data recorded include: production inventories and loss data, cleaning-in-place (CIP) chemicals, service and energy usage, maintenance, processing routes (number of tanks and shared equipment), number of start-ups and shut-downs, production planning and planning of shift work. This information is invaluable for the monitoring of waste (Figure 3.3).

3.4.3 Pasteurisation systems

The application of heat to certain foods for perservation dates back to the mid-1700s. By 1922, commercial pasteurisation began in Europe and North America

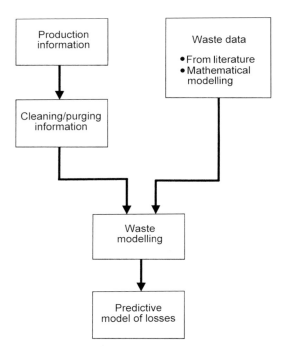

Figure 3.3 Typical waste-modelling work in a milk plant. Source: Chester (personal communication). Reproduced by courtesy of APV UK Co. Ltd, Crawley, UK.

(Harvey and Hill, 1967). The pasteurisation process was originally carried out by heating milk in a tank (batch system), but nowadays large volumes of milk are processed by using plate or tubular heat exchangers. The building blocks of a milk pasteurisation plant are common to different types of plant for the manufacture of many different kinds of market milk (see Figure 3.4). Typical products include:

- whole milk without homogenisation, usually packed in returnable glass bottles—the 'cream line' is regarded by consumers as evidence of legal quality
- fat standardised milk, ranging from virtually fat-free and ~1.5 to 3.5 g $100\,g^{-1}$—these standardised products are usually homogenised and thus have no 'cream-line'
- extended-shelf-life (ESL) pasteurised milk that has been subjected to bactofugation or microfiltration—this may also be fat standardised and homogenised

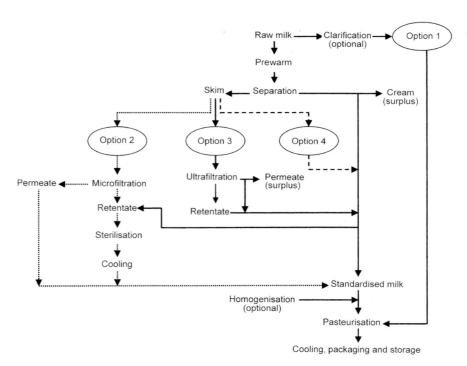

Figure 3.4 Flow chart illustrating possible routes for processing pasteurised milk. Option 1, whole milk without homogenisation; option 2, fat standardised milk and extended shelf-life (ESL) milk; option 3, fat and protein standardised milk; option 4, fat standardised milk only. Homogenisation is carried out either fully or partially and is used for bactofuge ESL milk.

- protein adjusted milk, in which the protein content has been adjusted by ultrafiltration of the skimmed milk before recombination with the cream to the desired final concentration.

A typical process line for pasteurised market milk (e.g. Figure 3.4, options 1 and 4) is shown in Figure 3.5. The main sections of a pasteurisation plant are: balance tank (item 1 in Figure 3.5), pump (item 2), flow controller (item 4), plate heat exchanger (PHE, item 5; e.g. regeneration, heating and cooling sections), holding tube (item 15), flow diversion valve (FDV, item 16), and a booster pump (item 14) which ensures that the pressure of the pasteurised milk after heat treatment exceeds that of the heat-transfer media to prevent contamination with raw milk in the regeneration or cooling sections. However, for pasteurised whole milk without fat standardisation (Figure 3.4, option 1), the milk bypasses the separator (item 6), standardisation unit and the homogeniser (item 13); instead, a centrifugal clarifier may be installed in the process line.

According to Bylund (1995), the important parameters of a PHE for pasteurising milk are:

- plant capacity
- temperature programme (4°C, heat to 72°C, cool to 4°C)

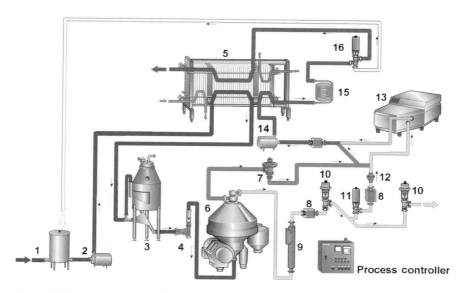

Figure 3.5 Illustration of a typical line for the production of milk to be sold as pasteurised milk. 1, balance tank; 2, product feed pump; 3, de-aerator; 4, flow controller; 5, plate heat exchanger (PHE); 6, separator; 7, constant pressure valve; 8, flow transmitter; 9, density transmitter; 10, regulating valve; 11, shut-off valve; 12, check valve; 13, homogeniser; 14, booster pump; 15, holding tube; 16, flow diversion valve (FDV). Reproduced by courtesy of Tetra Pak (Processing Systems Division) A/B, Lund, Sweden.

- regenerative efficiency (e.g. 94%)
- temperature of the heating medium (74–75°C)
- temperature of cooling agent (2°C) to cool milk to ca. 4°C

Thus, from these parameters the requirements for services (steam, water and/or chilled water) are readily calculated.

3.4.3.1 Control of heat exchangers

The heating medium in a PHE can be either hot water or saturated steam at atmospheric pressure. The former system is widely used, and Figure 3.6 shows details of the regulation of temperature in the heating section of a pasteuriser (Bylund, 1995).

The steam (generated by a boiler or steam generator at 0.6–0.7 MPa) indirectly heats the water, which, in turn, indirectly heats the milk in the PHE. The temperature difference between the hot water and the milk is 2–3°C. The temperature of the hot water is regulated by control of the steam into the circulated water by a proportional steam-regulating valve. The condensate in the steam line is removed via a steam trap. An expansion vessel (item 7 in Figure 3.6) compensates for the increase in the volume of water that takes place when it is heated. The system is usually fitted with safety valves, pressure gauges and temperature indicators (see Figure 3.6). A similar circulating system is used to control the cooling section; in this case, the cooling medium may be glycol.

In order to regulate the temperature (during heating and cooling) in a heat exchanger, temperature transmitters are located in the heating section and heating medium and in the product outlet and cooling medium. The feedback from these sensors is used to maintain a stable and continuous process (see Figure 3.7).

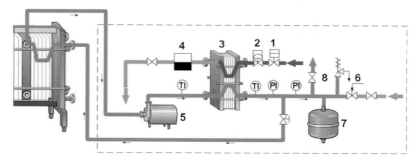

Figure 3.6 A flow chart illustrating the principle of a hot-water system connected to a pasteuriser. 1, Steam shut-off valve; 2, steam-regulating valve; 3, plate heat exchanger (PHE); 4, steam trap; 5, centrifugal pump; 6, water-regulating valve; 7, expansion vessel; 8, safety and ventilation pipe. TI, temperature indicator; PI, pressure indicator. Reproduced by courtesy of Tetra Pak (Processing Systems Division) A/B, Lund, Sweden.

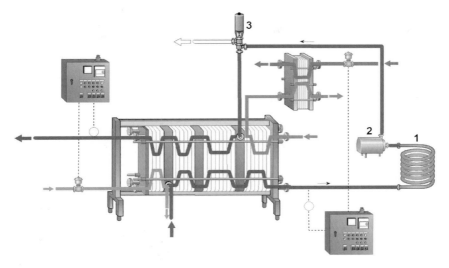

Figure 3.7 Automatic temperature controllers for heating and cooling systems in a plate heat exchanger. 1, Holding tube; 2, booster pump; 3, diversion valve. TT, temperature transmitter. Reproduced by courtesy of Tetra Pak (Processing Systems Division) A/B, Lund, Sweden.

Milk is usually heated 2–3°C above the pasteurisation temperature to compensate for the temperature drop during passage through the external holding tube. A temperature sensor transmits continuous signals to the temperature controller and to the recording chart in the control panel.

Another sensor is located after the holding tube. Should the temperature at this point fall below a preset value, a flow diversion valve (FDV; item 16 in Figure 3.5) diverts the milk back to the balance tank. The booster pump remains on, but other valves around the heat exchanger (not shown in Figure 3.5) operate automatically in the regeneration and cooling sections of the processed milk to maintain the pressure in the plant. Such process design ensures proper temperature balance, and as soon as the required heating conditions at the end of the holding tube are restored, the forward flow of the processed milk resumes (Bylund, 1995).

3.4.3.2 Control of fat standardisation
Milk fat standardisation is controlled automatically on-line (see Figure 3.5; see also Chapters 1 and 6). One example is the OL-7000 manufactured by On-Line Instrumentation Inc. (New York, USA; Figure 3.8). This type of unit is a third-generation design and ensures that the fat content is within ± 0.02 g 100 g^{-1} of the target level in the final product with 95% confidence (Anon., 1992). The milk is pre-warmed and separated into skimmed milk and cream. A unit measures the density of the cream and calculates the fat content. This information is relayed to the OL-7000 controller. In addition, a sample of milk is bled from the milk

stream after the homogeniser, and its fat content is measured in the OL-7000 device (see Figure 3.8). Fat analysis is by an improved method (Haugaard and Pettinati, 1959). The measured fat content is compared with a set point, and the difference is used to transmit a signal to an electrical valve positioned in the cream line to regulate the flow of cream for mixing with the skimmed milk. Hence, the throughput of the cream can be maintained by using an electrically-operated valve. The OL-7000 on-line fat standardisation system of market milk employs full-stream homogenisation of the product. A similar approach is used during the manufacture of yoghurt (see Chapter 6).

The processed milk may be either partially homogenised (Figure 3.5) or fully homogenised (Figure 3.8). In the former system, most of the skimmed milk is not homogenised, only the cream. This method of processing reduces total power consumption by 65% because smaller volumes pass through the homogeniser, and a smaller capacity homogeniser is required which could be a third of the output capacity of the plant (Bylund, 1995). Moreover, the level of homogenisation of the final product is enhanced if the fat content of the product passing through the homogeniser does not exceed $12 \text{ g } 100 \text{ g}^{-1}$.

3.4.3.3 Control of fat separator and/or clarifier
The clarification and separation of milk on a pasteurisation plant is carried out by using a high-speed centrifuge. Most centrifugal machines, such as the Tetra Pak models, are controlled automatically from start-up to shut-down (Gladwell, personal communication) and consist of a rotating bowl, containing a stack of conical discs to facilitate separation (for a fuller description of centrifugal

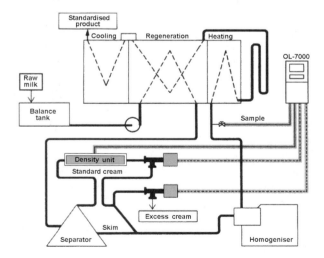

Figure 3.8 Automatic standardisation of the fat content in market milk. Reproduced by courtesy of On-Line Instrumentation Inc., New York, USA.

machines, see Section 5.2.1 and Section 8.3.2; for an illustration of a 'bactofuge', see Figure 8.1). Although the process can be carried out with cold milk, this is inefficient; it is better applied to hot product at 45–63°C. These processes are integrated into the process cycle of milk and may be carried out in the middle of or after regenerative heating and prior to final heating. In some instances, clarification is applied to cold milk at reception, before transfer to silos.

As a result of the centrifugal force applied to milk within the machine bowl and the disc stack, three phases are separated (sludge, skimmed milk and fat, i.e. cream). The low-density phase (fat) migrates to the centre of the disc stack whereas the intermediate-density phase (skimmed milk) is driven to the outer area of the disc stack. The sludge (high-density solid impurities) is thrown to the periphery of the separator and is collected in the sediment space. By controlling the respective flow rates of cream and skimmed milk out of the machine, and mixing the two phases in different proportions, standardised milk or cream is produced. This process can either be carried out automatically (by using a cream standardisation system; see Figures 3.5 and 3.8) or manually by a hand-operated flow-control valve (i.e. cream screw) on the cream outlet.

To maintain constant de-sludge volumes and consistent levels of fat in the cream, the pressure in the separator bowl must remain constant. To achieve this, a direct-acting, modulating valve, known as a constant pressure modulator (CPM), is installed in the outlet line for the skimmed milk. This valve reacts immediately to any change in the downstream pressure by either partially opening or closing to keep the required pressure in the bowl at a constant value. The centrifugal forces developed in the machine are very large because a typical bowl weighs 1.2 tonnes and rotates at a speed of 4000 to 7000 revolutions per minute (rpm). As a result, a sophisticated control system is required to take the machine from start-up to shut-down via a series of programme steps. To achieve this safely, the machine is supplied and installed as a complete system, able to 'stand alone', with its own control system. It is necessary to mount some of the control equipment adjacent to the separator. In addition, a set of peripheral equipment is supplied which is used to monitor and protect the machine as it is taken through its various start-up and shut-down programme steps.

The control system consists of a set of panels including: (a) an operator-interface electronic control panel, (b) auxiliary box and (c) starter panel (optional). A typical example is the electronic separator control system (SCS) (Gladwell, personal communication). This microprocessor control system can either be used in a local operator control mode or in a remote unit control mode. In the remote mode, the SCS becomes a 'slave-to-master' control system, and they are linked via a series of digital (i.e. volt-free switch contacts) signals or by a serial link (e.g. single-wire communication system). In either the local or the remote control modes, the same program steps are available with feedback, for example, when in the remote mode to confirm the plant status to the master control system. When used in the local control mode, the operator selects the various programmes (start, feed-off, standby, production or cleaning, and stop) via a keypad, complete with a display monitor. The display shows the operator the machine status plus its speed (rpm), power consumption, alarm faults and the status of the discharge count-down timers.

Milk clarification (i.e. de-sludging) can be controlled either manually or automatically via the panel. In automatic mode, these operations are carried out at pre-set intervals (i.e. the intervals are set up on plant commissioning). The de-sludge volume is monitored continuously via the power consumption. If the volume of the sludge is too large, the rotating bowl slows down and discharges excess product. To guard against out-of-balance forces, the machine may also be monitored for vibration. Excessive vibration will initiate a shut-down for safety reasons. Faults will, of course, trigger an audible or visual alarm, locally and remotely, to warn the operator or master control system that external action is required. Liquid flow to the machine is controlled at all times by the panel via an automatic inlet valve. The purpose of the valve is to prevent liquid flowing into the bowl until it is up to full speed and consequently closed.

The auxiliary box houses the water pressure regulators, water solenoid valves and pneumatic solenoid valves, all of which are controlled via the separator panel. These valves are used to control the various pneumatic service, product and cleaning valves on and around the machine. They are used to close the bowl, open the bowl, stop the flow of liquid entering the bowl, etc.

The starter panel houses the machine panel star or delta starter. Owing to the way the machine bowl is driven (and its substantial mass), when starting from rest it will take about 13 min for the machine bowl to reach 'delta' (its normal running speed). The power consumed by the motor to do this generates a significant amount of heat within it, which therefore limits the number of times it can be started within a given time period (Gladwell, personal communication).

3.4.3.4 Control of homogenisers

Homogenisers can be readily controlled: for example by operation of the motors (main motor, oil pump and hydraulic pump), by water valves to cool and lubricate the pistons and by a transmitter to regulate the pressure (Lund, personal communication). The operator video display unit (VDU) controls the process from a monitor in the control room. For example, an APV Nordic homogeniser is operated by pressing 'start' on the screen. A digital 24 VDC signal from a programmable logic controller (PLC) is sent to the power panel via a relay and operates the main motor and the oil pump. In parallel, the water valve is given a direct signal to open, and the homogeniser starts running. The homogenisation pressure is regulated by a signal transmitted to a hydraulic pump, normally in several 'steps' (i.e. 1, 2, 3 and so on) to build up the pressure. The operator chooses the appropriate step to use and this is entered on the VDU screen; the PLC automatically sends the signal to the homogeniser to run on that step.

LIQUID MILK

Feedback from the pressure transmitter controls the hydraulic solenoid valves to provide consistent homogenising pressures without the need for intervention by plant operators.

To stop the homogeniser, the operator has only to push the stop button on the VDU screen and all operating commands are cancelled. However, during the cleaning cycle of the plant, no signal is actuated to operate the hydraulic pump, to allow the homogeniser to run without pressure.

3.4.3.5 Control of integration processes

In large factories, each hour a significant volume of milk must pass through equipment of the sort shown in Figure 3.5. Hence, the design of a process plant (Bylund, 1995) must take into consideration the following fundamental physical aspects.

- The piping system must be matched to the appropriate flow rate because the product velocity varies in relation to the diameter of the pipe. For example, at a flow rate of $20000 \, l\,h^{-1}$ the milk velocity in a 5.1 and 7.6 cm diameter pipe is 2.75 and $1.25 \, m\,s^{-1}$, respectively.
- High velocities can cause friction, which may damage the product—for milk the upper limit of velocity is ca. $1.8 \, m\,s^{-1}$. In addition, an appropriate velocity is required to control the flow pattern (laminar or turbulent) of liquid in pipes and in parallel paths, such as in a PHE.
- Every part of a processing plant causes resistance in the milk when it is forced through the system; this results in a pressure drop which needs to be compensated for.

To ensure trouble-free operation, according to Bylund (1995) the following pieces of equipment are required to control the process: (a) transmitters for pressure and temperature measurements; (b) a controller; and (c) regulating valves (see Figure 3.9).

3.4.4 Extended-shelf-life milk

Production of ESL milk (Figure 3.4, option 2; Figures 3.10–3.12), which retains its quality for up to three weeks at temperatures of 5°C or lower, is achieved by combining efficient pasteurisation with centrifugal bactofuge or a microfiltration plant.

3.4.4.1 Bactofugation

The bactofugation process is based on the centrifugal separation of microorganisms. There are many possible configurations of a bactofugation plant (Bylund, 1995). Some examples are:

- a two-phase bactofuge with continuous discharge of bactofugate ('sludge')

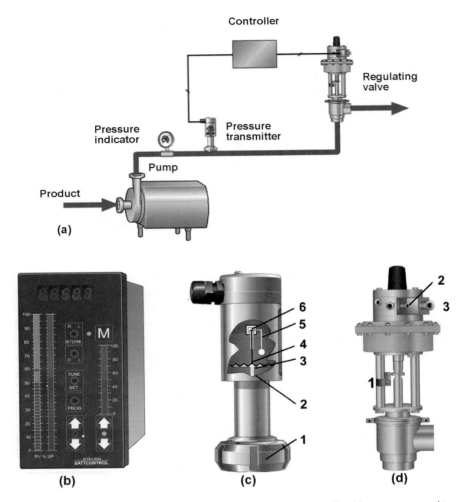

Figure 3.9 Illustration of (a) a typical control loop for pressure; (b) a controller; (c) a pressure transmitter (1, nut; 2, process pressure; 3, membrane; 4, capillary pipe; 5, reference pressure; 6, sensor); (d) pneumatic regulating valve (1, visual position indicator; 2, connection for electrical signal; 3, connection for compressed air). Reproduced by courtesy of Tetra Pak (Processing Systems Division) A/B, Lund, Sweden.

- a single-phase bactofuge with intermittent discharge of bactofugate
- double bactofugation with two single-phase bactofuges in series

Bactofugation of milk was developed for the processing of cheese milk to remove spore-forming bacteria (*Clostridium tyrobutyricum*) responsible for 'blowing' in high-pH varieties of cheese, such as Emmental, Gouda or Edam. Although two-phase bactofugation can remove up to 99% bacteria and spores, its application to extend the shelf-life of pasteurised milk is doubtful if the product is stored at 7°C or above (Bylund, 1995).

The process line of a two-phase bactofuge, with continuous discharge, for the treatment of milk is shown in Figure 3.10. Whole raw milk is pre-warmed, separated and the fat standardised. The milk then flows through a bactofuge which works under airtight conditions and produces a continuous flow of heavy phase (i.e. bacterial concentrate). The bactofugate or 'sludge' (equivalent up to $3 \, \text{ml} \, 100 \, \text{ml}^{-1}$ of the feed intake) is heat treated in an infusion steriliser at ca. 130°C for a few seconds. The hot bactofugate ('sludge') leaving the steriliser is partially cooled by mixing it with half the volume of treated milk before it is remixed with the rest of treated flow (Figure 3.10). Later, the milk is pasteurised in the PHE, as described in Section 3.4.3 (Bylund, 1995); alternatively, a single-phase bactofuge may be used with intermittent discharge and without sterilisation of bactofugate.

3.4.4.2 Microfiltration (MF)

The membrane pore size of a microfilter is ca. $1.4 \, \mu\text{m}$. Microfiltration can reduce the presence of bacteria and spores in milk up to 99.5–99.99 vol%. There are many microfiltration systems available on the market and, although the principles of bacterial separation are similar, the processing of milk differs from one equipment supplier to another. Different approaches are taken to the ESL process. For example, the Tetra Therm ESL process (Figure 3.11) is as follows: whole raw milk is pre-warmed to ca. 60°C in the PHE and separated in a cream separator. The skimmed milk is cooled to ca. 50°C, fed into the MF module under constant pressure (10-50 kPa) and fractionated into permeate ($95 \, \text{ml} \, 100 \, \text{ml}^{-1}$) and retentate ($5 \, \text{ml} \, 100 \, \text{ml}^{-1}$, $9\text{–}10 \, \text{g} \, 100 \, \text{g}^{-1}$ total solids). The

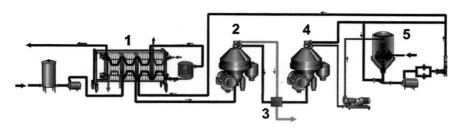

Figure 3.10 Two-phase bactofugation with continuous discharge and sterilisation of bactofugate. 1, Pasteuriser; 2, centrifugal separator; 3, automatic fat standardisation unit; 4, two-phase bactofuge; 5, infusion steriliser. Reproduced by courtesy of Tetra Pak (Processing Systems Division) A/B, Lund, Sweden.

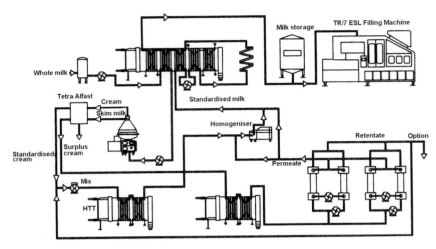

Figure 3.11 A milk pasteurisation processing line with Tetra Therm ESL. ESL, extended-shelf-life; HTT, high-temperature treatment. Reproduced by courtesy of Tetra Pak Filtration Systems A/S, Aarhus, Denmark.

retentate contains almost all the bacteria and yeast that were present in the milk. If an extra MF stage is used in series with the primary units, the retentate volume can be as low as 0.5 ml 100 ml^{-1} of the skim feed. The MF concept is based on a balanced transmembrane pressure achieved by one centrifugal pump for circulation of the retentate and another for circulation of the permeate (Harbo, personal communication; Wamsler, 1996).

The bacteria-rich retentate is either bled off to waste or is mixed with cream, sterilised at 120–130°C for 2–4 s, cooled to ca. 50°C and remixed with the microfiltered skimmed milk. Optionally, the retentate may be mixed back with incoming whole raw milk and reprocessed. The milk is homogenised using the partial stream system described earlier, fat content is standardised in the product, pasteurised at 72°C for 15–20 s, cooled to less than 4°C and aseptically packaged. An MF plant can be designed for different capacities, and the layout of a unit handling 10 000 l h^{-1} of raw milk is shown in Figure 3.11. The ESL milk may have a shelf-life of up to 45 days (i.e. if the retentate is bled off to waste) in an unopened container stored at less than 7°C. The mass balance of whole milk, skimmed milk, permeate, retentate and fat standardised ESL milk with use of an MF system is shown in Table 3.3.

In contrast, in the APV Nordic MF system, the retentate from the MF unit is mixed with the incoming raw milk to be separated and later microfiltered or alternatively is discarded. Figure 3.12 details such an MF process, and this method of manufacture provides the following advantages (Bøejgaard, personnal communication): (a) longer shelf-life (by a minimum of 50%); (b) longer-lasting

Table 3.3 Proximate compositional balance of the microfiltration or bactocatch process for the production of extended-shelf-life (ESL) milk

Product	Volume (l)	Constituents (g 100 g^{-1})		
		protein	fat	total solids
Raw milk	10 000	3.47	4.05	12.92
Separator:				
skimmed milk	9 000	3.60	0.05	9.26
cream[a]	720	2.16	40.00	45.50
Microfiltration:				
permeate	8 550	3.58	0.04	9.23
retentate	450	3.98	0.24	9.83
Cream plus retentate	1 170	2.86	24.71	31.72
Standardised milk	9 720	3.50	3.00	11.94

[a] Actual yield is 1000 l, but for standardisation purposes only 720 l is required; the balance (280 l) is surplus cream. Source: Harbo (personal communication). Reproduced by courtesy of Tetra Pak Filtration Systems A/S, Aarhus, Denmark.

fresher taste; (c) uniform milk quality; (d) healthier product (virtually bacteria free); and (e) a product that has a special niche market (i.e. of high quality and retailing at a higher price).

3.4.5 High-temperature pasteurisation

In some countries, the pasteurisation temperature of 72°C is raised to ca. 80°C for liquid milk processing. The holding time may range from a fraction of a second to a few seconds, and the processed milk is known as a high or 'flash' pasteurised product. The process line is similar to those described in Section 3.4.3, and the pre-set temperature to operate the FDV is slightly more than 80°C. In some countries, products from UHT systems are referred to (incorrectly) as high heat pasteurised products (see Section 3.4.7).

3.4.6 In-container sterilisation

Traditionally, milk was sterilised in glass bottles at 118–122°C for 30 min after being hermetically sealed. Such intense heat treatment changes the colour of milk (i.e. it becomes brown), and the product has a caramelised taste or cooked flavour. However, recent developments by Stork in collaboration with The Netherlands Institute for Dairy Research (NIZO) have resulted in improved products with use of a two-stage sterilisation system. The principle of the new process is first to heat treat the milk under UHT conditions (indirect), then package the milk in glass or plastic bottles and apply a second sterilisation stage at ca. 120°C for 11–15 min (see Figure 3.13).

The in-container sterilisation of milk (i.e. fat standardised and/or flavoured) may be described as follows. Clarified milk (at 5°C) is heated in a UHT Stork

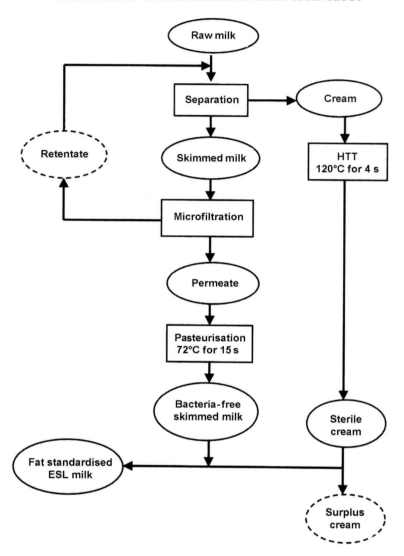

Figure 3.12 Microfiltration of market pasteurised milk. HTT, high-temperature treatment; ESL, extended-shelf-life. Source: Bøejgaard (personal communication). Reproduced by courtesy of APV Nordic, Aarhus, Denmark.

indirect tubular steriliser to 75°C (in the regeneration section), homogenised and heated to 120°C in the second part of the regeneration section. In the heating section, the product is further heated to 138–140°C for 2 s, and cooled to 30°C before bottling. The heat exchanger is of the double-tube or triple-tube

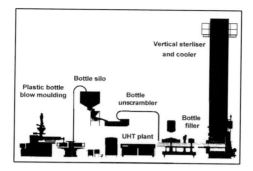

Figure 3.13 Two-stage in-plastic-bottle sterilisation plant. UHT, ultra high temperature. Reproduced by courtesy of Stork Food & Dairy Systems B.V., Amsterdam, The Netherlands.

configuration. The second heating stage takes place in a Stork hydromatic continuous steriliser, which is constructed as a vertical tower ca. 11–15 m high. The tower is divided into four sections where the filled bottles are transported through two endless chains, upwards and downwards through each section. The temperature profile of such a steriliser is as follows:

- the first section is a pre-heater, where the temperature is increased from 30°C to 85–95°C; this section acts as an inlet lock;
- the second section is the sterilising compartment; the product is heated by means of a steam–air mixture to 117–123°C for 10–12 min depending on the type of container used (see below), and achieves the correct sterilisation level (the F_0 value is 1);
- the third section is a pre-cooling unit, where the pressure is gradually reduced to cool the product, and, on the downward flow of the bottles, cooling continues by spraying water;
- the last section cools the processed milk to 35–40°C by spraying water on the bottles.

Temperature profiles for two-phase heating of milk for bottle sterilisation are shown in Figure 3.14.

Plastic bottles are usually made of high-density polyethylene (HDPE) which can melt and soften at temperatures between 124°C and 127°C, and hence the optimum sterilisation temperature of 120°C is applied. However, at this temperature the plastic container softens and is unable to withstand any internal pressure. Because air is enclosed in the headspace of the container, during the heating stage the total pressure is equal to the sum of the partial air and partial vapour pressures. Thus, if the sterilising section of the tower operates under saturated steam pressure only, the plastic containers will blow up or be badly deformed. Therefore, to minimise such an effect, a mixture of steam and air

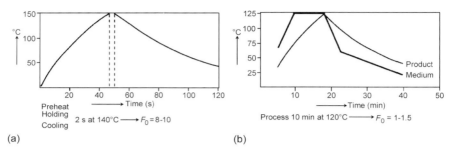

Figure 3.14 Illustration of temperature profiles (two-phase process) during the in-bottle sterilisation of milk: (a) first phase (in flow); (b) second phase (in-bottle). Reproduced by courtesy of Stork Food & Dairy Systems B.V., Amsterdam, The Netherlands.

is used to provide an overall pressure between 0.03 and 0.05 MPa higher than the pressure of saturated steam at the same temperature. This allows the air to expand inside the container during heating without distortion of the shape of the bottle. As a consequence, an increased height of hydromatic column is required for in-plastic-bottle sterilisation of milk.

Glass bottles are sealed with use of crown corks, and plastic bottles are heat-sealed with aluminium foil laminates. The bottles (plastic or glass) emerging from the steriliser are conveyed to a labelling unit, packaged in crates or cartons and, finally, placed in pallets. A highly mechanised post-heat-treatment production line may include:

- bottles (to be crated, stacked and placed in pallets)
- cardboard cases and/or trays (including erectors, robot packers, in-place packers, slide loaders, gluing and sealing units, and plastic wraparound systems)
- pallets (including a robot palletiser and pallet transporter)

3.4.7 Ultra high temperature (UHT)

The primary objective of any type of UHT processing method is to achieve so-called commercial sterility. Usually, the temperature used ranges between 135°C and 150°C and is applied for only a few seconds. The main options for UHT are indirect heating or direct heating by steam injection or infusion. In the former method, heat exchangers (plate, tubular or scraped-surface) are used for heating and cooling purposes. However, in the direct system, the UHT processing plants are equipped with: (a) a special steam injection nozzle through which steam is injected into the milk or steam infusion chamber (i.e. the product is introduced into a vessel filled with steam); and (b) an expansion chamber under vacuum to remove the equivalent amount of water to that added to the milk as a result of the steam injection. The steam infusion method may vary from one equipment manufacturer to another, and, according to Bøejgaard (personal communication), the APV Nordic UHT processes are categorised as steam infusion, high heat infusion and instant infusion systems (Anon., 1999).

All types of UHT plants (direct or indirect) are widely used for the manufacture of sterile market milk, including flavoured milks. Scraped-surface heat exchangers (SSHEs) and the instant infusion types are more suitable for processing highly viscous products containing large particles or for sticky products, such as whey protein concentrates or egg white; these plants are not discussed in this chapter. Nevertheless, all UHT plants are fully automated (Bylund, 1995) and consist of four main operational modes: (a) plant pre-sterilisation; (b) production; (c) aseptic intermediate cleaning (AIC); and (d) cleaning-in-place (CIP). Other aspects that have to be considered in a UHT plant to ensure the sterility of milk are as follows.

- Temperature sensors may be used similar to those used in a pasteurisation plant, but the transmitter and controller are different to provide a quicker response in a UHT plant. The heating control system, for example in a Tetra Pak UHT plant, is a cascade type, where the temperature of the hot water is monitored and controlled against a set point supplied by the hot product temperature sensor. This system gives a fast response to any temperature change in the raw milk supply to the plant and will also compensate for the loss of efficiency arising from fouling on the product heating surfaces. Fouling is caused by a gradual build-up of aggregated milk constituents on the heating surfaces. This build-up then acts as an insulator. The corresponding drop in product temperature is monitored and the signal controls the hot water set point. The overall accuracy of such a temperature sensor plus its electronic control should be ca. $\pm 1°C$. However, the accuracy of the individual temperature sensors should be equal to or better than $\pm 0.3°C$ (Gladwell, personal communication).
- An aseptic homogeniser is required for downstream processing of the milk.
- Plant pressure is higher (ca. 0.4 MPa) so that the milk does not boil at 135–150°C.
- Interlocks in the control programming of plant operation provide greater security against operator errors and tampering with the process; for example, it will be impossible to start production if the plant is not sterilised properly.

Different designs and types of UHT plant used to heat the milk will ultimately give different temperature profiles for the product during processing (Figure 3.15). Brief descriptions of the APV Nordic UHT systems are given in the next sections.

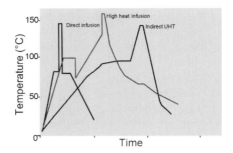

Figure 3.15 Temperature profiles of different ultra heat-treatment processes. UHT, ultra high temperature. Source: Anon. (1999). Reproduced by courtesy of APV Nordic A/S, Aarhus, Denmark.

3.4.7.1 Indirect ultra high temperature plant

Indirect heat exchangers (either plate or tubular types) heat the incoming raw milk from 5°C to 75°C initially in a product-to-product regeneration section. The product is then homogenised at a pressure up to 25 MPa and further heated to 90°C for a few seconds to stabilise the protein. The milk is then sterilised at 138°C in a holding tube for a few seconds (Figure 3.16). Finally, cooling takes place in two stages to less than 25°C, and the product leaves the plant directly to an aseptic filler or to an aseptic tank for intermediate storage. If any temperature drop occurs during production, the milk is diverted to be reprocessed, but in contrast to pasteurisers (see Section 3.4.3) the plant is cleaned and sterilised before the process is restarted; this precaution is also applicable to any type of UHT plant.

The flowchart of an indirect UHT tubular system is similar to that of the plate type, but there are significant differences in performance, as outlined in Table 3.4.

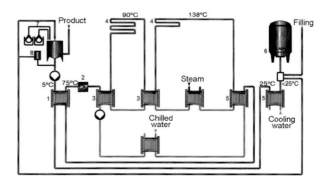

Figure 3.16 Flow diagram of a plate-type ultra high temperature system. 1, Product-to-product regeneration system; 2, homogeniser; 3, indirect heater; 4, holding tubes; 5, indirect cooler; 6, sterile tank; 7, clearing-in-place unit; 8, sterilising loop. Source: Anon., 1999. Reproduced by courtesy of APV Nordic A/S, Aarhus, Denmark.

Table 3.4 A comparison of the plate type and tubular type ultra high temperature systems

	Plate type	Tubular type
Energy recovery	High	Medium
Plant volume at 90% regeneration	Low	High
Product shear at equivalent heat transfer	Low	High
Heat transfer at equivalent surface	High	Medium
Tolerated pressure drop (MPa)	1.5	6.2
Running time (h)	12	20

Some indirect UHT plants, for example the Tetra Pak system, are designed for variable capacity ranging 50–100% of the nominal amount and are connected directly to the aseptic filler. However, in case one of the filling machines stops, and to avoid overprocessing of the milk, the heating section in the PHE is divided into two units. This is known as 'split' heating (Bylund, 1995). If milk flow drops by 50%, a valve is activated so that the heating medium passes outside the first heating section, where the temperature of the homogenised milk is maintained at 75°C until it reaches the second heating section, where the milk is heated to 138°C.

3.4.7.2 Direct UHT plant

The essence of a direct UHT plant is steam injection into the milk through a specially designed nozzle (e.g. the APV Nordic design shown in Figure 3.17 or the ring-nozzle steam injector marketed by Tetra Pak; Bylund, 1995). The raw milk at ca. 5°C is supplied from the balance tank and heated to 75°C in a plate or tubular heat exchanger. The milk pressure is increased to 0.4 MPa by a pump before steam is injected into the product. This instantly raises the temperature to 143°C. Immediately afterwards, the milk passes through a holding tube for a few seconds before it is flash cooled in a condensor equipped with a vacuum expansion chamber. The vacuum is controlled to balance removal of vapour from the product equivalent to the amount of steam injected into the milk. An aseptic centrifugal pump then feeds the ultra high temperature (UHT) treated milk at 75°C to a two-stage aseptic homogeniser before the milk is cooled to less than 25°C and is filled into an aseptic tank.

Another type of direct UHT system is the steam infusion steriliser (Figure 3.18). The incoming raw milk at 5°C is heated to 75°C in a conventional regeneration section before it is heated to 143°C in a steam infusion chamber. The milk is heated accurately and gently in the chamber with precise holding time. The latter aspect is achieved because the infusion system is designed with a special pump mounted directly below the infusion chamber. The pump ensures sufficient overpressure in the holding tube in order to achieve a single phase flow free from air and steam bubbles. At present, there is a tendency to process milk at a reduced residence time (i.e. few seconds) to minimise chemical degradation of

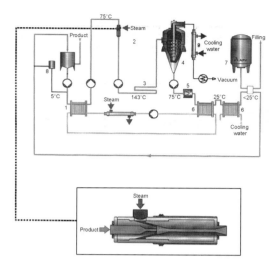

Figure 3.17 A typical illustration of direct steam injection ultra high temperature system. 1, Plate pre-heaters; 2, steam injection nozzle; 3, holding tube; 4, flash vessel; 5, aseptic homogeniser; 6, plate coolers; 7, aseptic tank; 8, sterilising loop; 9, condenser. Source: Anon. (1999). Reproduced by courtesy of APV Nordic A/S, Aarhus, Denmark.

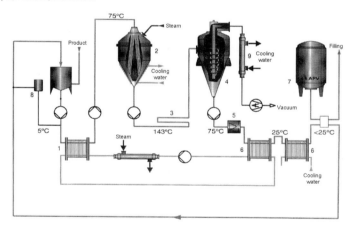

Figure 3.18 Flow diagram of a steam infusion ultra high temperature system. 1, Plate pre-heaters; 2, steam infusion chamber; 3, holding tube; 4, flash vessel; 5, aseptic homogeniser; 6, plate coolers; 7, aseptic tank; 8, sterilising loop; 9, condenser. Source: Anon. (1999). Reproduced by courtesy of APV Nordic A/S, Aarhus, Denmark.

the product. The concept is illustrated in Figure 3.19. Factors that can affect the residence time are: (a) dimension of the holding tube; (b) fluid flow behaviour (turbulent or laminar); (c) foaming; (d) air content; and (e) steam bubbles (Anon., 1999).

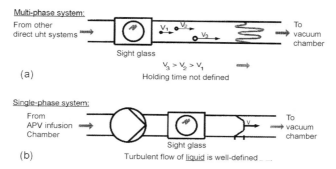

Figure 3.19 Flow behaviour of milk in the holding tube of an ultra high temperature (UHT) plant that can affect residence time: (a) holding tube without a centrifugal pump; (b) holding tube with a centrifugal pump. V_1–V_3, velocity in tube. Source: Anon. (1999). Reproduced by courtesy of APV Nordic A/S, Aarhus, Denmark.

After the infusion chamber, milk at 143°C passes through a holding tube for a few seconds before entering the aseptic flash vessel under vacuum to remove the added water. As a result, the milk temperature drops to 75°C. It is then homogenised (e.g. 20–25 MPa by using an aseptic model), followed by progressive cooling to less than 25°C before being packaged aseptically. An alternative process, which is also supplied by APV Nordic, is called the Pure-Lac™ infusion system. The plant design is similar to that shown in Figure 3.18 except the holding tube (item 3) is replaced with a PTT™ unit (i.e. a loop between the pump inlet and what was the end of the holding tube, which accurately controls the preset time–temperature combination of the processing system; Henyon, 1999; Kjoerulff, 2000). Furthermore, in the Pure-Lac™ system the milk is heated in the infusion chamber from 70 to 145°C in less than 1 s, and the final temperature is reached in 0.2 s. The heating rate is around 500–600°C s^{-1}. The cooling jacket located at the bottom of the infusion chamber eliminates fouling and possible foam by means of a constant condensate film. Fouling can also be avoided by ensuring that the initial Reynolds number of the milk jet is greater than 10 000 (de Jong et al., 1994).

The high-heat steam-infusion UHT system is a direct heating method to inactivate heat-resistant spore-formers in milk. The temperature required to achieve commercial sterility of the product ranges 145–150°C for 3–10 s. If a conventional steam infusion plant is used, energy consumption is excessive and the capital cost to heat the milk to 150°C for 10 s is prohibitive. The high-heat infusion system (marketed by APV) is shown in Figure 3.20. The milk is heated to 90°C for few seconds before entering the flash vessel (nonaseptic) under vacuum to remove an amount of water equivalent to the added steam in the steam infusion chamber *before* steam infusion. As a consequence, the temperature of the partially concentrated milk drops to 60°C, but is then followed by heating to 125°C in a tubular heat exchanger before entering the steam infusion chamber

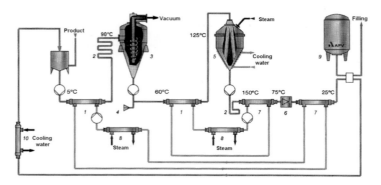

Figure 3.20 Flow diagram of a high-heat infusion ultra high temperature (UHT) system. 1, Tubular pre-heaters; 2, holding tube; 3, flash vessel (nonaseptic); 4, nonaseptic flavour dosing (option); 5, steam infusion chamber; 6, homogeniser (aseptic); 7, tubular coolers; 8, tubular heaters; 9, aseptic tank; 10, nonaseptic cooler. Source: Anon. (1999). Reproduced by courtesy of APV Nordic A/S, Aarhus, Denmark.

to heat the milk to 150°C. The milk then passes through a holding tube (e.g. up to a 10 s residence time), is cooled to 75°C, homogenised at 20–25 MPa (aseptic type) and finally cooled to less than 25°C before being filled into an aseptic tank.

Such plant design improves energy recovery in the regeneration sections to 75% compared with 40% and 80–85% with conventional steam infusion and indirect tubular systems, respectively. The time–temperature profiles of different UHT systems are compared in Figure 3.15, and the processing parameters of some UHT systems are shown in Table 3.5.

3.5 Recombination technology

It is estimated that around 75% of the world population lives in countries of limited milk production due to unfavourable climatic conditions and/or low income (Nielsen et al., 1999). Hence, these factors limit distribution of fresh milk and short shelf-life dairy products, even under refrigeration, because of the perishable nature of such products. Sterile and UHT dairy products overcome such problems, but undesirable changes in flavour and colour may occur. Recombination offers an alternative in these countries by using skimmed milk powder (SMP) and anhydrous milk fat (AMF) to manufacture a wide range of dairy products (IDF, 1990).

The process involves reconstituting SMP and blending AMF to yield a product with similar characteristics to those of fresh milk. A multitude of systems (batch or in-line) are available for recombination purposes (for further details refer to the reviews by Bylund, 1995; Nielsen et al., 1999; Tamime and

Table 3.5 Comparison of some processing parameters of different APV Nordic ultra high temperature (UHT) systems

	Indirect tubular models			Direct steam infusion	High heat infusion
	A	B	C		
Temperature profile (°C)	5 → 140 → 25	5 → 140 → 25	5 → 148 → 25	5 → 150 → 22	5 → 150 → 25
Holding time (s)	1	25	1	3	2
Bacteriological kill rate (F_0 value)	≤ 8	40	40	40	40
Regeneration (%)	85	85	80	45	up to 72
Operation time (h)	>16	>12	>8	>24	>24

Source: Bøejgaard (personal communication).

Robinson, 1999). A typical layout of milk recombination plant made by APV Nordic is shown in Figure 3.21. The unit for dissolution of powder is known as a Liquiverter and consists of a square and single-shell-type vessel supported by a stainless steel frame. It is equipped with a heavy-duty impeller disk driven by an electric motor at 1420 rpm. It is supplied with different sizes to provide capacities up to $30\,000\,l\,h^{-1}$. It is recommended to use low-heat skimmed milk powder and to rehydrate at 40–50°C for 20 min. The melted AMF is added to the recombined skim at a temperature of 40°C or above and is processed within 2 h of recombination because of risk of microbial growth.

An alternative method is to add the AMF in the recombined milk via a metering device just before entering the homogeniser (Bylund, 1995; Tamime and Robinson, 1999). The milk then passes through a duplex filter to remove undissolved or scorched particles before being pasteurised or ultra high temperature treatment. In the former (pasteurisation) system, the milk is pre-warmed to approximately 60°C in the PHE, de-aerated to remove air due to the recombination process, homogenised (i.e. first stage at 14–17.5 MPa and second stage at 3.5 MPa), heated to 72°C for 15 s, cooled to 4°C and packaged. However, sterilisation of the recombined milk (e.g. in-container at 120°C for 10 min or UHT at 137–150°C for a few seconds) is also possible, but the fault of age-thickening or gelation may occur in product stored at more than 27°C. This is a result of the survival of heat-resistant proteolytic enzymes (i.e. origin psychrotrophic microorganisms) during UHT processing and, to prevent such a fault, it is of paramount importance to produce powder from low-count milk around 24 hours old (Nielsen *et al.*, 1999).

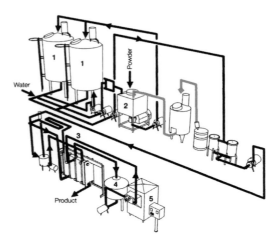

Figure 3.21 Layout of a milk-recombination plant. 1, Two mixing tanks; 2, Liquiverter; 3, plate heat exchanger; 4, de-aerator; 5, homogeniser. Source: Neilsen *et al.* (1999). Reproduced by courtesy of APV Nordic A/S, Aarhus, Denmark.

The process parameters for the control of the pasteurisation and sterilisation of market milk made from recombined dairy products are similar to those described earlier. However, in high-capacity plants, automatic powder handling is advisable to replace manual operation. Powder is received in bags (25 kg or 500–1200 kg), which are emptied and transported pneumatically to special silos; from there the powder is weighed and transferred to the Liquiverter to be rehydrated and processed (see Figure 3.21).

3.6 Packaging lines and storage

A number of different packaging options are available. The returnable glass bottle was the first form of packaging to be used with milk. At present, flexible plastic containers (i.e. 'pillows'), laminated paper board including a layer of aluminium foil, rigid plastic (bottles or cups) and metal cans are widely used for packaging market milks (pasteurised or sterilised). Although filling lines are highly automated, they are in fact semi-automated because certain packaging units (e.g. plastic caps, paper board and aluminium foil reels) have to be fed manually into the machine(s) to maintain a continuous run.

It is beyond the scope of this review to detail the automation of each type of packaging system, but specific aspects dealing with the packaging of milk and dairy products have recently been reported by the International Dairy Federation (IDF, 1995). It is safe to assume that volumetric and level filling are universal systems used in milk packaging lines.

The Tetra Brik Aseptic (TBA 1500S) unit is an example of an aseptic form–fill–seal cartoning machine including wraparound box and pallatising and is shown in Figure 3.22. The Tetra Brik Aseptic (TBA/8) filling machine is shown as item 1. All the functions, including the alarms, are controlled from a central processing unit which is mounted in the electrical cabinet (Matty, personal communication). The formation of the carton starts from a reel of paper board or aluminium foil coated on both sides with polyethylene; this material is referred to as a 'material web'. This passes through an automatic splicer and date print unit. The material web then passes to a pull-tab unit where a hole is punched; later, a plastic patch will be applied over this hole. A foil tab is applied to the outer sides and is heat sealed. At this stage, the temperature is accurately controlled with alarms that monitor: (a) a missing or misplaced patch or tab; and (b) temperature fluctuation. A plastic strip is attached to the right-hand edge of the reel by using pressure and hot air. Incidentally, such plastic strips ensure a proper seal on the opposite edge of the material at the longitudinal sealing stage or at the formation of the tube. Immediately after the plastic strip is applied, the web is dipped in a hydrogen peroxide (H_2O_2) bath and passes through rollers to remove any excess H_2O_2 before the flat paper board is made into a tube. Hot and sterile air evaporates H_2O_2 from the surface of the packaging material; the

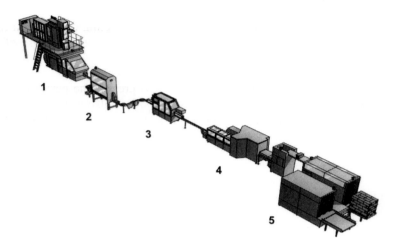

Figure 3.22 Tetra Brik Aseptic (TBA 1500S) complete packaging line including re-cap applicator and palletiser. 1, Filling machine (capacity 7500 lh^{-1}); 2, carton accumulator (666 packages per 8 min); 3, cap applicator; 4, cardboard packer, including system for applying a wraparound box and tray; 5, palletising system. Reproduced by courtesy of Tetra Pak Ltd, Uxbridge, UK.

process is monitored and controlled. At this stage, the sterile milk is delivered to the filling area via an aseptic product valve, with use of steam as a barrier in order to maintain sterile conditions. The forming of the unit container and the filling sequence are coupled to an electronic package-control system, but the filling method is electromechanical. A jaw system (i.e. a pair) is responsible for the formation of the package, particularly the tranversal seal, and is operated by using a high-frequency electrical system. Milk is then delivered to the container, the top is tranversally sealed (e.g. two seams are made) and cut to separate the carton from the tube. The mouth of the filling pipe is below the level of milk, to prevent frothing. The final fold unit (FFU) gives the carton its shape; at this stage, the side flaps are stuck down by using pressure and hot air.

The carton is discharged from the filling machine and passes, in sequence, to the carton accumulator unit, the plastic-cap applicator, cardboard packer, including a system for applying a wraparound box and tray, and the palletiser (see items 2–5, respectively, Figure 3.22); such a packaging line as provided by Tetra Pak can be fitted with an automated tool known as a packaging line monitoring system (PLMS) for measuring and analysing the performance of the filling line (Matty, personal communication). The data are automatically logged for the packaging line and are stored at the filling machine. The operator interacts with the multilanguage interface by means of manual data and fault-solving assistance. This system consists of an industrial programmable computer (IPC) known as Flexbox, which is located in the electrical cabinet, with an interface placed in the operator panel of the filling machine. A set of interfaces is also available for counting the cartons on the line and collecting information from the distribution equipment. Serial communication, together with use of the PLMS 'send' and 'receive' programs, allows the system to be connected to a remote PC (optional) for automatic transferring of collected data. The PLMS off-line analysis program is used to analyse and present the performance of the packaging line. However, an automatic printout is also available from the system, and remote access of the collected data is possible by means of the PLMS 'send' and 'receive' options, modem and telephone line (Matty, personal communication).

Ultimately, the (bottled) packaged milk (pasteurised or sterilised), held in stacked crates, a metal trolley and/or in pallets is transported to the store (at a temperature of less than 4°C or at 20°C). In large plants, automated robotic handling systems are widely used to store and retrieve the packaged milk products. The system is based on first-in, first-out principles; also, the retrieval of products is based on customers orders. This system is also used for fermented milk, and a description of such a robotic system is detailed in Chapter 6.

3.7 Statistical process control

The data-gathering processes in different dairies have much in common (see Chapter 1) and information may be generated from: (a) milk reception and purchases of added ingredients, including packaging materials; (b) production areas; (c) packaging and storage; (d) cleaning-in-place; (e) services; and (f) administration and marketing. It is evident that analysis of such information helps to manage and run the business efficiently: process management can also be improved, however, by use of statistical process control.

Modern automated processes record and store large quantities of data. This information is usually transient. It is examined by the plant operators on a day-to-day basis but is then archived and, unless a problem occurs with a product, may never be referred to again. Nevertheless, these measurements often contain useful information that can be unravelled by simple statistical procedures. This treatment of data is called statistical process control and, for a detailed exposition, the reader is referred to Oakland and Followell (1992).

In practice, the data are used to estimate two components—the mean and the standard deviations (σ) of the measurement as a function of time. The standard deviations are used to compute the warning limits ($\pm 2\sigma$), and the 'out-of-control' or action limits ($\pm 3\sigma$). A practical application of these principles is in the control of homogenisation efficiency. Simulated data are used to illustrate this process. Over a period of time, homogeniser valves are subject to wear, and early detection of this avoids deterioration in product quality. Close inspection of the individual measurements from day to day does not readily identify trends (Figure 3.23; runs 1–50). In this case, the specific surface area (SSA) of the

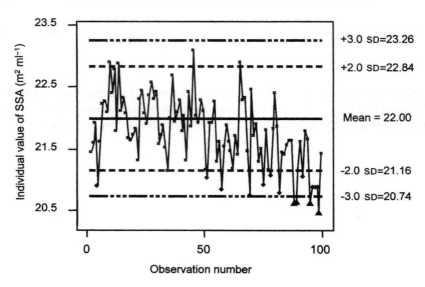

Figure 3.23 Control chart of specific surface area (SSA) of fat globules in homogenised milk. SD, standard deviation; ♦, warning signal attained; ▲, out of control; - - -, warning level; — - - -—, action limit.

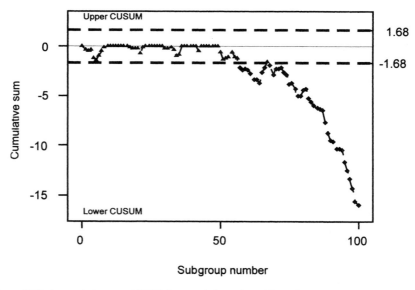

Figure 3.24 A cumulative sum (CUSUM) control chart of specific surface area (SSA) of fat globules in homogenised milk.

fat globules in very efficiently homogenised milk has a mean value, over 50 consecutive runs, of 22.00 m^2 ml^{-1}, and a standard deviation of 0.42 m^2 ml^{-1}. Over the period shown, the process is 'in control', that is, variations are random. In four cases (over the first 50 runs), warning signals were attained, but the action limits were never attained. Also shown in Figure 3.23 are the data for the next 50 simulated runs. A slow deterioration of homogeniser efficiency can be seen to be taking place (see Figure 3.23). Two alarm points are set: process out of control ($\pm 3\sigma$), and 9 consecutive data points fall lower than the centre (mean) line. As can be seen in Figure 3.23, an alarm occurs at runs 62, 63 and 78 and very regularly after run 88. In addition, by run 73 warning limits start to occur regularly, and by run 88 the process is clearly out of control. The control chart shown in Figure 3.23 is not very sensitive, and when deterioration of the efficiency of a process is anticipated a different approach can be adopted. For example, the cumulative sum (CUSUM) chart (Figure 3.24) gives a much clearer signal. These plots are the cumulative sums of the deviations of each sample value from the mean value. It is clear that from run 58 there is a drift downwards in the SSA, and by run 70 action needs to be taken to avoid deterioration of product quality. This might take the form of refacing the existing homogeniser valves or by replacing them with new valves.

Control charts are a powerful tool and can identify process drift or out-of-control conditions. Data processing can be carried out by using readily available and user-friendly software packages. The examples shown here were analysed by means of Minitab (version 13, Minitab Ltd; e-mail: sales@minitab.co.uk).

Acknowledgement

The Hannah Research Institute (D. Donald Muir) and the Scottish Agricultural College (Adnan Y. Tamime) receive funding from the Scottish Executive Rural Affairs Department (SERAD).

References

Anon. (1992) On-line standardisation of milk. *Milk Industry*, **94**(2) 22-6.
Anon. (1999) *Technology Update: Long Life Dairy, Food and Beverage Products*, Technical Bulletin 0999, APV Nordic (Unit Systems), Silkeborg, Denmark.
Burton, H. (1988) *Ultra-high-temperature Processing of Milk and Milk Products*, Elsevier Applied Science, London.
Bylund, G. (1995) *Dairy Processing Handbook*, Tetra Pak Processing Systems A/B, Lund, Sweden.
Fox, P.F. (ed.) (1989) Monograph on heat-induced changes in milk. *Bull. Int. Dairy Fed.*, **238**.
Fox, P.F. (ed.) (1991) *Food Enzymology*, Vol. 1, Elsevier Science Publishers, London.
Fox, P.F. (ed.) (1995) Heat-induced changes in milk. *Special Issue Int. Dairy Fed.*, **9501**.
Harvey, W.C. and Hill, H. (1967) *Milk: Production and Control*, 4th edn, H.K. Lewis, London, pp. 363-86.
Haugaard, G. and Pettinati, J.D. (1959) Photometric milk fat determination. *J. Dairy Sci.*, **42** 1255-75.

Henyon, D.K. (1999) Extended-shelf-life milks in North America: a perspective. *Int. J. Dairy Tech.*, **52** 95-101.
Holdsworth, S.D. (1992) *Aseptic Processing and Packaging of Milk*, Elsevier Science Publishers, London.
IDF (International Dairy Federation) (1990) Recombination of milk and milk products. *Special Issue, Int. Dairy Fed.*, **9001**.
IDF (International Dairy Federation) (1995) Technical guide for the packaging of milk and milk products. *Bull. Int. Dairy Fed.*, **300**.
de Jong, P., Waalewijn, R. and van der Linden, H.J.L.J. (1994) Performance of a steam infusion plant for heating milk. *Netherlands Milk and Dairy J.*, **48** 181-99.
Kjoerulff, G. (2000) Innovative UHT systems for process safety and product quality. *Scandinavian Dairy Information*, **14**(1) 18-20.
Nielsen, W.K., Bøejgaard, S.E. and Vesterby, P. (1999) *Technology Update: Recombined Dairy Products*, technical bulletin 0499, APV Nordic (Dairy), Aarhus, Denmark.
Oakland, J.S. and Followell, R.F. (1992) *Statistical Process Control: A Practical Guide*, 2nd edn, Butterworth Heinemann, Oxford.
Tamime, A.Y. and Robinson, R.K. (1999) *Yoghurt Science and Technology*, 2nd edn, Woodhead Publishing, Cambridge, pp. 61-71.
Walstra, P., Geurts, T.J., Noomen, A., Jellema, A. and van Boekel, M.A.J.S. (1999) *Dairy Technology: Principles of Milk Properties and Processing*, Marcel Dekker, New York.
Wamsler, C. (1996) Liquid milk with extended shelf-life. *Scandinavian Dairy Information*, **10**(4) 15-7.

4 Concentrated and dried dairy products

P. de Jong and R.E.M. Verdurmen

4.1 Introduction

Market milk, and fresh dairy products are highly perishable, and it is desirable to preserve their nutritional components for later consumption. To extend the limited shelf-life of milk products, concentration and drying are among the most important techniques. Another advantage of water-removing techniques is the decrease in cost of storage and transportation by reduction of product volume. The disadvantage, however, is that energy consumption is high; no other process in the dairy industry has such a high energy demand per tonne of finished product. This is because approximately 90 g 100 g^{-1} of milk is water, and all that water must be removed by heat. The removal of water usually takes place in two stages. The first stage is concentration by vacuum evaporation, and the second stage is drying; 90 g 100 g^{-1} of the water is removed in the evaporator and only 9–10 g 100 g^{-1} in the spray dryer. However, the energy required per kilogram of water evaporated in the dryer is 15–20 times the energy required per kilogram of water removed in the evaporator.

Besides the processing of milk products, an important criterion for preservation by concentration and drying is the quality of the (recombined) product. Modern technologies are focused on minimising the loss of nutritive value and on improvement of microbiological quality. Nowadays, optimal design by means of predictive processes, product models and advanced automation are the ingredients to produce high-quality products for the food market.

4.1.1 Evaporation

In the dairy industry, falling-film evaporators are commonly used. The heat exchange surface consists of a bundle of vertical tubes known as 'calandria'. A schematic illustration of a modern falling-film evaporator plant for concentration of milk products is shown in Figure 4.1. In practice, a large number of different evaporator configurations are used in the industry. For example, the number of evaporation stages (or 'effects') varies from one to seven. The actual configuration depends on the desired properties of the concentrate and on the state-of-the-art at the time of installation of the evaporation plant.

In order to obtain a high thermal efficiency, the products are sometimes heated first in spiral tubes placed in the condenser and the evaporators. Before the milk enters the evaporator effects, it is pre-heated to a temperature above the

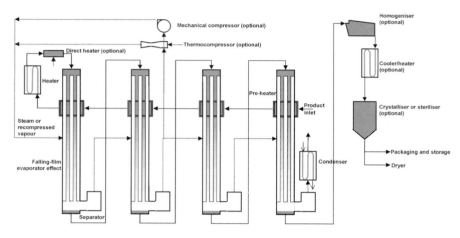

Figure 4.1 Schematic representation of industrial configurations of falling-film evaporators.

temperature of the first effect. In general, pre-heating has a significant impact on the properties and quality of the concentrate and powder produced. To meet certain quality standards, it may be necessary to use a direct heating system (e.g. steam injection, steam infusion) to apply a high-temperature short-time treatment (HTST).

The milk entering the falling-film evaporator is distributed (e.g. by nozzle or distribution plate; Figures 4.2 and 4.3) over the bundle of evaporation tubes. The liquid 'falls' as a film through the inside of the tube. On the outside of the tube steam is condensing. The use of water evaporation, which usually takes place below 70–80°C, is based on the physical law that the boiling point of a liquid is lowered when that liquid is exposed to a pressure below atmospheric pressure. The vapour is separated from the product in a separator placed at the base of each evaporator effect and is used as the heating medium for the next effect. From the last effect, the vapour goes to the condenser. The boiling temperatures vary from 70–80°C in the first effect to 40–50°C in the last effect.

It is common knowledge that by increasing the number of vacuum units (effects) the energy consumption is decreased (Bouman *et al.*, 1988). For example, in the case of four effects 1 kg of steam results in 3 kg water evaporation. Thermal vapour recompression (TVR) will also decrease the consumption of steam. In the thermocompressor, steam is introduced through a nozzle creating a steam jet in the mixing chamber whereby vapour from the separator is sucked into the mixing chamber. Therefore, the application of a thermocompressor (Figure 4.1) results in an energy consumption of 1 kg steam per 5 kg water evaporation.

A more energy-efficient way of achieving recompression is by the application of mechanical vapour recompression (MVR) in which, in contrast to TVR, all the vapour is recompressed. Usually, the MVR evaporator consists of only one or

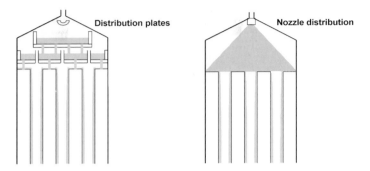

Figure 4.2 Liquid distribution systems used at the top of an evaporator effect.

Figure 4.3 The distribution of milk in a falling-film evaporator (demonstration model, NIZO food research).

two effects, and the boiling temperature can be chosen depending on the desired product properties. Apart from the steam used for start-up, an MVR evaporator requires no steam and no cooling water. Modern MVR evaporators use a heavy-duty fan instead of a relatively complex compressor. This has resulted in diminishing investment costs and nowadays most of the new evaporators use MVR.

Depending on the desired product properties, the concentrate from the last effect is homogenised, heat-treated and/or crystallised. When the concentrated milk is to be used for powder production, it is transported to a balance tank.

4.1.2 Drying

In the dairy industry, conversion of concentrate into powder is brought about mainly by spray drying. Spray drying is a relatively gentle process that has replaced the cheaper, but also the more product-denaturing, drum or roller dryers. Moreover, spray drying makes it possible to manufacture powder with varying specifications for different applications and quality standards.

A schematic illustration of a multi-stage dryer is shown in Figure 4.4. In practice, a spray dryer can consist of one, two or three stages. In the first stage, the pre-heated product (at a temperature of less than 100°C) is sprayed by atomisation into a chamber filled with circulating hot air (Figure 4.5). The outlet temperature of the air ranges between 150°C and 250°C. By atomisation, the concentrate is converted into droplets of 10–200 μm. In the dairy industry, two types of atomisation system are used: a stationary pressure nozzle and a rotating atomiser. Droplets flow down inside a tower and adsorb the heat necessary to evaporate the moisture, which is removed by hot air. Depending on the dimensions of the tower, the residence time of the powder particles is in the order of 5–30 s. The dried powder falls to the bottom of the dryer, and is transported to the next drying stage or to a packaging system. The exhaust air is removed through an outlet duct, and passes through cyclones and filters where small powder particles (fines) are removed. The fines can be recycled to the top of the dryer or to other drying stages. The result is an agglomerated powder.

The first processing stage is performed in the dryer tower. For subsequent stages fluid-bed dryers are generally used, both internal to the drying tower and external. In the fluid-bed dryer a powder layer of a defined height is formed by flowing hot air (at a temperature lower than that of the tower). In certain cases

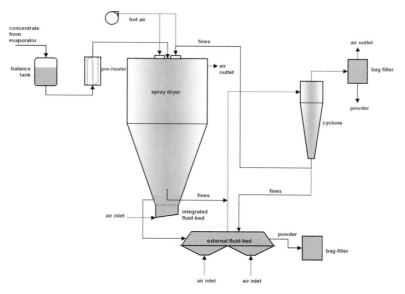

Figure 4.4 Schematic illustration of industrial configurations of spray dryers (single-, two- and three-stage).

Figure 4.5 The cone of an industrial spray-dryer (stork) with mechanical hammers to remove deposits.

a fluid-bed dryer is used to achieve product transformations, for example, to achieve lactose crystallisation in whey powder. In the final part of the external fluid-bed dryer, the powder is cooled to the packaging and storage temperature.

4.2 Product and process technology

4.2.1 Evaporated and dried products

The technology (process operation, packaging and storage) applied in the manufacture of evaporated and dried products is dependent on the type of product. The main types are:

- condensed milk
- sweetened condensed milk
- milk powder
- whey powder
- high-protein powders
- buttermilk powder
- caseinate powder

4.2.1.1 Condensed milk

For production of condensed milk, evaporated milk (i.e. the concentrate from the evaporator) is first homogenised in at least a two-stage homogeniser (12.5–25 MPa). The homogenisation conditions are determined mainly by the minimum stability of the protein and the emulsion. Inferior stability results in coagulation of the fat globules during sterilisation. After homogenisation, the product is cooled to 5–15°C. To control the stability of the condensed milk, a sample is sterilised (15–20 min at 110–125°C). The stability is adjusted by adding stabilising agents. The product is then packaged in bottles or cans and sterilised. Finally, the product is cooled and transported to a storage room. An alternative to in-bottle sterilisation is an ultra high temperature (UHT) (approximately 2–5 s at 138–142°C) before packaging. In this case, an aseptic filler system is required.

Important quality parameters for condensed milk are: colour, viscosity, taste, pH level and microbiological quality.

4.2.1.2 Sweetened condensed milk

Sweetened condensed milk is produced by adding sugar to the milk either before pre-heating and evaporation or during evaporation. In the former method, the sugar will be involved in the protein denaturation reaction and will have a greater impact on the viscosity of the stored product. After the evaporation stage, the product is cooled to about 30°C. During cooling, crystallisation of lactose occurs. This is a critical stage of the process because crystals with diameters greater than 10 μm may result in a sandy texture. The mean diameter of the crystals formed during cooling can be controlled by maintaining the shear stress above a certain level; this may be achieved by mixing, for example. Another means of controlling crystallisation is to seed the condensed milk with lactose crystals. Finally, the product is usually filled into cans and then stored.

Important quality parameters for sweetened condensed milk are size of lactose crystals, colour, viscosity (including age thickening), taste, pH level and microbiological quality.

4.2.1.3 Milk powder

Milk powder properties are governed by the pre-heating stage before evaporation. For example, with skimmed-milk powder, the minimum heat load is determined by the activity of phosphatase, whereas for whole milk powder the pre-heating is aimed at the inactivation of the lipase, which demands a higher heat load. In the evaporator, the milk is concentrated to a dry matter content of 45–53 g 100 g^{-1}. The configuration of the spray dryer is largely dependent on the product's desired properties, such as solubility. For example, instant milk powder should have excellent solubility properties, and, to obtain such an effect, a two- or three-stage dryer using fluid-beds is necessary for the production of powder consisting of large and porous agglomerates. After cooling, the powder is generally packed in multi-layer paper bags with a polyethylene liner. This type of package is permeable to oxygen but impermeable to moisture. In the case of whole milk powder, oxidation may occur during storage, and it is recommended that the product be packaged in an atmosphere flushed with inert gas.

Important quality parameters for milk powder from condensed milk are: moisture content, heat classification (low, medium or high, which is determined by the degree of protein denaturation required, insolubility index, the presence of scorched particles, colour, bulk density and microbiological quality.

4.2.1.4 Whey and high-protein powders

Whey and high-protein powders should have their proteins in an undenatured form and all the lactose in a crystalline form. To achieve these properties, preheating before evaporation should be gentle, and the lactose should be converted into crystals before drying. To achieve a nonhygroscopic and nonsticky powder, β-lactose must as far as possible be transformed (muta-rotated) into α-lactose. α-Lactose crystallises as a monohydrate, which is a nonhygroscopic form. The optimal temperature for muta-rotation is about 30°C. To achieve 85% crystallisation of the lactose, the concentrate from the evaporator is crystallised by cooling in special tanks. An agitator in the tank controls the degree of crystallisation and the viscosity of the concentrate. In the dryer, the outlet temperature should be below the sticking point of the powder (60–80°C). To obtain high-quality whey powders, a secondary crystallisation stage takes place after spray drying in the external fluid-bed dryer at a temperature of 50–60°C.

Important quality parameters for whey powder are moisture content, heat classification (low, medium or high), determined by the degree of protein denaturation, solubility (or insolubility index), package volume, lactose crystal size, and microbiological quality.

4.2.1.5 Buttermilk powder

Buttermilk powder products can be classified as sweet or sour. The sweet type is processed like skimmed milk powder, but sour buttermilk (a by-product of sour cream butter) has a high level of lactic acid and can be evaporated to a maximum of 28 g 100 g^{-1} of solids as the concentrate becomes very viscous. To reduce the viscosity, the configuration of the evaporator is designed in such a way that the last effect operates at a high temperature. Also, spray-drying is difficult because of the stickiness and hygroscopicity of the product. To avoid large amounts of deposits in the dryer a low temperature should be used, with frequent cleaning of the apparatus.

4.2.1.6 Caseinate powder

Caseinate powder is made from acid casein, which is derived from skimmed milk and blended with water and NaOH. The liquid sodium caseinate has a dry-matter content of approximately 20 g 100 g^{-1} before it is spray-dried to a moisture content of about 5 g 100 g^{-1}. Caseinate is a water-soluble protein; the main problem in the processing of this product is the exponential increase of the viscosity at a dry-matter content higher than 20 g 100 g^{-1} and with a pre-heating temperature above 90°C. As a consequence, operation costs are relatively high.

4.2.2 Process design and operation

The process design and operation of evaporators, including spray dryers, are governed by the deposition of product components. These may be proteins, minerals, fat and microorganisms. Since there is a clear relationship between deposition, equipment design, product quality and production costs, mechanisation and automation in the dairy industry are highly related to the various type(s) of deposit.

4.2.2.1 Microbial fouling

In evaporation and drying, the method and time of production are highly dependent on the bacterial count in the raw materials and the final bacterial count in the evaporated or dried milk product. Negative effects of a bacterial load (e.g. *Streptococcus thermophilis*, *Bacillus stearothermophilus* subsp. *calidolactis*, *Bacillus licheniformis*, *Thermus thermophilus*) include: an inferior-quality product (in terms of stability and safety) and nitrite formation in the evaporator.

The main cause of increase of bacterial count with operation time is the adherence of bacteria and spores to the surface walls of the equipment when operated at temperatures below 80°C (de Jong *et al.*, 1999). A coverage of less than 2% may have a large impact on product quality (Bouman *et al.*, 1982). Figure 4.6 illustrates the mechanism of adherence, growth and release of bacteria from equipment surfaces.

A complete evaporation and drying plant can be considered as a cascade of model reactors. In each reactor the growth, adherence and/or destruction of bacteria can occur. Two mass equations form the basis of the proposed model: one for the wall to which bacteria may adhere and a second for the liquid. The bacterial growth as a function of the operating time, t, at position, x, on the wall in a tubular plug flow reactor is defined by the transfer equation as:

$$\frac{dn_w}{dt} = \mu_T n_w (1 - \beta) + k_a C \qquad (4.1)$$

where, n_w is the wall coverage [colony forming units (cfu) m^{-2}], μ_T is the bacterial growth rate at temperature T (1 s^{-1}), β is the fraction of generated bacteria which is released into the bulk and k_a is the adhesion constant (m s^{-1}).

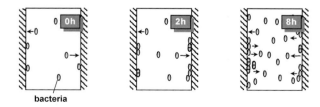

Figure 4.6 Adherence, growth and release of bacteria: one of the main causes of contamination in evaporators.

The local bulk concentration, C, (cfu m^{-3}) at operating time, t, follows from the component (bacteria) equation:

$$\frac{dC}{dx} = \frac{\pi D}{\phi}(\beta \mu_T n_w - k_a C) + \frac{\pi D^2}{4\phi}(\mu_T - k_d)C \qquad (4.2)$$

where, ϕ is the milk flow (m^3 s^{-1}), k_d is the destruction constant (1 s^{-1}) and D the hydraulic diameter of the reactor (m). In the case of a tank reactor, the concentration is independent of position x:

$$\frac{dC}{dt} = \frac{4\phi}{\pi D^2 L}(C_{in} - C) + \frac{4}{D}(\beta \mu_T n_w - k_a C) + C(\mu_T - k_d) \qquad (4.3)$$

where, L is the liquid level in the tank and C_{in} is the inlet bacteria concentration. Since the wall temperature is a function of position x in the heat exchanger, the differential equations (1), (2) and (3) must be solved numerically in parallel.

Figure 4.7 shows the effect of adherence of microorganisms to the walls of equipment on product contamination (industrial conditions). The predicted concentration of thermophilic bacteria in whey after pre-heating the milk before it enters the evaporator is shown as a function of operating time. In the case where no adherence or growth occurs (the lower line, consisting of short dashes in Figure 4.7), the outlet concentration will be nearly two log$_{10}$ cycles lower than the concentration in the raw material as result of pasteurisation. If the growth of bacteria in the whey (product) phase is taken into account, the upper line (consisting of longer dashes in Figure 4.7) will give the actual situation. Figure 4.7, however, shows clearly that the adherence of bacteria is a major factor in describing the actual increase of the bacterial load with operating time.

Important design factors determining the bacterial concentration arising from adherence are the temperature profile and the surface/volume ratio of the installation, and the local shear stress in the equipment. Predictive models may be applied with good results to minimise bacterial contamination (de Jong *et al.*, 1999).

4.2.2.2 Protein and mineral fouling

The operation time of equipment at temperatures above 80°C is determined largely by the deposition of proteins and minerals. Up to a heating temperature of 115°C, the fouling layer is relatively soft and consists mainly of proteins (50–70 g 100 g^{-1}). At higher temperatures, the fouling layer becomes harder, more compact and granular and the mineral content increases. In the case of whey, the total solids content, as well as the temperature, determines the composition of the fouling layer. At higher concentrations, relatively more calcium phosphates and calcium citrates become insoluble, which increases mineral precipitation during the formation of the deposit layer (Schraml and Kessler, 1996).

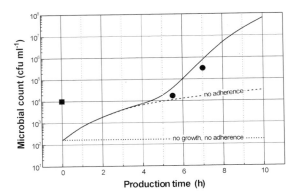

Figure 4.7 Concentration of *Streptococcus thermophilus* in the product as a function of the operating time of downstream processing of whey: effect of local adherence and growth in processing equipment. cfu, colony forming units. Production line (pasteuriser, buffertanks, fermenter, centrifuge, cooler). ■, Raw milk; ●, Product.

In general, protein and mineral fouling can be described by an adsorption reaction:

$$J_{x,t} = k''(C_{x,t})^{1.2} \tag{4.4}$$

where, $J_{x,t}$ is the local flux (kg m^{-2} s^{-1}) of food components to the wall and $C_{x,t}$ is the local bulk concentration of the key component related to fouling (e.g. whey proteins; de Jong, 1997). The mechanism of protein and mineral fouling is shown in Figure 4.8. The amount of fouling is obtained by integrating the flux over the operating time and surface area of the equipment walls. This amount can be related to the costs due to cleaning, change-over (rinsing losses), depreciation, energy, operator, pollution and product losses (de Jong, 1996).

Some important measurements to minimise the amount of fouling in equipment are:

- optimisation of the time–temperature profile
- lowering the surface temperature of the equipment wall of the heat exchangers by changing water flow rates and/or configuration of the plates or tubes
- increasing the velocity of the product (to more than 2 m s^{-1})

As in the case of microbiological fouling, protein and mineral fouling can be efficiently reduced by use of predictive models (de Jong, 1996).

4.2.2.3 Design of evaporators
For the concentration of milk and milk products use is generally made of energy-saving evaporators, the multistage arrangement of which leads to a low specific steam consumption. However, it is known that in these evaporators

CONCENTRATED AND DRIED DAIRY PRODUCTS

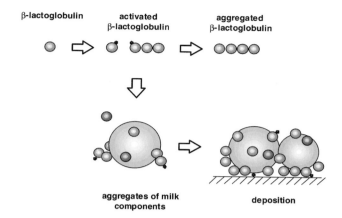

Figure 4.8 Schematic representation of the mechanism of protein/mineral fouling: grey circles, β-lactoglobulin; small black circles, free sulphydryl group; large light grey circles, casein micelles; dark grey circles, other milk components, such as calcium.

a considerable amount of product is lost as a result of fouling of the evaporator tubes. As a rule, the design of falling-film evaporators is based on heat and mass equations and, in most cases, on empirical relations with a limited number of variables, such as total solids content, boiling temperatures and minimum wetting rates. The heat transfer and pressure drop in the evaporator tubes are complex phenomena, and detailed correlations are not applied. Jebson and Iyer (1991) described the performances of falling-film evaporators used in the New Zealand dairy industry. They concluded that the viscosity of the product and the momentum of the vapours passing down the tubes are the main factors controlling the heat-transfer coefficient. However, they also reported a considerable scatter in the values for the heat-transfer coefficients. To design falling-film evaporators, it is necessary to know which factors control the heat transfer and which control the pressure drop in the evaporator tubes. The optimal diameter and length of the tubes can be determined if the changes in evaporating conditions (e.g. boiling temperature, vapour flow rate, wetting rate and physical properties) along the tubes are taken into account (Bouman *et al.*, 1993).

The temperature difference between steam and product $(T_s - T_p)$ used in computing the heat transfer in falling-film evaporators is frequently an arbitrary figure, since it is difficult to determine the temperature of the boiling liquid at all positions of the evaporator tube. Generally, the heat transfer, Q, is given by:

$$Q = kA(T_s - T_p) \tag{4.5}$$

where, A is the heat transfer area and k is the overall heat-transfer coefficient, given by:

$$k = \left(\frac{1}{\alpha_p} + \frac{\delta_w}{\lambda_w} + \frac{1}{\alpha_s}\right)^{-1} \qquad (4.6)$$

where, α_p and α_s are the heat transfer coefficients at the product and steam side of the tube, respectively, δ_w is the wall thickness and λ_w is the thermal conductivity of the wall. The temperature difference between steam and boiling liquid will be smaller at the top of the tube than at the bottom owing to pressure losses along the tube. With the change in boiling temperature, the thermal driving force and consequently the heat transfer coefficient, α_p, will also vary along the tube.

Figures 4.9a and b show the evaporation behaviour of skimmed milk at a low and a high heat flux, respectively. In Figure 4.9a, convective boiling occurs, and in Figure 4.9b nucleate boiling occurs. Both evaporation regimes are important in film evaporators for milk and milk products. Convective boiling occurs when the temperature differences across the film are small (less than 0.5 K for milk; less than 5 K for water); evaporation only takes place at the liquid–vapour interface of the film. In the case of nucleate boiling, vapour bubbles are formed at the metal surface. The boiling phenomena has a clear impact on the heat-transfer equations for milk and milk products. The general form is:

$$\alpha_p = a \times q^b \times m^c \times \eta^{-d} \qquad (4.7)$$

where, q is the heat flux, m is the wetting rate, η is the dynamic viscosity, and a, b, c and d are coefficients. The friction factor, f, for vapour flow in evaporator tubes also shows a relation with the boiling phenomena (r, s, t and u):

$$4f = r \times q^s \times m^t \times (N_{Re})^u \qquad (4.8)$$

where, N_{Re} is the Reynolds number of gas flow (Bouman et al., 1993).

The designer of a falling-film evaporator must determine the heating surface area at each stage of the evaporator, taking into account the desired wetting rate needed to prevent fouling. The wetting rate can be increased by installing longer tubes or by dividing the stage into two or more passes in series (Gray, 1981). Recirculation, as a method of increasing the wetting rate, cannot be considered for milk, on bacteriological grounds.

The effect of tube length and diameter on the heating surface area and the wetting rate for the first effect of a six-effect evaporator for whole milk is shown in Figure 4.10, which is based on a model calculation that incorporates the effect of pressure drop on heat transfer in evaporator tubes (Bouman et al., 1993). The horizontal lines indicate the change of the wetting rate in the tubes from top to bottom. For a desired minimum wetting rate, the tube length and diameter can be determined. The required surface area decreases with

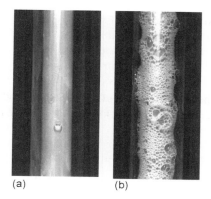

Figure 4.9 Skimmed milk boiling at 70°C; wetting rate, 400 kg m^{-1} h^{-1}. (a) Convective boiling, $q = 0.8$ kW m^{-2}; (b) nucleate boiling, $q = 6.2$ kW m^{-2}. q, heat flux.

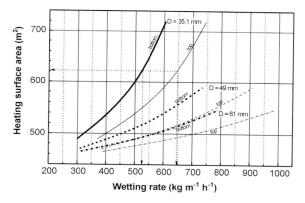

Figure 4.10 Heating surface area and wetting rate (at bottom and top of evaporator tubes) in the first effect of a six-effect evaporator in relation to tube diameter, D. Feed capacity, 30 tonnes per hour; steam temperature, 73°C; boiling temperature, 70°C.

increasing tube diameter (as there will be less pressure drop). The wetting rate and the temperature difference between condensing steam and boiling product are interdependent variables. From the point of view of energy consumption, such temperature differences, accompanied by lower wetting rates, will offer prospects for decreasing the energy consumption by increasing the number of stages (effects). However, high wetting rates are favourable to prevent fouling. In this context, it is worth noting that the liquid must be distributed uniformly at the feed inlet to ensure a sufficient supply of liquid to each evaporator tube; otherwise, deposit formation in individual tubes will occur, causing a decrease in the running time of the evaporator and an increase in product losses.

4.2.2.4 Designing spray dryers

The quality of milk powders is determined by a variety of properties depending on the specific application. In general, the final moisture content, the insolubility index and the bulk density are of primary importance. Nowadays, the main challenges in the production of powders are the development of specific 'functional' products and the reduction of processing costs. Therefore, the production capacity of available installations is maximised and the process conditions are directed towards minimal fouling of equipment, minimal product losses and reduction of energy consumption. Also, on-line product quality control is implemented as far as possible. For these purposes, models of the drying process can be very helpful.

The earning capacity of spray dryers can be improved further by increasing both the total solids content in the feed and the inlet temperature of the drying air. However, the formation of insoluble material, which, especially in dairy products is boosted by these process modifications, is often a limiting factor for product quality. The amount of insoluble material formed is largely determined by the temperature curve and the total solids content of the drying particles. To achieve a better understanding of the formation of insoluble material, a model of drying describing the behaviour of the individual particles during drying is needed, coupled with a model for the kinetics of the formation of insoluble material.

In the drying process, the flow pattern of the air depends on the geometry of the dryer and the location and design of the air inlet and outlet channels. The trajectories described by the particles depend not only on the airflow pattern but also on the position and method of atomisation. Nowadays, several universal software packages making use of computational fluid dynamics (CFD) techniques are commercially available.

In Figure 4.11a.i, the temperatures and air humidity in a working dryer are shown as grey contours; in Figure 4.11a.ii, the simulated flow pattern of the air and the particle trajectories are illustrated (Straatsma et al., 1999a). The rotary speed of the atomiser is low: droplets leave the atomiser with a velocity of 58 m s^{-1}. The air inlet is placed at the centre of the top of the dryer and the airflow is directed straight downwards. In the dryer there is a main circulation airflow, which is downwards at the centre axis and upwards at the outer side. Part of the airflow leaves the dryer at the outlet, which is placed at the upper end on the outside. The air which reaches the outlet is cooled to about $90°C$. Over the course of years, the capacity of the dryer and the rotary speed of the atomiser have been increased. In this modified dryer, the momentum exchange from the atomised spray droplets to the air has a strong influence on the flow pattern, as shown in Figure 4.11b.ii. Here, the results of the simulation with a higher rotary speed (initial droplet velocity 150 m s^{-1}) are shown. The main airflow circulation has reversed compared with Figure 4.11a.ii and there is a circulation of hot air at the upper part of the dryer. The temperature of the outlet air is now

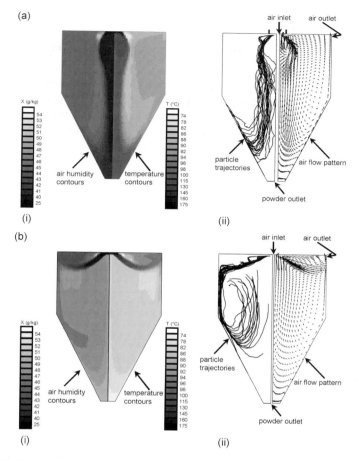

Figure 4.11 Computational simulation (NIZO-DrySim) of an industrial spray dryer. (a) Initial droplet velocity, 58 m s^{-1}: (i) air humidity and temperature contours; (ii) simulated airflow pattern. (b) Initial droplet velocity, 150 m s^{-1}: (i) air humidity and temperature contours; (ii) simulated airflow pattern.

much higher (Figure 4.11b.i). The actual droplet velocity in the industrial dryer was slightly higher than the predicted minimum droplet velocity of 150 m s^{-1} needed to cause a reversal of the main airflow circulation, so the operators were correct: at the actual rotary speeds, reversal of the main airflow circulation and a shortcut of hot air from inlet to outlet may be expected. The dryer concerned has now been reconstructed for high-pressure nozzle atomisation.

Another working example is provided by the simulation of the operation of an industrial spray dryer for the production of a powder with a high lactose content (Straatsma *et al.*, 1999a). The dryer was equipped with one central and several noncentral air inlet channels with nozzle atomisers. This dryer had

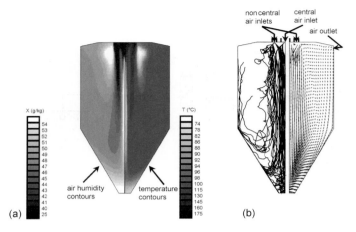

Figure 4.12 Computational fluid dynamics simulation (NIZO-DrySim) of an industrial spray dryer with a fouling problem: (a) air humidity and temperature contours; (b) simulated airflow pattern.

serious fouling problems in the bottom part (cone). The simulation is shown in Figure 4.12b and indicates clearly that the powder particles collide with the cone wall when the airflow reverses in the conical part of the dryer. Under these conditions the particles were not dried enough and were very sticky. Furthermore, the simulations indicated that the thermal load of the particles atomised in the central air inlet was much higher than that of the particles atomised in the noncentral air inlets owing to the temperature distribution in the dryer (Figure 4.12a). According to new simulations the fouling problem would be reduced by adapting the cone angle; however, this would be an expensive operation. The problem has now been solved by reconstructing the air inlets, but only the noncentral air inlets, placing then further from the centre. The diameters of the inlet channels have also been adapted, resulting in different initial air velocities.

4.2.2.5 Cleaning

A large part (40–80%) of the operating costs of evaporators and dryers results from fouling and cleaning of the equipment. For the cleaning procedure, automated cleaning-in-place (CIP) installations are used (see Figure 4.13). The evaporator is cleaned by liquid flow, both under evaporating and nonevaporating conditions. The spray dryer is commonly cleaned by use of nozzles. Critical factors for an optimal cleaning procedure are rinsing times, switching between the different steps of cleaning, the temperature and the separation of the different liquid phases.

In general, evaporators are cleaned in multiple processing steps by using caustic and acid cleaning. Caustic (lye) solution is used to break down the organic deposit (protein and fat) and acid is used to remove the mineral deposit. The

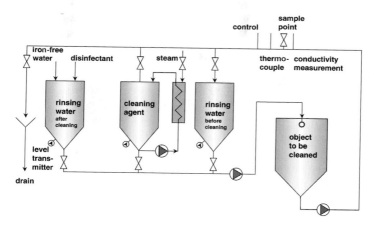

Figure 4.13 Scheme of a cleaning-in-place installation used in the dairy industry.

sequence of caustic and acid cleaning is determined by the composition of the deposits. In the case of mineral fouling, the procedure starts with cleaning with acid. If the majority of the deposits are proteins and fat, the cleaning programme starts by circulating the lye solution. Generally, more expensive single-phase cleaning agents (e.g. NIZO K500®) require shorter cleaning times. Depending on the actual situation (e.g. annual capacity of the plant), this may result in a reduction of cleaning costs and production costs.

4.3 Quality control

In general, two types of control strategies can be distinguished. The first strategy is focused on maintaining constant values of process conditions, such as temperatures, flows and pressures. The selected values of the conditions are based on experience gained over the years. The idea is that constant process conditions will result in constant product quality. However, in practice, 'constant' may be understood as 'average value with a certain standard deviation'. Moreover, a small standard deviation in a process parameter may cause a large standard deviation in a product parameter (e.g. the bacterial count).

The second control strategy is based on indirect or direct control of the desired product quality parameters. In this case, the set points of the process conditions are determined by mathematical models which describe the interaction between process and product properties. At present, the models used are black-box models with little or no physico-chemical background. In the near future, it is expected that physico-chemical predictive models will also be applied to model-based control.

4.3.1 *Control of process conditions*

In the dairy industry, the evaporator and the dryer are controlled separately. Both the control of the evaporator and the control of the dryer are directed at control of moisture content. The occurrence of changes in dry matter content of the feed at a subsequent drying stage is one of the major sources of disturbance in the drying process. To obtain a high-quality powder, a constant dry matter content in the concentrate produced in the evaporator is therefore a necessity. From an energy-saving point of view, it is advantageous to remove as much water as possible at the evaporation stage. However, this possibility is limited by the increase in viscosity that occurs as the dry matter content increases, which will eventually hinder the subsequent drying process. Less variation in the dry matter content of the concentrate enables this limit to be very closely approached and thus contributes to energy efficiency.

In closely-coupled production lines, the production rate of the evaporator and the dryer should be balanced. Most drying processes are either controlled by the feed rate or require a constant feed rate. The concentrate flow from the evaporator should closely match this rate in order to avoid the need for large buffers.

4.3.1.1 *Evaporator control*

The control of evaporators is focused mainly on a constant flow rate and constant dry matter content in the concentrate. Where multiple-effect falling-film evaporators are to be controlled, evaporators should be viewed as complex interacting systems. The commonly used conventional control technology, such as single-loop proportional–integral–derivative (PID) controllers, will therefore perform poorly compared with multivariable controllers. Modern multivariable robust control design methods make it possible to design compensators that optimise performance objectives where there is uncertainty about exact plant behaviour. Central in this approach is the process, or predictive, model. One part models the effect of control inputs on process outputs, and another part models the rate of conversion of product (by measuring the dry matter content of the feed) to process outputs (Figure 4.14).

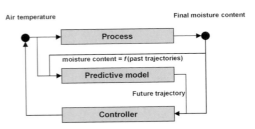

Figure 4.14 Example of a model-based control scheme. f (past trajectories), function of past trajectories.

Two approaches have been followed (van Wijck *et al.*, 1994) to obtain models describing process dynamic behaviour: (a) a physical model constructed from first principles; and (b) a black-box model which has been fitted to experimental data with system identification techniques. The first approach is more flexible and robust for handling changes in the design and process operation. The advantage of the latter approach is that it requires less knowledge about the design and process operation of the evaporator.

4.3.1.2 Spray dryer

The main issue of the automatic control of spray dryers is to achieve a constant level of moisture in the powder. As illustrated in Figure 4.15, a reduced standard deviation in the moisture content minimises operating costs.

As with the evaporator, two types of control strategies can be distinguished. The classical control strategy is based on the PID concept (Stapper, 1979). Modern dryers use the moisture content determined by infrared, resistance or capacitance measurements for use as a control variable. The variables used are thermal flux (air temperature, gas flow rate) or concentrate flow rate. The choice of the product flow rate as the variable to be manipulated is generally cheaper but it should be avoided in the case of continuous flow production because it modifies the production capacity locally and creates larger buffers.

Particularly in drying processing there is a trend to use more and more predictive models in the control strategy (Figure 4.14). In most cases, these models are first-principle or neural network models (respectively, white-box and black-box approaches). Only a few first-principle approaches have been described in the literature (Alderlieste *et al.*, 1984; Chen, 1994; Delemarre, 1994; Pérez-Correa and Farías, 1995). A comparison of both approaches is given in Table 4.1. Since the neural network approach is still considered to be a

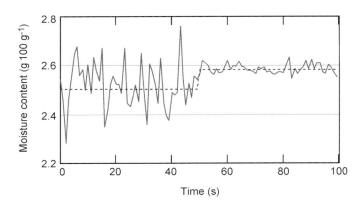

Figure 4.15 The effect of a well-defined automatic controller on operating costs.

Table 4.1 Comparison of two approaches for model-based control of dryers

First-principle approach	Neural network approach
It has relatively high predictive power outside the operation point (location-independent)	It has less predictive power outside the operation point (location-dependent)
Process knowledge is needed	No process knowledge is needed
Custom-made dryer models are needed	Standard software tools are available (process-independent)
Few measuring points are needed	Many measuring points are needed for model training

specialised field, manufactures of dryer installations have initiated cooperation with specialised software houses.

4.3.1.3 Carbon monoxide detection

An important issue in spray dryer control is the detection of smouldering milk powder, as powder deposits in a production plant can start a fire. Before the fire starts, the first step is the smouldering of powder, which produces a significant amount of carbon monoxide (CO). It has been shown that this increase of CO can be detected by a sensitive CO analyser (Steenbergen *et al.*, 1991). Figure 4.16 shows a measuring system, which comprises an air sample line from the inlet of the dryer, an air treating unit for cleaning and drying the air issuing from the outlet of the dryer, a CO analyser and, if necessary, a control system for consecutive performance of the measuring cycle. Nowadays, an increasing number of CO detection systems are operating on an industrial scale and these are linked to the dryer control systems to perform automatic shut-down, if necessary.

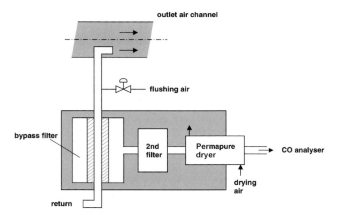

Figure 4.16 The carbon monoxide measuring system for dryers.

4.3.2 Control of product properties

The automated control of industrial concentration and drying installation is generally directed at only a few product properties, such as the viscosity and dry matter content of the concentrate, and the moisture content of the powder. However, there are a number of other important product properties which determine the product quality. Some of the most important aspects are as follows.

4.3.2.1 Microbiological specifications

The microbiological quality of the concentrate and powder is very often defined by a heat classification, such as low, medium or high. These classifications are based on protein denaturation and are thus an intrinsically incorrect definition of microbiological quality. The same is valid for the F_0-value, which is based on the average temperature dependency of microbial deactivation. Although both have their practical use, it is better to select a number of so-called key microorganisms (e.g. *Bacillus stearothermophilus* spores) that determine the microbiological quality and to focus measurements on them. In the case the actual temperature-time profile of the concentrate and powder are well measured and a process model is available, the inactivation of these microorganisms can also be estimated by using inactivation kinetics of the microorganism (de Jong, 1996).

4.3.2.2 Heat stability

Heat stability, defined by the heat coagulation time, is one of the most important quality parameters in the production of concentrated milk. Heat stability is related to a complex of different factors, such as protein composition and concentration, calcium activity, pH, fat content and pre-heat-treatment. Pre-heating results in a certain amount of whey protein associated with κ-casein at the surface of the casein micelles (de Jong and van der Linden, 1998). In Figure 4.17, a typical heat coagulation curve is shown as a function of the initial pH of the milk. At a pH of less than 6.5, pre-heating increases the heat stability of the milk concentrate, since at a pH of more than 6.6, κ-casein dissociates from the micelle, so the effect of pre-heating on the heat stability is decreased (Kessler, 1996). In order to increase the heat coagulation time of evaporated milk, stabilising salts can be added. Salts such as disodium phosphate and trisodium citrate reduce the calcium ion activity.

4.3.2.3 Viscosity and age thickening

In the manufacture of (sweetened) condensed milk, the control of viscosity of the product is of major importance. As well as the consistency aspect, the viscosity must be high enough to prevent sedimentation and creaming of the fat but low enough to be able to allow processing. During storage the viscosity may increase; this process is called age thickening. The exact nature of the physico-chemical processes that cause age thickening are poorly understood. As with heat stability, the way in which the product is pre-heated has a large impact on changes in viscosity.

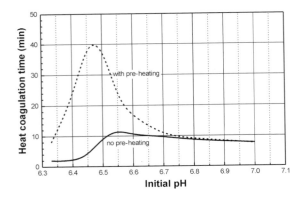

Figure 4.17 Heat coagulation time (120°C, skimmed milk concentrate) as a function of pH and pre-heating. Source: van Boekel (1993).

4.3.2.4 Solubility

Several tests have been developed which give a measure of the solubility of milk and whey powder. The most well-defined method is the International Dairy Federation's method for the determination of the insolubility index (IDF, 1988). More qualitative tests are, for example, determination of the number of white flecks or specks. Straatsma *et al.* (1999b) developed a kinetic model that predicts the insolubility index as a function of temperature and particle diameter.

4.3.2.5 Integrated quality control

With the availability of predictive models partly based on first principles and partly on the chemical engineering approach (see Section 4.3.1), it becomes

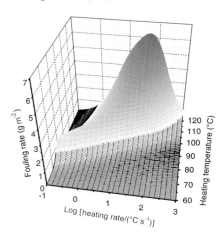

Figure 4.18 Simplified example of an objective function used for process optimisation.

possible to control the product quality directly instead of the process conditions. Both the off-line and on-line optimisation of process operation and control are based on two factors: quality and economics. To optimise the entire production line, a so-called objective function, $F(u, x)$, must be minimised:

$$F(u, x) = \alpha c_q(u, x) + \beta c_o(u) \tag{4.9}$$

where, u is a vector of process control variables (e.g. temperature, flow), x is a vector of desired product properties related to food quality and safety, and β is a weight factor. The value of c_q is dependent on the outcomes of the predictive models for contamination and transformation of food components; the value of c_o is related to operating costs. By using weight factors, the optimisation approach accounts for the relative importance of the individual changes in product properties and operating costs. Figure 4.18 provides a simplified example of an objective function used for the minimisation of the fouling rate of a heating system. By implementation of the objective function into the control scheme a high-level automation system arises.

References

Alderlieste, P.J., Fransen, J.J. and van Boxtel, A.J.B. (1984) Control of moisture content of milk powder. *Voedingsmiddelentechnologie*, **17** (6) 21-3.

Bouman, S., Lund, D.B., Driessen, F.M. and Schmidt, D.G. (1982) Growth of thermoresistant streptococci and deposition of milk components on plates of heat exchangers during long operating times. *J. Food Protection*, **45** 806-12.

Bouman, S., Brinkman, D.W., de Jong, P. and Waalewijn, R. (1988) Multistage evaporation in the dairy industry: energy savings, product losses and cleaning, in *Preconcentration and Drying of Food Materials*, (Ed. S. de Bruin) Elsevier, Amsterdam, pp. 51-60.

Bouman, S., Waalewijn, R., de Jong, P. and van der Linden, H.J.L.J. (1993) Design of falling-film evaporators in the dairy industry. *J. Soc. Dairy Technol.*, **46** 100-6.

Chen, X.D. (1994) Towards a comprehensive model-based control of milk drying processes. *Drying Technology*, **12** 1105-30.

de Jong, P. (1996) *Modelling and Optimization of Thermal Processes in the Dairy Industry*, PhD Thesis, University of Technology, Montfoort, The Netherlands.

de Jong, P. (1997) Impact and control of fouling in milk processing. *Trends in Food Science and Technology*, **8** 401-5.

de Jong, P. and van der Linden, H.J.L.J. (1998) Polymerisation model for prediction of heat-induced protein denaturation and viscosity changes in milk. *Journal of Agricultural and Food Chemistry*, **46** 2136-42.

de Jong, P., Bouman, S., Langeveld, L.P.M. and Kiezebrink, E.H. (1999) Predictive model for the adherence, growth and release of thermoresistant streptococci in production chains, in *Fouling and Cleaning in Food Processing '98* (Eds. D.I. Wilson, P.J. Fryer and A.P.M. Hastings), European Commission, Brussels, pp. 118-24.

Delemarre, V. (1994) New control strategy of spray dryers reduces operating costs. *Voedingsmiddelentechnologie*, **17** (3) 20-1.

Gray, R.M. (1981) Technology of skimmed milk evaporation. *J. Soc. Dairy Technol.*, **34** 53-5.

IDF (1988) Dried milk and dried milk products: determination of the insolubility index, standard 129A, International Dairy Federation, Brussels.

Jebson, R.S. and Iyer, M. (1991) Performance of falling film evaporators. *Journal Dairy Research*, **58** 29-38.

Kessler, H.G. (1996) *Food and Biotechnology*, Kessler, Munich.

Knipschildt, M.E. (1986) Drying of milk and milk products, in *Modern Dairy Technology, Volume 1: Advances in Milk Processing* (Ed. R.K. Robinson), Elsevier Applied Science, New York, pp. 131-234.

Pérez-Correa, J.R. and Farías, F. (1995) Modelling and control of a spray dryer: a simulation study. *Food Control*, **6** 219-27.

Schraml, J.E. and Kessler, H.G. (1996) Effect of concentration on fouling of whey. *Milchwissenschaft*, **51** 151-4.

Stapper, H.L. (1979) Control of an evaporator and spray dryer using feedback and ratio feedforward controllers. *New Zealand J. Dairy Sci. Technol.*, **14** 241-57.

Steenbergen, A.E., van Houwelingen, G. and Straatsma, J. (1991) System for early detection of fire in a spray dryer. *J. Soc. Dairy Technol.*, **44**, 76-9.

Straatsma, J., van Houwelingen, G., Steenbergen, A.E. and de Jong, P. (1999a) Spray drying of food products: 1. Simulation model. *J. Food Engin.*, **42** 67-72.

Straatsma, J., van Houwelingen, G., Steenbergen, A.E. and de Jong, P. (1999b) Spray drying of food products: 2. Prediction of insolubility index. *J. Food Engin.*, **42** 73-7.

van Boekel, M.A.J.S. (1993) Mechanisms of heat coagulation of milk products, in Protein and Fat Globule Modifications by Heat Treatment, Homogenisation and Other Technological Means for High Quality Dairy Products. *Special Issue, Int. Dairy Fed.*, **9303**, 205-15.

van Wijck, M.P.C.M., Quaak, P. and van Haren, J.J. (1994) Multivariable supervisory control of a four-effect falling film evaporator. *Food Control*, **5**, 83-9.

5 High fat content dairy products

H.M.P. Ranjith and K.K. Rajah

5.1 Introduction

The biosynthesis of milk fat, in common with other dairy milk components, is initiated in the glandular epithelium of the mammary gland. Once formed, milk fat, in the form of globules, appears to be surrounded by a thin membrane and gradually increases in size. The secretory cells exude these globules at the apical cell boundary into the lumen, with the membrane still intact. The globule size is typically in the 0.1–10 μm range. These fat globules are uniformly dispersed and intermingled with other milk components to form an 'oil-in-water' emulsion. This milk fat emulsion is unique owing to the presence of membrane material. The intricate structure of the membrane is complicated and the exact form is uncertain (Figure 5.1).

Inside, the fat globule comprises a mixture of high and low melting fats and therefore exists as an admixture containing liquid and fat crystals at chilled temperatures. The functional properties and nutritional significance of milk fat have been well researched and documented by many scientists. Equally, its effect on human health has been the subject of many seminars and conferences in recent years. Controversy still surrounds the long-term effects of milk fat in the human diet, yet some of the major dairy products used in food manufacture contain significant quantities of milk fat (see Figure 5.2). Consequently, some of the best known consumer food and snacks in many countries contain these high milk fat ingredients.

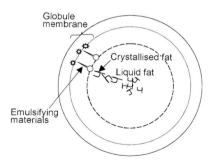

Figure 5.1 Indication of basic layers of the milk fat globule.

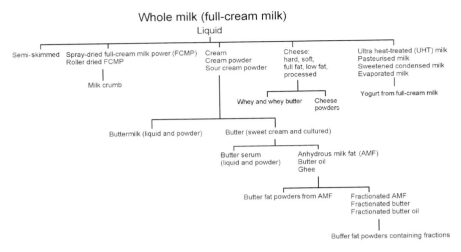

Figure 5.2 High fat content dairy products. Reproduced courtesy of Klondyke Management Consultants, Essex, UK.

5.1.1 Properties of milk fat

Milk fat is predominantly (more than $98 \text{ g } 100 \text{ g}^{-1}$) triglycerides (also known as triacylglycerols). Although only about 15 fatty acids account for more than $95 \text{ g } 100 \text{ g}^{-1}$ of those present in triglycerides, more than 400 fatty acids are found in milk fat. Uniquely, significant amounts of the easily digested short-chain and medium-chain fatty acids are present (which can be useful to those with special dietary needs) in addition to the long-chain fatty acids (usually found in edible animal and vegetable fats). About $5 \text{ g } 100 \text{ g}^{-1}$ occur naturally as trans-fatty acids. A number of the fatty acids are also associated with the distinctive flavour of butter, particularly saturated short-chain butyric acid (i.e. C4:0). Small amounts of other glycerides and fatty materials are also present, such as diglycerides, monoglycerides, free fatty acids, phospholipids, cerebrosides and sterols (cholesterol and cholesterol esters).

In mechanised operations, some of the unique chemical and physical properties of fatty acids (and therefore of triglycerides), such as the degree of unsaturation (i.e. number of double bonds present), density, viscosity and melting point, will influence operating conditions. For instance, milk fat from milk produced by pasture-fed cows contains relatively more unsaturated fatty acids and so will have a lower melting point (and be less firm when present in stored butter) and therefore require relatively less mechanical working during blending.

A number of conditions are known to cause variation in the type and quantity of fatty acids present: (a) age of the cow; (b) stage of lactation; (c) health; (d) welfare; and (e) feed. The feed is held to be the main contributing factor for what is known as the 'seasonality effect' (i.e. the firmness of milk fat, which varies with the seasons) because when cows are fed on fresh pasture (typically during the nonwinter period), there is an increase in the presence of unsaturated fatty acids (particularly C18:1), which results in the fat being relatively soft. With commercial solid feed (made with use of hydrogenated fats) during the winter months, the levels of saturated fatty acids in milk fat increase and, correspondingly, the melting point and the solid fat content (SFC) values are also high. Therefore, one hears also of 'summer' and 'winter' butter. The SFC, which is sometimes also referred to as the solid fat index (SFI), is shown in Figure 5.3 for UK summer and winter milk fats; the main differences are summarised in Table 5.1.

The SFC is a measurement of the solid/liquid ratio in the fat at a given temperature and is measured by pulse nuclear magnetic resonance (NMR) spectrophotometry. Readings are usually taken at temperatures between 0°C and 45°C at intervals of 5°C. The value at each temperature (N_T) is expressed in general terms as:

$$N_T = x \tag{5.1}$$

where, x is the percentage of fat which is in crystalline form.

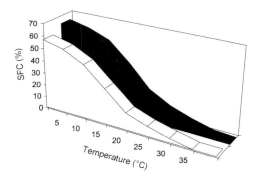

Figure 5.3 Solid fat content (SFC) of summer (unshaded chart) and winter (shaded chart) milk fat in the UK.

Table 5.1 The seasonality effect in milk from UK Friesian cows: a summary of the main differences between summer and winter milk fats

Property	Summer	Winter
Flavour	Rich in flavour precursors	Lower flavour level
Typical level of carotene (mg kg^{-1})	6–8	4–6
Vitamin A and E content	Rich	Lower levels
Tendency to oxidate	More stable	Stable
Degree of saturation	Increase in unsaturation	Increase in saturation
Melting point (°C)	28–32	30–35

Source: Klondyke Management consultants, Essex, UK.

5.1.2 Melting and crystallisation

The behaviour of milk fat is characterised by its ability to melt or solidify. It becomes a complete liquid at 40°C and crystallises to a complete solid at temperatures lower than −40°C. At intermediate temperatures products contain a liquid/solid mixture. In the main, mechanisation is used to control this solid/liquid ratio, which determines the rheological properties and hence the functionality of the fat. For instance, cream, which is treated to the Alnarp process (where it is put through a cool–heat–cool cycle to control closely the type of crystals formed) can be churned to a softer, spreadable butter. This process is used in commercial operations to produce a softer butter for spreading or to soften the much harder 'winter' butters.

The solid/liquid ratio is therefore determined by both the composition of the triglycerides present, and the conditions under which the cream or fat is treated as it is cooled. In an alternative process, butter which has not been produced through either of these routes can be worked mechanically and will be softer as a result of this 'work softening'.

In all cases, however, the softer products are said to be in a metastable crystalline state. This means that if the temperature of the butter is raised repeatedly to the melting point of the metastable form, typically by moving it between refrigerator and a warm bakery, the improved spreadability at low temperatures will be lost, since the fat will revert back to its stable (firmer) form and more of the butter will be in the solid phase.

More than one crystalline form can be produced during crystallisation of milk fat. van Beresteyn (1972) described these forms as α, β', β, with melting points at about 22, 30 and 36°C, respectively. This behaviour is referred to as polymorphism. The tendency to develop one or another crystal form is termed the crystal habit (Table 5.2).

Table 5.2 Crystallisation properties of oils and fats: β and β' crystalline forms

β	β'
Soybean	Cottonseed
Rapeseed	Herring
Sunflower	Menhaden
Palm kernel	Palm
Coconut oil	Milk fat[a]
Lard	Modified lard

[a] Cream, butter, anhydrous milk fat.

5.2 High fat content emulsions: oil-in-water type

In an oil-in-water emulsion, the fat or the oil is uniformly distributed in a continuous aqueous phase. The original method used to recover fat from milk was to apply gravity in order to separate particles either by sedimentation or by flotation. For sedimentation, the liquid is allowed to stand for a period of time until the heavy solids and particles have settled to the bottom of the container, after which the clear liquid is carefully decanted, allowing the solids to remain at the bottom. In flotation, the particles to be removed are of lower density than the surrounding liquid phase; these particles will rise to the surface of the liquid and can then be harvested by skimming the surface. The flotation technique is the traditional method for the separation of cream from milk.

The dairy industry has pioneered the mechanisation of the fat recovery process from milk by making use of the effect of a centrifugal force on milk and its constituents. The design of milk separators was based on other industrial centrifugal separation techniques; the chemical industry has been using centrifugal separation of mixtures for a number of years, and the dairy sector was able to learn from such experience. The principles of centrifugal separation are described below.

5.2.1 Centrifugal separation

Centrifugal force is generated when materials are rotated; the magnitude of the force depends on the radius and speed of rotation and the composition (or density) of the centrifugal material. The denser liquid or phase moves to the outer wall of the rotating unit (or bowl) and the lighter liquid is displaced to an inner annulus. The early design of centrifugal separators consisted of an elongated hollow bowl mounted vertically with one inlet and two outlets. The bowl was mounted on a spindle, which was driven by a geared motor arrangement.

The early designs suffered from low throughput and low separation efficiency. Fat particles had to travel a long distance in the bowl full of milk to get separated and before coming to rest. In later designs, the separation efficiency was increased by inserting a number of conical discs stacked on top of each other within the bowl, and milk was introduced through distribution holes in the disc stack. This design ensured that during separation the solid particles, dirt and sludge moved to the outer wall of the bowl, whereas the lighter fat particles or cream moved inwards to the inner section of the bowl.

The exits for the cream and skimmed fractions are located at the head of the upper section of the bowl. The cream exit port collects cream in the vicinity of the centre shaft, whereas the skim port collects skimmed milk away from the centre shaft. The milk separator bowl rotates at a very high speed, approximately 6000–9000 revolutions per minute (rpm) and therefore each unit must be constructed to a high degree of precision.

The separators may take one of three basic designs: open, semi-enclosed or hermetic. In the open design, milk is introduced through a float on top of the rotating bowl. The separated fractions, skimmed milk and cream, are allowed to overflow from the top of the bowl and are collected in corners or spouts. The semi-enclosed design is similar to the open design but the outlet ports are fitted with paring discs which act as cream and skim pumps. In the hermetic seal design, the separator is fitted with seals to isolate the milk, cream and skim from the atmosphere. Milk is fed into the bowl from the bottom of a hollow spindle via a feed pump.

The following steps are important in the separation of milk:

- milk is warmed to above 38°C prior to separation
- after separation, the skimmed milk and cream fractions are heat treated (to kill pathogens) and immediately cooled to below 10°C
- preferably, a positive pump is used to pump the cream
- milk may also be separated after being pasteurised and subsequently cooled to 40°C—in this case, the two streams (skimmed milk and cream) are cooled immediately to below 10°C, and the cream can be standardised by mixing with pasteurised whole milk or skimmed milk

The efficiency of milk separation also depends on certain other factors, such as the milk itself, the mechanical performance of the separator and the design of the process. Large fat globules in the original raw milk are desirable for efficient separation, whereas large numbers of small fat globules have the tendency to cause loss of fat from the cream phase into the skimmed milk. It is important to minimise the mechanical damage to fat globules from pumps and agitators and from factors arising from the design of the pipe line itself.

The separator alone can produce poor-quality cream and lead to fat losses if the speed of rotation, assembly of the internal bowl components and the throughput are not optimised. Regular maintenance and checks are essential for efficient mechanical performance. Bowl discs and the bowl itself should be clear of sludge and other built-up material to prevent blockage of flow paths. Also, the milk process pipeline linking the separator must be designed with care to provide optimum conditions for centrifugal separation.

The efficiency is affected by the pumps, by the feed and discharge pressures in the circuit and by the heating method. A suitable pump should be selected to match the throughput of the separator. Oversized pumps tend to damage the fat globules in the milk and also have a tendency to suck air through joints in the product inlet side of the pump. such air in the circuit is detrimental to the separation process.

The pressure in the circuit must not fluctuate during separation as the milk inlet, skim outlet and cream outlet pressures all affect the final fat percentage in the cream and skim. Severe heating conditions, such as steam heating in batch pasteurisation, tend to damage fat globules.

5.2.2 Control of fat content in creams

In the dairy industry, cream standardisation is carried out routinely and the method employed is known as 'Pearson's square'. A typical example of the use

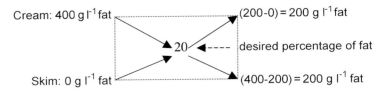

Figure 5.4 Cream standardisation: Pearson's square.

of this method is given in Figure 5.4. It works on the basis that a 50:50 mixture of 40 g 100 g^{-1} fat cream and skimmed milk will produce cream with 20 g 100 g^{-1} fat; the fat content in skimmed milk is usually less than 0.05 g 100 g^{-1}.

In most separators, the exit ports for cream and skimmed milk are provided with flow adjustments so that the required fat content can be achieved. However, this is only a rough adjustment to provide a fat content of approximately the correct magnitude. In the semi-open designs, paring discs located at the top of the disc stack gather the cream and pump it to the exit port. The volume of cream exiting the separator is controlled by a throttle valve. An increase in the cream discharge volume has a tendency to reduce the fat content of the cream. Therefore, the rate of discharge corresponds to a given fat content in the cream fraction. A similar procedure is followed to throttle back the discharge of the skim fraction so that the desired relative volumes exit the separator and the correct fat content is achieved in the cream.

The cream and skim discharge mechanisms have been further improved to facilitate flow monitors in the form of a flow indicator in the cream exit line and a pressure indicator in the skim exit line. For a specified milk separator, the system pressure in the skim outlet corresponds to the designed throughput of the separator. The cream discharge volume is adjusted to achieve the desired fat content. The cream flow can be estimated as follows. Let us assume the separator is designed to run at 2000 l h^{-1}. If the fat content of the milk is 40 g l^{-1}, then 80 l of fat alone pass through the separator every hour [(2000 l h^{-1}) × (4 l/100 l)]. If the desired fat content in the cream is 400 g l^{-1}, then 80 l of fat will be present in 200 l of cream. This is an approximate value and the cream flow can be throttled back until the flow indicator corresponds to the desired reading. In a hermetic separator, the skim flow may be controlled by an automatic pressure-modulating valve so that the exit pressure is kept at the desired value.

In modern milk separators, an on-line fat control system or standardisation unit exists as part of the ancillary processing instrumentation. Such a system enables the dairy to prepare the basic products, such as skimmed milk, standardised milk and creams. The basic components in an automation unit for high fat content products are the density transmitter, flow transmitter, an electrical or electronic control system and valves. The flow diagram shown in Figure 5.5

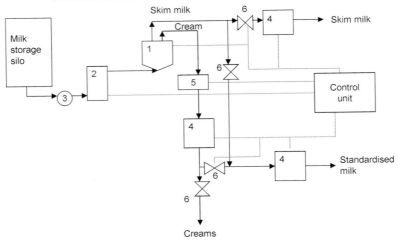

Figure 5.5 Automatic control of milk separation to produce milks and creams. 1, Separator; 2, heat exchanger; 3, milk pump; 4, flow transmitters; 5, density transmitter; 6, control valves; ———, wires associated with the control unit; ———, other connections.

indicates the basic principles and units used in handling raw milk to produce skimmed milk, standardised milk and creams.

The flow transmitter (item 4, Figure 5.5) may be of a type based on electromagnetic principles, whereby a voltage is applied to generate a constant magnetic field. The product flows through the magnetic field and at right angles to it, generating a detectable current. The magnitude of the current generated is directly related to the mass of product passing through it in a defined period. The control unit in the flow transmitter maintains a steady magnetic field, and the current generated by the movement of the product is amplified and converted into a suitable display signal. This signal is used to adjust the flow rate of separated milk fractions to produce the desired products.

The density transmitter (item 5, Figure 5.5) is a widely-used instrument in modern automated systems. It is capable of detecting small changes in the density of cream caused by changes in fat content. As the fat content in cream increases its density becomes lower (i.e. the fat content in milk varies inversely with density). This is due to the oil phase in the oil-in-water emulsion being lighter than the aqueous phase. Measurement of the density is affected by the entrained air in milk and cream. Therefore, if the raw milk is aerated, the cream will be lighter because the centrifugal force in the separator will direct the gas bubbles into the fat phase. It is also important to ensure that atmospheric air is not sucked into the milk by the pumping arrangement or because of an inadequate milk supply reaching the separator. In some installations, a de-aerator is included to ensure the entrained air is reduced to a constant level for daily operations. Such a de-aeration unit (not shown in Figure 5.5) will also eliminate any errors arising from the presence of gas bubbles in the measurement of the volumetric flow of the milk and cream. It is equally important to measure the density at a constant temperature or in the range recommended by the equipment suppliers.

The density transmitter is also equipped with a temperature sensor to allow necessary corrective adjustments to be made when there are slight changes in the incoming milk to the separator. For example, if the temperature at which the density is measured is higher than desirable, the density value transmitted is lower and, similarly, if the temperature is lower than desirable, the value transmitted will be higher. The signals (temperature and density) from the transmitter reach the control unit for computation of the values and comparison against pre-set values. This triggers the control unit to take the necessary steps to implement the required action and also activate other signals, if appropriate.

The basic components in a density transmitter are an inner stainless steel tube (its outside diameter closely matching that of the dairy process circuit pipe) and an outer tube to house the coils of the electrical circuits. The electrical circuits are for input and output signals, and the information from them is analysed by the control unit to activate the necessary signals.

5.2.3 Cleaning of milk separators

A separator contains a large number of parts assembled in a precise configuration. However, this intricate arrangement makes the internal parts difficult to clean. The bench-top type and the small separators are initially rinsed with hot water while rotating to remove milk, cream and skimmed milk inside the bowl. The bowl is then dismantled and each part is washed manually with use of a suitable detergent. It is important to use the correct strength of detergent to remove the fat and other sludge deposits completely. Mechanical dishwashers may be used for cleaning of mechanical separator bowl parts.

Large milk separators are designed with 'self-cleaning' and/or 'de-sludging' facilities. The rotating bowl is made in two halves; the bottom half slides down to open discharge ports on the sidewall to eject sludge. The de-sludging periods are predetermined to be activated at regular intervals. Hydraulic pressure is used to achieve the downward movement of the bottom half of the bowl. These separators are cleaned by using an existing dairy cleaning-in-place (CIP) system. The bowl is usually dismantled for inspection once every 3–4 months, but this is done more frequently if problems are encountered.

5.2.4 Description of creams

A typical definition for cream is found in the UK Dairy Product (Hygiene) Regulations (1995) in which 'cream' is defined as 'that part of milk rich in fat which has been separated by skimming or otherwise and which is intended for sale for human consumption'.

To place this within context, milk, being an (oil-in-water) emulsion, has a fat content of, on average, $3.9\,g\,100\,g^{-1}$. The fat is present as minute globules, of average diameter $3\text{–}10\,\mu m$, with some 15 billion such globules being present in a teaspoonful (5 ml) of milk.

Table 5.3 shows the specifications of different types of cream processed in the UK.

Table 5.3 Cream categories in the UK: minimum fat content ($g\,100\,g^{-1}$)

Type	Fat content
Half cream	12
Single cream	18
Whipping cream	35
Double cream	48
Clotted cream	55
Sterilised cream	23

5.2.5 Processing of cream

The two products from milk separation, skimmed milk and cream, each require heat treatment to comply with legal requirements as well as hygiene standards. The heat treatment may be pasteurisation, or ultra high temperature (UHT), or sterilisation. Legal definitions for heat-treatment processes in the UK specify the following minimum time–temperature combinations:

- pasteurisation, 72°C for 15 s
- UHT, 135°C for 1 s
- sterilisation, 108°C for 45 min

An additional heat treatment regime called thermisation is used by the dairy industry to improve the keeping quality of milk on the production site pending its final use. This method may use a final temperature of 63 or 65°C for 15–20 s and does not destroy pathogenic microorganisms.

The thermal preservation of cream is similar to that of milk, involving heat treatment to the desired temperature, holding at that temperature for a predetermined period (the holding time) and then cooling to a lower temperature for packing.

5.2.5.1 Batch heat treatment

Small volumes of cream (up to about 400 l) may be pasteurised by using a batch method. Here, a jacketed tank or a tank with an internal coil is used; the tank is fitted with a slow-moving agitator (20–30 rpm). Warm water at 63–65°C is circulated around the jacket or coil and a temperature probe is inserted into the cream to monitor the product temperature. The cream temperature is also recorded on a chart, as required by the legal guidelines. Upon reaching the pasteurisation temperature (63°C), the product is held at that temperature for a minimum period of 30 min before cooling to about 10°C. To commence cooling, cold water is circulated through the jacket or coil. Alternatively, the cream is pumped through a small heat exchanger for continuous and rapid cooling to below 10°C.

The batch system offers flexibility and economy since product wastage is also minimised. However, automation is limited and, with the absence of regeneration, energy efficiency is also poor. Traditionally, this type of processing was used to manufacture clotted cream, with typical conditions being the use of a temperature of 96–98°C for 40–45 min.

5.2.5.2 High temperature for short time processing

High temperature for short time (HTST) processing is the most popular method used for the production of fresh cream. The heat exchange system can be made by using a series of thin plates or by using a tubular arrangement.

Heat transfer area calculations for plates and tubes have been described by Kessler (1981), Lewis (1987) and Fellows (1996). For plate systems, 98% regeneration efficiency may be possible with modern pasteurisers. The aim is to re-use the energy from the hot product (milk or cream) to heat the incoming cold product. The regeneration efficiency, e^R, in a heat exchange system may be calculated as follows:

$$e^R = \frac{E^{\text{with}}}{E^{\text{without}}} \times 100 \tag{5.2}$$

where, E^{with} is the total energy used with use of regeneration, and E^{without} is the total energy used without use of regeneration. To improve heat exchange performance, design features such as a narrow gap between plates, turbulent flow, resistance to chemical detergents and ability to withstand operating temperatures and pressures are included.

In a HTST plate pasteurisation plant, several passes are provided for heating as well as cooling duties based on the flow rate and on the physical properties of the product, such as specific heat, thermal conductivity and viscosity. Heating energy for most commercial plants is supplied in the form of steam. Tubular heat exchangers are also used for pasteurisation duty. Some small commercial plants and laboratory heat exchangers have been designed to use electrical energy for heating. A laboratory tubular pasteuriser is shown in Figure 5.6.

The basic requirements for a HTST commercial pasteurisation plant include the following:

- a product pump usually of a centrifugal type, although a positive type is desirable for creams
- a flow control and monitor

Figure 5.6 A laboratory tubular pasteuriser.

- a product-holding section to complying with regulations and hygiene requirements
- a flow diversion value (FDV) located in the product-holding section
- automatic temperature control and a temperature-recording unit
- an event recorder to indicate when the plant diverts product

Operation of the product heat-treatment programme should include safety interlocks to ensure that the plant enters production mode only after CIP and sterilisation cycles have been completed. The pressure differential across various sections in the heat exchanger (especially in plate systems) must be maintained at a minimum safety level to prevent cross-contamination of the heat-treated product by raw product. If the pressure monitoring and recording system is not installed as a standard feature, the plant should be pressure tested annually and certificated.

In most tubular heat-exchange systems the energy can be recovered only from the heating and cooling medium (i.e. hot water and cooling water). Hot product cannot be cooled by using cold incoming product to exchange heat. This is because the inside surfaces of the service side of the tubular heat-exchange unit do not satisfy the standard of finish required for food applications.

5.2.5.3 Ultra high temperature treatment

Ultra high temperature (UHT) treatment is designed to produce a commercially sterile product by a continuous flow method; the product is then filled into an aseptic container. The legislation governing the UHT process may vary according to country. In the UK, Council Directive 92/46/EC (OJEC, 1992) and the Dairy Products (Hygiene) Regulations (1995) stipulates that, while milk should receive a minimum of 135°C for not less than 1 s to be designated as UHT milk, for UHT cream, the UK regulations require a heat treatment of 140°C for not less than 2 s or equivalent. However, typically, commercial processes use continuous heat treatment in the range 135–150°C for 6–2 s, and then cooling to 15–25°C prior to aseptic filling.

Definitions of UHT methods are specific to the precise nature of the heat treatment. The methods commonly used are (Ellborg and Tragardh, 1985):

- indirect, with use of steam or super-heated water but without direct contact with the product
- direct, with steam or electrical heating:
  - using steam can be further catagorised as steam injection, where steam is introduced into product or steam infusion, where product is introduced to steam
  - electrical heating (e.g. ohmic heating), with use of electrical resistance

A comprehensive account of UHT processing has been documented by Burton (1988). The heat treatment given to a product was originally quantified in terms of lethality values based on the work carried out in the food canning industry. For example, the lethality value given to food products is described in terms of F_0 values with reference to the death rate of the organism *Clostridium botulinum*. An F_0 of 1 is given when a product receives a heat treatment of 121.1°C for one minute. A simplified formula to derive the F_0 values in high-temperature processes is as follows:

$$F_0 = 10^{\left(\frac{T-121.1}{z}\right)} \times t \qquad (5.3)$$

where, T is the temperature of the process in degrees centigrade, t is the time in minutes, and z is the change in temperature (in °C) required for thermal death time to transverse one $\log_{10}$ cycle.

Kessler and Horrak (1981) described alternative dimensionless values, B^* and C^*, to quantify the lethal effects of a heat treatment. A B^* value of 1 refers to a heat treatment where the spores of *Bacillus stearothermophillus* were reduced by 10^9 $\log_{10}$ cycles. A C^* value of 1 refers to a heat treatment where vitamin B_1 (thiamine) content is reduced by 3% of the original concentration. For milk and cream processing it is necessary to achieve the highest B^* value possible and the lowest C^* value possible. These values can be calculated by using graphical methods, as described by Kessler (1981) and by Kessler and Horrak (1981). Modern UHT processing plants are designed to achieve the optimum lethal rates for a particular product. Various temperature and time profiles for commercial UHT plants are given in Figure 5.7. A complete UHT processing and aseptic filling installation is shown in Figure 5.8.

Steam barriers are used to isolate various units (e.g. the process plant from the aseptic buffer tank). One drawback in aseptic installations using steam protective

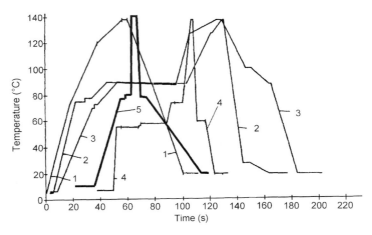

Figure 5.7 Temperature–time profiles for some commercial ultra high temperature plants. 1, Indirect plate, 85% regeneration; 2, indirect plate, 95% regeneration; 3, indirect tubular, 60% regeneration; 4, indirect plate, 70% regeneration; 5, direct steam injection. Reproduced by courtesy of Milk Marque Worcester, UK.

barriers is deposit formation in the valve clusters arising from stagnant product in that area as a result of frequent stoppages. Deposit formation can be reduced by lowering the steam pressure (e.g. 20 kPa) or through modifications to allow condensation to build up in the barrier without any danger to the integrity of the system (Carlin, personal communication).

5.2.5.4 In-container sterilisation

The market for in-container sterilisation in the UK and in the rest of Europe is small. Here the cream is prepared by standardising to 23 g 100 g^{-1} fat. Permitted stabilisers are then added and homogenised at (15–20 MPa pressure) before filling into tin cans. Filled and seamed cans are then retorted at 115–120°C for about 15–20 min depending on the can size and temperature used. The retorting may be carried out in a batch system or in a continuous process. In the 1995 regulations, the minimum requirement for sterilisation of cream was 108°C for at least 45 min or equivalent.

5.2.6 Factors affecting cream quality

When good-quality cream is made available for processing it can deteriorate. The following are the main factors involved:

- shearing effect (fat globules can be damaged as a result of mechanical action)
- heat-treatment temperature profile
- homogenisation conditions
- storage temperature

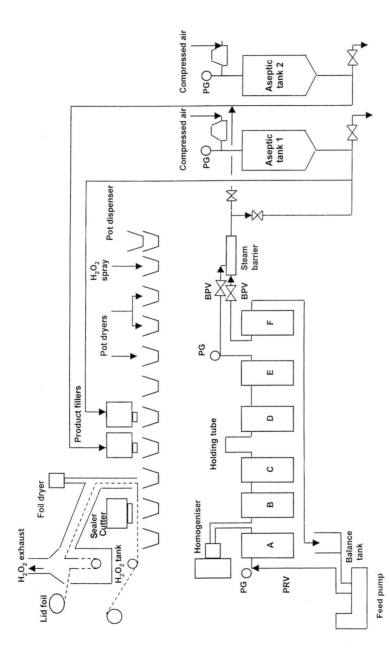

Figure 5.8 A schematic diagram of a continuous aseptic processing and packaging system. A, Pre-heat 1; B, pre-heat 2; C, ultra high temperature unit; D, cooler 1; E, cooler 2; F, sterilising cooler; BPV, back pressure valve; PRV, pressure relief valve; PG, pressure gauge.

Shear damage arising from pumping methods and flow restriction causes disruption to the fat globule membrane. After separation of the skimmed milk, the cream fraction should receive gentle handling to minimise the extent of damage to fat globule membranes. In some installations, centrifugal pumps were used to pump warm cream (40–60°C) so that the damage caused was relatively low compared with that caused when the cream was pumped at a temperature below 10°C. However, in modern installations, positive pumps are used as a standard for handling creams as they cause less damage to the fat globule membrane. Such pumps are fitted with speed control devices, such as hydraulic systems, mechanical gears or frequency inverters. Therefore, the pump speed can be adjusted to the desired product flow rate to minimise the shear forces acting on the fat globules. The processing installation for cream requires uninterrupted flow with minimal use of orifice plates or bends.

Once the fat globule membrane is damaged, some of the fat exits the fat globule and floats to the surface. Damage to the fat globule membrane also increases clumping and cluster formation, which leads to destabilisation of the emulsion. The degree of fat globule damage can be assessed by testing for free fat by using the solvent extraction and centrifugal methods described by Fink and Kessler (1983). A high percentage of free fat in cream reflects poorly processed cream. It is also important to reduce the free fat in whipping cream as it tends to increase serum separation in the foam as well as to reduce its overrun and firmness, making it less than ideal for bakery applications. Also, the oxidation of free fat is rapid, and, if free fat is present, rancidity is quick to develop before the end of the normal shelf-life of the cream.

A certain proportion of fat globule damage, together with globule size adjustment and crystal formation plays an important role in the production of thick cream. The milk separation temperature also influences the thickening process. For example, a separation temperature of 40°C increases the viscosity compared with a separation temperature of 65°C (Rothwell *et al.*, 1989).

A dairy homogeniser is a high-pressure positive pump designed to develop pressures up to about 40 MPa, but in some operations it may be higher. Three or more pistons are used to develop such high pressures against a homogenising valve. Details of homogeniser components have been reported by Phipps (1985) and Bylund (1995).

The homogenisation process reduces the size of fat globules; typically, they have a diameter in the range of 0.5–1 μm at 15–20 MPa pressure. The homogenising head usually consists of two stages to develop the pressure. Homogenisation is achieved in the first stage; the second stage helps to keep a constant pressure in the first stage and to break up clusters formed immediately after the first stage. In early equipment designs, the homogenisation pressures were adjusted manually, but in modern units this is done with use of hydraulic adjustment devices.

Homogenisation efficiency is affected by temperature, the flow rate of the product and the design of the homogenising valve (Phipps, 1985). In most commercial installations, milk and creams are homogenised at temperatures above 50°C to ensure that the fat is in liquid form.

In addition to having a particle size reduction capability, the homogeniser is a metering unit which controls the flow rate of the process plant.

5.3 Processing recommendations for high fat content products

Butter, anhydrous milk fat (AMF), Ghee and fractionated milk fats are all products derived from milk and cream (Figure 5.2). The best practice for processing cream destined for buttermaking is as follows.

1. *Starting material*: one should use only fresh cream. Even holding pasteurised cream at low temperatures can result in microbial growth. The heat-stable proteolytic or lipolytic enzymes which survive processing can cause off-flavours to develop in stored butter (Russell, 1973).
2. *Pumping and piping*: positive pumps cause less damage to the fat globules, although centrifugal pumps can be used if correctly sized. Avoidance of features which can cause excessive turbulence from high pressure drops, such as sharp bends, will minimise product damage.
3. *Temperature*: to minimise microbiological and fat globule deterioration, the separated cream should be held and transported or piped below 6°C. The evidence shows that at temperatures between 10–35°C cream is more susceptible to fat globule damage (Mulder and Walstra, 1974; Te Waiti and Fryer, 1975).
4. *Neutralisation*: this is best avoided. However, when poor handling and management of cream leads to the production of lactic acid, automatic dosing of alkali for neutralisation is possible (although this practice is rare these days).
5. *Taint removal*: cows which graze on fresh pasture sometimes feed on weeds which can taint milk (McDowall, 1953). Equally, aromas in the milking parlour or shed can also lead to similar problems. The removal of taints is particularly important for the production of sweet cream butter, but it is also important for lactic butter. Vacuum pasteurisation of cream eliminates most taints. However, when taint levels are very high, a steam distillation apparatus is used (Scott, 1954). Jebson (1994) has described a number of items of equipment used for cream handling and has also compared flash pasteurisation and vacreation used in taint removal.

5.3.1 *Properties required of high fat emulsions for table spreads*

The terms texture, consistency and plasticity describe the physical qualities of fat. Texture primarily refers to its structure and can be smooth, floury, grainy, granular, sandy, coarse or lumpy. Consistency is temperature dependent, and may be, a smooth, even plastic, state varying from soft, medium, firm or tough to hard and brittle. As it is possible to encounter smooth as well as grainy and coarse plastic products, plasticity is associated with both texture and consistency.

The characteristic feature of a plastic fat is its ability to retain its shape under slight pressure, such as encountered during rolling, mixing or spreading. Rheologically, therefore, plastic fats possess a yield stress. This is made possible by three inherent properties, all of which can be controlled and manipulated during mechanised processing. First, the fat contains two phases, one of which must be a solid and the other a liquid (or at least they must behave as such). Second, the dispersion of the solid phase is sufficiently fine for the entire mass to be effectively held together by internal cohesive force (i.e. the size of the solid particles should be sufficiently small so that the force of gravity on each is negligible in relation to the adhesion of the particle to the mass), and the interstitial spaces must be small enough to prevent the liquid phase seeping from the material. Third, in keeping with the concept of plasticity developed by Bingham (1922), the balance between the phases present must be such that the solid particles should be small enough to facilitate mass flow without causing obstructions or by forming a rigidly interlocking crystalline structure.

The spreadability of butter, margarine and other yellow fat spreads is a term often used in connection with their solid state; however, it is a function of consistency, and the shear of the material on which the fat is spread (e.g. bread) also has a strong influence.

Texture, consistency and spreadability are important for their physiological and psychological effects on flavour and general palatability. Such effects are essentially decided by glyceride composition, state of emulsion and process conditions used during crystallisation of the fat.

Larsson (1966) found that complex triacylglycerol mixtures, such as margarine, exhibit four polymorphic crystal forms: α, β', β_2 and β_1, in increasing order of stability. The effect of cooling on milk fat during buttermaking results in several changes in polymorphic form (Table 5.4).

Table 5.4 Polymorphism in milk fat

Cooling	Polymorphic crystalline form	Melting point (°C)
Rapidly to 5°C	α	17–22
To 15–18°C	β'	28–31
To ≥ 21°C	β	35–38

5.3.1.1 Emulsions for table spreads

Controlled crystallisation of fat in an emulsion contributes a solid crystal network to stabilise the emulsion. However, this alone is insufficient. Since an emulsion is a mixture of two immiscible fluids, the presence of an emulsifier at the interface of the two fluids is required to reduce interfacial tension. Typically, this is provided by mono- and diacylglycerols, the base materials for emulsifiers which are used in water-in-oil emulsions such as margarines.

Two types of butter are produced through churning: sweet cream and lactic butter (Table 5.5). The latter process was developed by The Netherlands Dairy Research Institute and is a method for producing an aromatic soured butter from sweet cream without acidifying it (Frielink, 1977). Butter churned from cream does not require additional emulsifiers (see Table 5.5). However, butter that is made spreadable by addition of vegetable oils requires some addition of emulsifiers as well as colour. Products that are lower in fat content (i.e. at 65–75 g $100\,g^{-1}$ fat) begin to require additional ingredients to hold both fluids together and also to enhance eating quality so that they resemble butter closely. In low fat products (with less than 40 g $100\,g^{-1}$ fat), the aqueous phase must be fortified to increase its viscosity to close to that of fats to prevent emulsion breakdown and to ensure its organoleptic quality is not impaired. Sometimes fat replacers, such as carbohydrates or microparticulate whey proteins, are also used.

5.3.1.2 Preparation of emulsion for buttermaking

Oil-in-water emulsions such as cream (Jebsen, 1994) are concentrated to 35–40 g $100\,g^{-1}$ fat by expelling a considerable amount of water. The cream is then 'ripened' or aged (i.e. held overnight or for at least 8 h at a temperature below 4°C). This helps the milk fat in the globules to undergo partial crystallisation, which aids churning.

During churning the emulsifiers, which stabilise the original oil-in-water emulsions, are inactivated by aeration, particularly the milk fat globule membrane. This makes it possible for phase inversion (i.e. for the fat to flow out of the globules and form a continuous phase). Thus, a water-in-oil emulsion is formed, as found in butter. However, unlike margarine-type emulsions, the emulsion in butter is complex as it also contains some oil-in-water droplets and is therefore a mixed emulsion.

5.3.1.3 Preparation of emulsion for dairy spreads

With respect to recombined butter or margarines and lower-fat water-in-oil emulsions, the first step is to introduce an aqueous phase into a fat phase. Hence, a pre-emulsion is sometimes required. Before scraped-surface cooling was widely available the emulsion had to be first cooled by ice and then worked or kneaded to produce the final product. The introduction of scraped-surface heat exchangers (SSHEs) has significantly improved the mechanical stabilisation

Table 5.5 Summary of ingredients, technology and processing conditions used in the manufacture of yellow fats spreads

	PRODUCTS			
	Butter (sweet cream and lactic) 80–82 g 100 g^{-1} fat	Margarine/blended spread 80–82 g 100 g^{-1} fat	Blended spread 65–75 g 100 g^{-1} fat	Low-fat spread typically 40 g 100 g^{-1} fat
PROCESS	Churning (cream to be preformed emulsion, only a single stage is therefore needed)	Emulsification followed by churning or emulsification followed by SSC and texturisation	As for margarine	Emulsification followed by SSC and texturisation
INGREDIENTS (in the fat phase)	Cream (uncultured or cultured)	Cream vegetable oil, hydrogenated vegetable oil	Butter, butter oil, vegetable oil, hydrogenated vegetable oil	Butter oil, vegetable oil, hydrogenated vegetable oil, fat replacers (optional)
EMULSION TYPE	Oil-in-water	Oil-in-water for churning water-in-oil for SSHE	As for margarine	Water-in-oil for SSHE
OTHER INGREDIENTS	Water Salt Lactic culture 'Starter distillate'	Water Salt Emulsifier (E471) Colour (E160a)	Water Salt Emulsifiers (E471, E322) Colour (E160a, E160b, E100) Vitamins A/D Whey solids Flavouring	Water Salt Emulsifiers (E471, E322) Gelatin Stabilisers (E401, E412, EDTA) Colour (E160a) Vitamins A/D Preservative (E202) Skimmed milk powder Buttermilk powder Lactic acid Aroma

SSC, Scraped-surface cooling, SSHE, Scraped-surface heat exchanger (see page 139).

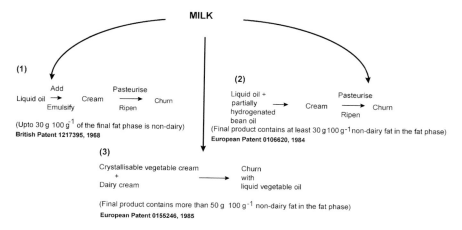

Figure 5.9 Techniques for preparing emulsions for manufacturing table spreads.

of the emulsion. This has been achieved mainly through efficient and fine dispersion of water in the fat phase (Rajah, 1992), which has also resulted in a reduction in the amount of emulsifier required.

Moran (1994) has reviewed the techniques available for blending fats. Three inventions best describe the variations in the manner of adding oil (Figure 5.9). According to Andersen and William (1965) the properties and functional benefits of a fine emulsion are as follows:

- A fine emulsion contributes to the plasticity of the product.
- A stable 80 g 100 g^{-1} fat emulsion contains 1 g water with an estimated 10–20 million water droplets; 95% of these water droplets have a diameter in the range 1–5 μm, 4% a diameter in the range 5–10 μm and 1% a diameter of around 20 μm.
- The shelf-life from microbiological spoilage of the product is improved by addition of emulsifiers, since an emulsion is more resistant to bacterial attack because in a small droplet any bacteria cannot multiply and soon die. [Note: bacteria found in spreads typically grow to about 15 μm, which makes it important to aim for droplet diameters below this (see above)]. However, as lower-fat emulsions are introduced, the increase in the amount of water results in a greater quantity of larger diameter droplets forming, and some may even have diameters of up to 80 μm.
- Flavour perceptions from added water-soluble flavours may be masked by too fine a dispersion. Although some good oil-soluble flavours are available, it is believed that only water can wet the taste buds. In addition, too fine an emulsion may also be absorbed without being broken on the palate and therefore may never really be tasted.

5.3.2 Butter manufacture

5.3.2.1 Traditional methods

The butter churn dates from before the 1480s. Early designs consisted primarily of a wooden barrel that rotated on its horizontal axis. To aid churning, these barrels were fitted with internal side baffles which lifted and dropped the cream to create the necessary working action.

At the turn of the twentieth century, as rail transport linked rural areas to the towns, buttermaking increased in scale from farm-scale dairies to industrial-scale creameries to meet increasing demand. The batch-scale units underwent significant modification and upgrading. Three notable changes were as follows:

- batch sizes were scaled up to typically 6 tonnes per batch
- the need for quality led to the use of more hygienic and easily sterilisable construction materials (e.g. magnesium aluminium alloys or sand-blasted stainless steel)
- automation was introduced with use of motor-driven churning

5.3.2.2 Continuous buttermakers

Continuous buttermaking techniques and processes were developed (McDowall, 1953) and introduced in the 1930s. Outstanding among these was the Fritz buttermaking process. The mechanisms involved have been reviewed by Walstra and Jenness (1984). Here, fresh ripened cream is continuously introduced into a churning cylinder which generates high shear conditions. This accomplishes aeration and foam formation within minutes. The foam then collapses, by a mechanism which is equally important in batch churning but takes substantially longer to complete: 20–40 min, depending on process conditions. Fat is released from the globules and droplets coalesce to form grains of butter. This mechanism applies to both batch and continuous processes. Contemporary Fritz-type continuous buttermaking equipment has a number of standard features:

- cream feed pump
- high-speed churning section
- section for separating buttermilk and butter grains
- buttermilk drain
- vacuum section
- working section
- salting and dosing (in the form of brine)
- final cone

Since final moisture levels need to remain within the $16 \, g \, 100 \, g^{-1}$ legal maximum, the size of the grains is important to allow good drainage of the aqueous phase (buttermilk). To aid removal of moisture, continuous buttermakers are designed with twin augers in the working section which squeeze out more of the buttermilk. The grains are worked and kneaded together, mixed and sheared as the augers force the butter through a series of perforated plates alternating with rotating impellers. The result is a homogeneous water-in-oil emulsion. The efficiency of the process can therefore be controlled by the design of the perforated plates and impellers, as well as by auger speeds.

Batch churned butter contains 3–5 g $100\,g^{-1}$ air and has a more open but sticky consistency. To improve product quality, the extruded butter from continuous buttermaking is passed through a vacuum chamber in which air is removed. This results in a more close texture and waxy appearance which is preferred over batch-churned butter. However, further mechanical disruption of globular fat also takes place in the vacuum chamber and working sections, which affects and controls the butter consistency and moisture content. These stages must be controlled closely so as to prevent underworking, which can lead to a coarse crumbly consistency, or overworking which results in a soft creamy texture. Both can cause formation of large moisture droplets resulting in free moisture in the butter (Hughes *et al.*, 1976).

5.3.2.3 Developments in buttermaking
Butter is also manufactured from high-fat cream (Munro, 1986). Typically, creams with more than 75 g $100\,g^{-1}$ fat undergo phase inversion and are then cooled and texturised mechanically following one of two routes. First, in the Tetra Pak equipment (i.e. in the New Way and Meleshin process) the cream is destabilised during cooling and mechanical working. Second, in the Cherry Burrell (Gold 'n' Flow) and Creamery Package processes, phase inversion takes place before the cooling and working stage.

An improvement has been claimed over both routes. In the Ammix process developed in New Zealand, scraped-surface cooling and pin working is used to convert an emulsion of butter oil, cream and brine into butter (Truong and Munro, 1983).

5.3.2.4 Texturisation of low-fat spreads and spreadable butter
The development of spreadable butter and dairy spreads (which contain some vegetable oil) using buttermaking technology is a relatively recent innovation. However, the addition of nondairy fat into butter was first reported over 100 years ago. It is now common practice for these spreads to be produced using scraped-surface cooling and texturisation. During the 1990s, these products overtook margarine as the main spread for bread.

Spreads containing mixtures of dairy and nondairy fats, known in the USA as butterines, have been churned in Wisconsin since 1967, containing 40 g $100\,g^{-1}$ milk fat and 40 g $100\,g^{-1}$ hydrogenated vegetable oil (Graf, 1972). In Europe, the Swedish product, Bregott, which was launched in 1976, is acknowledged to have started this phase of development. It was originally manufactured by adding 15–25 g $100\,g^{-1}$ vegetable oil (as percentage of total fat) to cream before batch churning. This technique was later readapted to Fritz continuous buttermaking by installation of an in-line static mixer. This enabled the cold vegetable oil to be blended with the cream as it is pumped to the continuous buttermaker.

In a further development in the UK, in 1983, a mixed-fat product, Clover, was successfully launched containing more than $50 \, g \, 100 \, g^{-1}$ of the fat phase as nondairy fat. Again, the product was manufactured by churning 'cream' prepared with vegetable oils to produce a spreadable butter-like product. A similar product, manufactured by Sköne, was launched in Sweden with use of scraped-surface technology.

5.3.2.5 Scraped-surface technology

Scraped-surface cooling and texturisation of fatty emulsions (i.e. blends of oils and dairy fats) has been adopted by the dairy industry worldwide to manufacture fresh butter, recombined butter, dairy spreads, and a variety of products based on anhydrous milk fat and milk fat fractions. It has enabled the industry to expand the use of dairy products globally and also to arrest the decline in butter consumption in many Western countries.

The tubular cooler, or 'votator' cylinder, was first introduced by the Girdler Corporation for margarine manufacture in 1938. In principle, the tube had the scraping surface in its inner wall. Inside, a coaxial rotor with an approximate diameter of 20 mm less than that of the tube was fitted with scraper blades which scraped the walls of the nickel tube under centrifugal force at speeds in the range of 400–700 rpm.

The votator as a heat exchanger has been reviewed from a chemical engineering standpoint by Hosking (1962). Details of major developments in plant and equipment up to the 1960s have been discussed by Andersen and William (1965). Nevertheless, in principle, at the start of the twenty-first century technology remains similar to the original Girdler design of 1938.

A number of important developments have shaped the growth and widespread use of scraped-surface technology since 1938. These include: improved seal arrangements, higher pressure ratings, more effective scraper blade designs and use of multiple cooling tubes and working sections for lower-fat products.

An increase in the variety of oils available for yellow fat products [the most notable for the dairy industry being fractionated anhydrous milk fat (milk fat fractions)], required technology which was flexible. During recent years, owing to consumer demand, other oils which have grown in importance are typically low erucic acid rapeseed oil (LEAR), sunflower oil, olive oil, palm oil and its numerous fractions, hydrogenated fish oil (in Denmark, nonhydrogenated fish oil has been used in spreads). A large proportion of these are used in spreads and bakery applications, both being product areas which require good quality textured fats.

The increased use of fat-modification techniques (e.g. hydrogenation, interesterification and fractionation) have provided new product opportunities for the yellow fats market (i.e. oils and fats with modified physical and chemical characteristics).

Increased consumer awareness of the need for a balanced diet led to the development of polyunsaturated fatty acid (PUFA) margarines in the late 1970s and early 1980s. This was followed closely by a move towards lower-fat margarines and dairy spreads and very-low-fat spreads. All of these required close attention to process conditions and, in some cases, use of specialist equipment (e.g. for the blending of vegetable oils with dairy cream prior to churning).

Increase in demand for high melting pastry margarine and fats in Europe and the Far East since the mid-1970s has led manufacturers to the more hygienic and efficient scraped-surface process in preference to the traditional chill drum method, which is also relatively more labour intensive. Equally, high-pressure (up to 16 MPa) tubular coolers are now standard in many of these operations.

A comprehensive treatment of margarine manufacture is found in a book by Andersen and Williams's (1965) and a recent review is available from Hoffman (1989). Moran (1994) has reported on the preparation of lower-fat emulsions. Details of buttermaking were described by McDowall (1953), and advances have been reviewed by Munro (1986) and Jebsen (1994).

5.3.3 Anhydrous milk fat

Commercial and factory operations are designed to manufacture AMF from one of three routes: from milk, cream or butter (Figure 5.10). There are a number of implications for each.

A butter-powder factory design typically comprises plant and equipment capable of milk reception, treatment, and processing into butter and liquid skim (the latter into skimmed milk powder if necessary).

5.3.3.1 Anhydrous milk fat from milk

Separation of milk into cream and skimmed milk form the initial step. The cream is retained for buttermaking or processing into AMF; the skimmed milk will need an outlet (e.g. spray drying, Feta cheese manufacture), and appropriate plant will be required to deal with this. Typically, the alternatives for skimmed milk are for retail sale into the liquid market, for spray drying into powder and for the manufacture of caseinates and or fermented milk products.

5.3.3.2 Anhydrous milk fat from cream

Cream can be sourced externally or produced from milk, as described above. During AMF manufacture, the serum leaving the heavy-phase outlet contains 1–2 g 100 g^{-1} fat, and this can be passed into the buttermilk separator for recovery. However, with efficient separation, some operations recover skimmed milk from the initial stages of fat concentration and buttermilk serum after phase inversion takes place. Both require further handling, and hence separate plant and equipment, if they are to be used for processing in-house or for sale as liquid products.

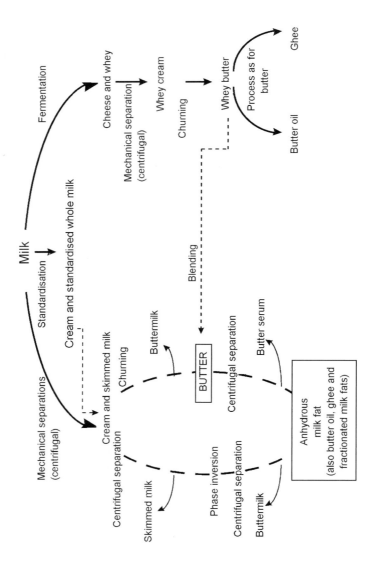

Figure 5.10 Process routes for the manufacture of anhydrous milk fat from milk, cream and butter. Reproduced courtesy of Klondyke Management Consultants, Essex, UK.

5.3.3.3 Anhydrous milk fat from butter

The butter serum produced, as a by-product when salted butter is used as the feed, will be far too salty for food use. Equally, butter of poor microbiological quality will result in butter serum which is unsafe. However, when unsalted butter is used by smaller AMF manufacturers, the serum volumes produced may be too small for regular supplies to be available or relied upon for use by industrial customers or in any high-volume retail products manufactured in-house. An illustration of the processing of butter into AMF is provided in Figure 5.11.

5.3.3.4 Sources of cream

The main source of cream is from whole milk which is procured mainly for buttermaking. However, there are other sources from which cream is obtained: surplus cream from low-fat or standardised market milk, or cheese whey. The latter source of cream is made into whey butter. It has been estimated that in 1995 in the UK 2000 tonnes of whey butter was manufactured from cheese whey, its final use being conversion to Ghee for export. The actual tonnage of whey butter from farmhouse cheesemaking that went into the retail sector as household table spread is not recorded.

5.3.3.5 Practical considerations

When making AMF with either cream or butter, it is important to take into account the need to adapt processing conditions when different feedstocks are used. Phase inverted cream at 80 g 100 g^{-1} fat or sweet cream butter tend to form an emulsion layer containing fat globular material, which on close analysis is found not to have been converted and, therefore, still to contain milk proteins and phospholipids.

Lactic butter (also referred to as cultured butter) or sweet cream butter with its serum pH adjusted to 4.5 may be used. In this case the proteins become denatured at this pH level and so, instead of an emulsion layer, a precipitate or sedimented solids are evident.

A critical stage in the conversion of cream to AMF is reached when it transforms from an oil-in-water into a water-in-oil emulsion (Figure 5.12). This process is dependent on efficiency of separation. Complete phase inversion is not possible. Illingworth and Bissell (1994) highlighted the major factors that influence the efficiency of phase inversion in order of importance:

- homogenisation pressure
- fat content of the cream
- free fat content of the cream
- temperature

They also proposed that acidity in cream does not have any significant effect.

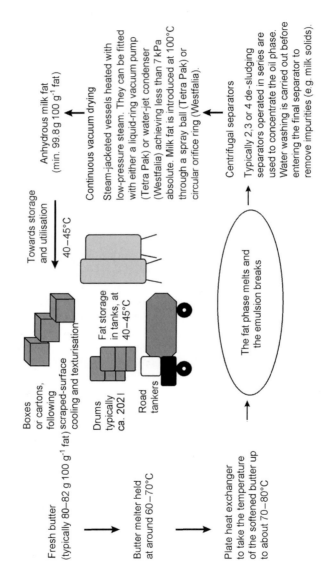

Figure 5.11 Processing of butter into anhydrous milk fat.

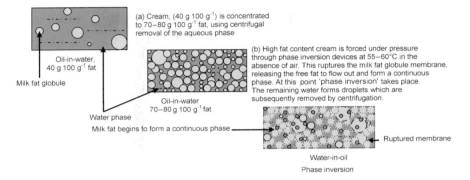

Figure 5.12 Mechanism for the phase inversion of an oil-in-water to a water-in-oil emulsion.

The shelf-life of AMF is considerably greater than that for butter. It is reported that AMF produced from good quality butter and stored under nitrogen at $-12°C$ will keep for 3–4 years with no off-flavour developing. Also, in recent years, large manufacturers of pastry who have traditionally used butter now find it more convenient to use AMF, which is more easily transported in bulk tankers, stored, distributed and portioned. The transport and storage of AMF in melted form in heated tanks is now common practice. AMF can be stored at 40–50°C for 2–3 weeks with no impairment of quality (Timmen, 1981).

5.3.4 Ghee

Ghee is clarified butter popular in the Middle East and Asia, particularly in the Indian subcontinent. In warmer climates, it is not possible to hold butter for long periods without deterioration of the product. Therefore, the removal of water by heating butter in open vessels is common practice, resulting in well-developed 'buttery' flavours. These flavours are not only from the fat itself, but also from the milk solids which are present. Additionally, the smoke from open wood fires is absorbed into the oil, which contributes to unique flavours depending on local practice and custom. Some processors even burn the dried leaves from specific plants to create flavour. When the butter oil is collected and stored, slow cooling and crystallisation leads to the formation of a grainy physical structure. This is a typical characteristic of Ghee.

5.3.5 Butterschmalz

Another traditional product with low residual moisture, called 'Butterschmalz' (butter lard), is made in South Germany, Austria and Switzerland. It is prepared from 'low-grade' butter (e.g. too high or low a moisture content, poor body) by removal of the serum or by boiling off the water. It is used mainly for baking and cooking (Timmen, 1981).

5.4 Fractionation of milk fat

The major thrust in milk fat fractionation came during the mid-1960s. Early studies looked at solvent as well as detergent fractionation, but neither process was developed further because of problems with residual solvents and flavour. Owing to the strong desire of the international dairy community to keep butter a 'natural' product, resources were soon focused on fractionation by crystallisation from the melt (i.e. by dry fractionation). Although initial interest centred around using the rotary drum Stockdale-type vacuum filter, commercial-scale dry fractionation of milk fat really only took off with the development of the Tirtiaux Florentine™ continuous-bed vacuum filter (see Figure 5.13).

During the mid-1980s, a stationary-bed 'closed' vacuum filter, the Vacuband, was evaluated in semi-commercial-scale investigations (Rajah, 1988). This filter gave a significant improvement in yield of olein (about 80%) over the Tirtiaux Florentine™ filter (about 70%). During the same period, the development of the membrane filter led to similar or better yields and, consequently, at least two plants have been put into commercial operation within the dairy sector. The introduction of the high-pressure membrane filter enabled high yields of olein.

The design of the crystalliser vessels and stirrers plays an important role in milk fat crystallisation and the quality of crystals formed. Also important are the effects of agitation and cooling rates.

Milk fat fractions in the melting point range 6–46°C have been produced by using the Vacuband filter. Analyses carried out [e.g. triglyceride carbon number (TCN) analysis, fatty acid methyl ester (FAME) analysis and solid fat content (SFC) analysis] show that the triacylglycerol groups C:38 and C:50 are two of the largest occurring fatty acids and are dominant in the olein and stearin fractions, respectively. Moreover, a useful means of distinguishing between milk fat and fractions is the oleic acid/stearic acid ratio, which also correlates well with slip point. The SFC value determined at 20°C (N20 value) is a valuable indicator of the physical character of milk fat and its suitability for application in specific food products. In England and Wales the range for the N20 value for summer–winter milk fat is 13–22%. Fractionation is used to produce milk fats with more consistent melting profiles (N20 values).

When milk fat is fractionated, the main changes evident are in relation to flavour (taste and aroma), colour and physical properties (melting point and solidification point). The fractions need to be specially prepared for use in any one of a range of food applications, which include a number of major food sectors [namely, bakery (biscuit, cake, pastries and buttercream), chocolate, yellow spreads, sauces, mayonnaise, cream and ice-cream].

Supercritical carbon dioxide (CO_2) extraction has been investigated. The technique facilitates more efficient separation of fractions based on molecular weight. The liquid fraction is enriched in short- and medium-chain triacylglycerols and fatty acids, while the solid fraction is abundant in both saturated

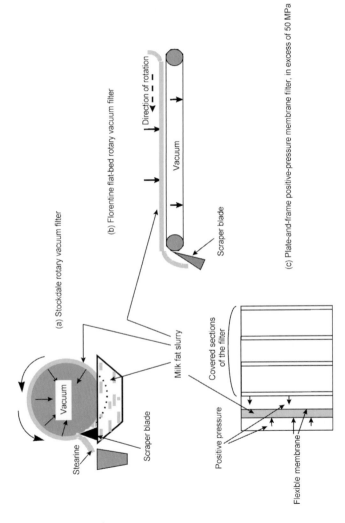

Figure 5.13 Mechanised milk fat fraction: principles of three major separation systems.

and unsaturated long-chain fatty acids. This offers fractions with distinctive differences in chemical and physical properties for use in a variety of food applications.

Short-path distillation as a fractionation technique is superior to supercritical CO_2 extraction. In principle, it consists of evaporating molecules into a substantially gas-free space (i.e. a vacuum) and is controlled by the rate at which the molecules escape from the heated surface of the distilling liquid and are received by the cooled condenser surface. It has been used to recover volatile compounds of milk fat and cholesterol from butter oil; the development of dietetic butter enriched in short- and medium-chain triacylglycerols for dyspeptic patients is also possible.

References

Andersen, A.J.C. and William, P.N. (1965) *Margarine*, Pergamman Press, Oxford.
Bingham, E.C. (1922) *Fluidity and Plasticity*, McGraw-Hill, New York.
Burton, H. (1988) *Ultra-high-temperature Processing of Milk and Milk Products*, Elsevier Applied Science, London.
Bylund, G. (1995) *Dairy Processing Handbook*, TetraPak Processing Systems A/B, Lund, Sweden.
Dairy Product (Hygiene) Regulations (1995) SI No. 1086, HMSO, London.
Ellborg, A. and Tragardh, C. (1985) Hydrodynamic considerations in continuous flow sterilisation, in *Aseptic Processing and Packaging of Foods*, Proceedings of a Symposium, Lund University, Tylasand, Sweden, pp. 85-99.
Fellows, P.J. (1996) *Food Processing Technology: Principles and Practice*, Woodhead Publishing, Cambridge.
Fink, A. and Kessler, H.G. (1983) Determination of free fat in cream for assessment of damage caused to fat globules by processing. *Milchwissenschaft*, **38** 330-4.
Frielink, J.G. (1977) Method for producing aromatic soured butter without acidifying it. *Netherlands Patent* 7513464.
Hoffman, G. (1989) *Chemistry and Technology of Edible Oils and Fat in High Fat Products*, Academic Press, New York.
Hosking, A.P. (1962) Chemical engineering. *Trans. Inst. Chem. Eng.*, **40** A97.
Hughes, I.R., Jebsen, R.S. and Tuttieth, P.T. (1976) Moisture control in continuous buttermaking. *NZ J. Dairy Sci. Technol.*, **13** 29-36.
Illingworth, D. and Bissell, T.G. (1994) Anhydrous milkfat products and applications in recombination, in *Fat in Food Products* (Eds. D.P.J. Moran and K.K. Rajah), Blackie Academic & Professional, London, pp. 111-54.
Jebsen, R.S. (1994) Butter and allied products, in *Fat in Food Products* (Eds. D.P.J. Moran and K.K. Rajah) Blackie Academic & Professional, London, pp. 69-106.
Johannson, M.J.S. (1980) *UK Patent* 1 582 806.
Kessler, H.G. (1981) *Food Engineering and Dairy Technology* (translated by M. Wotzilka), Verlag, A. Kessler, Freisling, Germany, pp. 173-83.
Kessler, H.G. and Horrak, P. (1981) Objective evaluation of UHT milk heating by standardization of bacteriological and chemical effects. *Milchwissenschaft*, **36** 129-33.
Lewis, M.J. (1987) *Physical Properties of Food and Food Processing Systems*, Ellis Horwood, Chichester, pp. 246-75.
McDowall, F.H. (1953) *The Buttermaking Manual*, Vol. 1, New Zealand University Press, Wellington.
Milk Marketing Board (1984) European Patent 0106620.

Moran, D.P.J. (1994) Fats in spreadable products, in *Fats in Food Products* (Eds. D.P.J. Moran and K.K. Rajah), Blackie Academic & Professional, London, pp. 155-207.

Mulder, H. and Walstra, P. (1974) *The Milkfat Globule*. Commonwealth Agricultural Bureau, Farnham Royal, UK, pp. 101-28.

Munro, D.S. (1986) Alternative processes, *Bull. Int. Dairy Federation*, **204** 17-9.

OJEC (1992) Council Directive 92/46/EC. *Official Journal of the European Communities*, Annex C, Ch. 1. No. 268/24.

Olsson, I.T.H. (1968) British Patent 1217395.

Phipps, L.W. (1985) High pressure dairy homogeniser. *Tech. Bull.*, **6** National Institute for Research in Dairying (NIRD), Reading, UK.

Rajah, K.K. (1988) *Fractionation of Milkfat*, PhD Thesis, Department of Food Science, University of Reading, UK.

Rajah, K.K. (1992) Milkfat emulsion behaviour in scraped-surface heat exchangers. *Lipid Technology* (Nov.–Dec.) 129-37.

Rothwell, J., Jackson, A.C. and Faulks, B. (1989) Modification of the control of cream viscosity, in *Cream Processing Manual* (Ed. J. Rothwell), Society of Dairy Technology, Huntingdon, pp. 87-97.

Russell, R.W. (1973) The effect of cream handling conditions on butter quality. *NZ J. Dairy Sci. Technol.*, **8** 124-6.

Scott, J.K. (1954) The steam stripping of taints from liquids: an analysis of processing methods with particular reference to the Vacreator. *J. Dairy Res.*, **21** 354-69.

Svenska Mejeriernas (1985) European Patent 0155246.

Te Whaiti, I.E. and Fryer, T.F. (1975) Factors that determine the gelling of cream. *NZ J. Dairy Sci. Technol.*, **10** 2-7.

Timmen, H. (1981) Pure butterfat, a source of fat in food. *AOCS Monograph*, **8** 89-100.

Walstra, P. and Jenness, R. (1984) *Dairy Chemistry and Physics*, John Wiley, New York.

6 Yoghurt and other fermented milks
A.Y. Tamime, R.K. Robinson and E. Latrille

6.1 Background

Yoghurt and various fermented milks have been produced and consumed throughout the Middle East, India and Eastern Europe for centuries, but most of the countries with large English-, French- or Spanish-speaking populations have tended to regard cheese as the most important dairy fermentation. From an historical perspective, this view has much to commend it, for the low moisture content of hard-pressed cheeses, such as Cheddar, Cantal or Machego, has meant that products could be transported and stored under comparatively primitive conditions and yet remain fit for human consumption (Robinson, 1995); even the brined varieties, such as Feta or Kashkaval, can withstand summer temperatures of 25–30°C (Robinson and Tamime, 1991). By contrast, a soft cheese or fermented milk would normally have had to be eaten on the day of production; a pattern of purchase and consumption that was suitable only for rural communities.

However, the advent of the twentieth century saw two events that changed the situation in favour of fermented milks, namely the introduction of small mechanical refrigerators to many suburban households who were then able to buy and store fresh foods, and the publication of a book linking longevity among the hill tribes of Bulgaria with the consumption of yoghurt (Metchnikoff, 1910). In essence, Metchnikoff suggested that one aspect of the ageing process in humans involved the passage of noxious compounds from the intestine to the blood stream and that these chemicals arose from the action of undesirable, putrefactive bacteria in the lower ileum and colon. If the activity of these bacteria could be suppressed, then the adverse effects of their metabolic products would no longer be manifest and the individual might anticipate a longer and healthier life. Such an hypothesis sounded perfectly reasonable, and rumours of the health-giving properties of yoghurt aroused considerable interest, as did stories about the medicinal properties of fermented milks from Russia—kefir and koumiss.

The negative side of this scenario appears to have been the quality of the fermented milks available, and it is likely that most yoghurt was consumed for its mystical properties rather than for pleasure. In the absence of scientific evidence to support the claims of potential longevity, interest in the fermented milks waned and it was not until the 1950s that yoghurt regained a place in the marketplace. In the event, it was the addition of fruit that proved the turning point, for consumption became a pleasure and any thoughts of benefits to health were almost incidental. Over this period, yoghurt, and fruit yoghurt in particular, became the major fermented milks produced by the dairy industry, and volumes of production are still rising; an increase in per capita annual consumption of fermented milks—including yoghurt—over the past decade has been reported (Tamime and Robinson, 1999). This massive surge in sales is mainly a reflection of the fact that yoghurt is perceived as a safe, healthy and satisfying 'snack', but it is interesting that, in a sense, consumer attitudes to yoghurt have almost gone full-cycle with the so-called 'bio-yoghurts' being hailed as offering protection to the human body against cancer, failures of the immune system and other less dramatic ailments.

6.2 Classification of fermented milks

6.2.1 Mesophilic microfloras

Species of bacterium, such as *Lactococcus lactis* subsp. *lactis*, *Lactococcus lactis* subsp. *cremoris*, *Lactococcus lactis* subsp. *lactis* biovar. *diacetylactis* or *Leuconostoc mesenteroides* subsp. *cremoris*, that grow at temperatures of 20–22°C are found in many cheeses, but also in Scandinavian cultured milks (Kurmann *et al.*, 1992; Tamime and Marshall, 1997). The tentative correlation between a cool climate and the natural selection of mesophilic starter cultures appears again with respect to Kefir because the mixture of microorganisms that produce this acidic, mildly alcoholic (0.3–0.8% alcohol) drink (Robinson and Tamime, 1990) are best grown around 25–26°C; a summer temperature that would be typical of the Caucasian plains where Kefir is believed to have originated.

In North America, immigrants from Northern Europe have made buttermilk quite popular, and one characteristic of all these products is their distinctive 'buttery' flavour that results from high levels of diacetyl released by the cultures (Marshall and Tamime, 1997). In North Africa, a product similar to buttermilk goes under the name of Leben but, in general, the market for fermented milks is dominated by those produced with thermophilic cultures.

6.2.2 Thermophilic and/or therapeutic microfloras

This group consists of bacterial cultures with temperature optima of 37–45°C. Some products are manufactured with just one species of bacterium, and in Europe and North America *Lactobacillus acidophilus* is a popular choice (Sellars, 1991). The other popular product made with only one species (*Lactobacillus paracasei* biovar. *shirota*) in the culture is yakult, which is marketed as offering specific prophylactic and therapeutic benefits. *Lactobacillus rhamnosus*, *Lactobacillus reuteri* and *Lactobacillus paracasei* subsp. *paracasei* are further members of the same genus and, in due course, these latter species may well feature in 'health-promoting' dairy products.

However, the more popular products made with thermophilic cultures contain more than one species, and the most important, both historically and commercially, is yoghurt, a product that should have over one million viable cells per millilitre of both *Streptococcus thermophilus* and *Lactobacillus delbrueckii* subsp. *bulgaricus*. These key organisms are normal inhabitants of milk, and their interaction during growth gives yoghurt its unique organoleptic properties; indeed, in many countries, retail cartons may, by convention or by law, only be labelled as 'yoghurt' if *Lb. delbrueckii* subsp. *bulgaricus* has been used in the fermentation. Terms such as 'mild yoghurt' or 'bio-yoghurt' may cause confusion among consumers but are permitted in some countries to describe a product that is 'yoghurt-like' in texture. However, the prefix implies that the culture may be different from normal yoghurt, and currently the International Dairy Federation (IDF) Group D39 Fermented Milk Products is preparing a bulletin to encompass definitions of such products. Nevertheless, products that are heat treated after fermentation should be regarded as 'desserts'.

In addition to *Lb. acidophilus* and similar *Lactobacillus* spp., other bacteria that are known to play a distinct role in the ecology of the human gut are being introduced into fermented milk products. These latter cultures may include a number of species that are present in the large intestine, for example, *Bifidobacterium bifidum*, *Bifidobacterium breve*, *Bifidobacterium lactis*, *Bifidobacterium longum*, *Bifidobacterium adolescentis* and *Bifidobacterium infantis*. Many 'health-promoting' products now contain a mixed flora of *Lb. acidophilus*, *Bifidobacterium* spp., *S. thermophilus* and *Lb. delbrueckii* subsp. *bulgaricus*. The first two organisms provide the main 'health benefits', whereas the second two give the milk the flavour and physical characteristics expected by the consumer of a yoghurt-like product; the proven therapeutic or prophylactic properties of *S. thermophilus* and *Lb. delbrueckii* subsp. *bulgaricus* tend to be more limited.

It has also been proposed that *Enterococcus faecium* or, perhaps, *Entercoccus faecalis* could be employed as additional therapeutic cultures, mainly on the grounds that these species are normal inhabitants of the human intestine (Tunail, 2000). However, this proposal has proved somewhat controversial because some strains of *Ent. faecalis* can behave as mild pathogens in susceptible individuals, and because many strains of entercocci carry genes that confer resistance to penicillin and there is a risk that this genetic material might be transferred to other species of bacteria in the intestine. If one or more of these latter species became infective, then control of the disease could prove a problem.

6.2.3 *Microfloras including yeasts and lactic acid bacteria*

Fermented drinks produced with a mixed flora of yeasts and bacteria have been popular in Eastern Europe for hundreds of years, and the best known examples are Kefir and Koumiss. Traditionally, Koumiss is made from mares' milk, which has limited geographical availability, but Kefir, based on cows' milk, is now produced in limited quantities in both Western Europe and North America (Tamime and Robinson, 1999).

Although the component species of the microflora of Kefir can be grown separately for addition to the process milk, traditional Kefir is made with so-called 'Kefir grains'. These grains are small granules (0.5–1.0 cm in diameter) of a polysaccharide, kefiran, to which a mixed population of millions of yeasts and bacteria adhere, with yeasts towards the centre of the granule and the bacteria around the outside. What is remarkable about these grains is that the complex structure is self-replicating, and new grains that form during the Kefir fermentation are similar to the parent grains that were added originally; also, the microfloras that compose the grains vary from factory to factory. Nevertheless, some species of yeast or bacterium occur quite frequently (Robinson and Tamime, 1990; Tamime and Marshall, 1997). Some microflora that have been isolated from Kefir grains are shown in Table 6.1.

Koumiss has a similarly ill-defined microflora but, in this case, the process milk is inoculated with a batch of product made on the previous day, and the balance between the different genera is governed by environmental factors (e.g. quality of milk and exact temperature of incubation).

Other closely-related products, which are popular in Eastern Europe, are acidophilus-yeast milk and acidophiline. These fermented milk products have been developed for the treatment of certain intestinal and other diseases. The *Lactobacillus* spp. have a high antimicrobial activity against undesirable microfloras in the intestinal tract.

6.2.4 Microfloras including moulds and lactic acid bacteria

A unique cultured milk product, which is similar to Nordic or Scandinavian fermented milks, is widely produced in Finland and is known as viili. The

Table 6.1 Some selected microfloras of Kefir grains

Yeast	Bacteria
Torulaspora delbrueckii	*Lactobacillus kefir*
Zygosaccharomyces florentinus	*Lactobacillus brevis*
Kluyveromyces marxianus var. *fragilis*	*Lactobacillus acidophilus*
Saccharomyces cerevisiae	*Lactobacillus cellobiosus*
Candida kefyr (*kefir*)	*Lactobacillus helveticus*
Kluyveromyces marianus var. *lactis*	*Lactobacillus delbrueckii* subsp. *bulgaricus*
Kluyveromyces marxianus var. *marxianus*	*Lactococass lactis* subsp. *cremoris*
	Enterococcus durans
	Leuconostoc mesenteroides subsp. *dextranicum*
	Acetobacter aceti
	Acetobacter rasens

Source: Robinson and Tamime (1990); Tamime and Marshall (1997).

starter culture consists of mixed strains of *Lactococcus* spp., *Leuconostoc mesenteroides* subsp. *cremoris* and the mould, *Geotrichum candidum*. The milk base is not homogenised but is heated to 83°C for 20–25 min, cooled to 20°C, inoculated with the starter culture, packed and incubated at 18–20°C. During the fermentation stage, the fat rises to the surface where the mould grows to give a velvet-like appearance to the product, whereas the lactic acid bacteria produce acid to gel the milk (Tamime and Marshall, 1997).

Although markedly different with respect to the cultures and, in many cases, organoleptic properties, the manufacturing processess for all these milks have many features in common. For this reason, attention can be focused on the stages of production that are, at least to some degree, common to all fermented milks.

6.3 Manufacture of fermented milks

Yoghurt and similar fermented milks have been produced in warmer regions around the Mediterranean for centuries, but the more widespread popularity of the products throughout Europe, North America and elsewhere is comparatively recent. These newer markets are dominated by two types of retail product. One variant has a firm, gel-like strucuture (set yoghurt), whereas the other has the consistency of 'double cream' and the background flavour is usually modified by the addition of fruit or flavours and sugar (stirred yoghurt) (Tamime and Robinson, 1999). Despite the apparently contrasted nature of the end-products, the manufacturing procedures for both types of fermented milks are broadly similar and an outline of the overall process is shown in Figure 6.1. The quality of these products is governed by a multitude of factors, such as the compositional and microbiological qualities of the raw and added ingredients, the preparation and processing of the milk base, and the handling of the coagulum after the fermentation stage. The scientific background of fermented milk production is briefly reviewed below, followed by a detailed account of the automated systems that are currently used in large dairies.

6.3.1 Raw materials

The raw material for the production of most fermented milks is cows' milk, although the milk from other mammals, such as the horse, sheep, camel or buffalo, is equally amenable to fermentation. Goats' milk can also be employed, but, owing to the high level of β-casein and the poor functional properties of caprine α_{s1}-casein, the coagulum formed during the fermentation stage is soft and the end-product may lack the attractive 'mouth feel' of sheeps' or cows' milk products. Although milk fat can be present or absent according to taste and/or market demand, the critical feature of the milk, in the present context, is the level

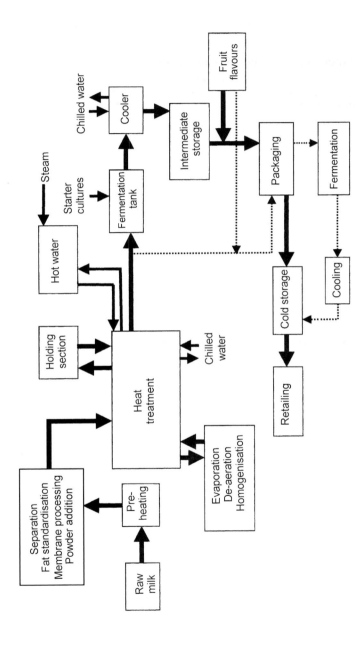

Figure 6.1 Flow diagram showing the manufacturing stages of fermented milk products. Note: dotted lines illustrate the manufacturing stages of set-type products.

of solids–not-fat (SNF). In cows' milk, the level (g $100\,g^{-1}$) is 8.5–9.0, of which around 4.5 is lactose, 3.3 protein (2.6 casein and 0.7 whey proteins) and 0.7 mineral salts. Each of these components is vital for the production of a fermented product: the lactose provides an energy source for the starter bacteria (Section 6.3.4), and the protein, together with minerals such as calcium and phosphorus, gives rise to the basic structure of the gel. For drinking yoghurts, Kefir, Koumiss or Laban, the basic level of SNF is sufficient to given an acceptable fluid product even though, in some industralised countries stabilisers, or other ingredients are added to generate additional viscosity. However, the levels present in liquid milk are not sufficient to produce a satisfactory set or stirred product for, in these cases, the intention is to eat the fermented milk with a spoon. Consequently, the first step in manufacture is to raise the level of SNF in the incoming liquid milk.

6.3.2   Fortification of the milk

Traditionally, the solids content of milk was raised by gently heating the milk in an open pan over a fire, and evaporation of the water was sufficient to achieve the desired effect. Nowadays, three systems are available: the addition of powders [e.g. skimmed milk (including retentates)], whey (including whey protein concentrates), caseinate or buttermilk, to liquid milk; the evaporation of water from liquid milk under vacuum; or the removal of water from liquid milk by membrane filtration (e.g. ultrafiltration or reverse osmosis).

Whether the milk base is skimmed, semi-skimmed or full-cream depends on the fat content desired in the retail product for, although $0.05\,g\,100\,g^{-1}$ may be attractive for the diet-conscious, around $1.0–1.5\,g\,100\,g^{-1}$ gives the fermented milk a more attractive 'mouth feel'. In practice, the choice of method of fortification tends to depend upon the process plant available but, as long as the process milk has around $4.0–5.0\,g\,100\,g^{-1}$ protein and a total solids content of $13–14\,g\,100\,g^{-1}$, it will provide a suitable base for most commercial fermented milks; for some niche markets for natural yoghurts, total solids levels of $16–18\,g\,100\,g^{-1}$ may be used.

Although fermented products sold without fruits or flavourings are usually based entirely on milk, some countries permit the use of stabilisers to achieve an acceptable consistency in stirred fruit yoghurts or similar retail items. Many of these stabilisers are complex carbohydrates (Rao and Steffe, 1992), and the incorporation of starch or modified starch or one of the vegetable gums, such as guar or locust bean gums, at a level of $0.5–0.75\,g\,100\,g^{-1}$ can give a smooth texture to a product without the need for high levels of milk solids (Tamime and Robinson, 1999). Guar gum, locust bean gum, as well as the carrageenans and cellulose derivatives, are long-chain polysaccharides composed of regular arrangements of monosaccharide units, and it is this structure that gives them the ability to bind water. It is also significant that the more complex molecules cannot be attacked by digestive enzymes in the human body and hence they may contribute to human nutrition by providing a 'bulking agent' for the contents of the intestine and, by stimulating intestinal peristalsis, help to prevent some types of colonic malfunction.

In order to achieve full hydration of these complex molecules and, at the same time eliminate any contaminant yeasts or moulds, stabilisers are usually added to the milk before the heat-treatment stage.

6.3.3 Heat treatment of the milk

Once the desired level of SNF has been achieved, the milk may be homogenised to ensure full incorporation of any dry ingredients and to break down the fat globules to a uniform size of around 1 μm. This size reduction is essential to prevent the separation of cream during the production of full-fat yoghurts, for example, but homogenisation also improves the consistency of all stirred products (Robinson and Tamime, 1993; Tamime and Robinson, 1999).

In most production plants, homogenisation takes place before the heating step so that any coliforms or other vegetative microorganisms that might be contaminating the homogeniser are destroyed. However, with some drinking products homogenisation may be delayed.

The heating stage involves passing the milk through a heat exchanger in order to raise the temperature to 90–95°C with a residence time in the holding tube of 2–5 min. Alternatively, the milk may be heated in the main process vessel to 80–85°C and then held for 30 min, but this approach does not lend itself to automation. Whatever the choice of treatment, this step is essential to:

- give an end-product with the desired textural properties; in particular, heating or holding alters the physio-chemical properties of the caseins and denatures the whey proteins, so that β-lactoglobulin may become attached to the κ-casein; this linkage improves the texture of viscosity of the final product;
- cause some breakdown of the whey proteins to give products that stimulate the activity of the starter culture;
- expel oxygen from the process milk for, as both the mesophilic and the thermophilic starter bacteria are microaerophilic, this de-aeration provides the correct environment for rapid growth; and
- kill any non-sporing pathogens that may be present.

In order to achieve these aims, the critical feature is the holding time, for the application of high temperatures, [e.g. ultra-high-temperature (UHT) treatments at 140°C for around 2 s or more (Section 6.4.3.4)] does not give rise to a 'spoonable' product with attractive physical properties. For fluid fermented milks, such as acidophilus milk, an acceptable alternative to UHT is to use a temperature of 90–95°C with an extended residence time (Salji, 1993).

6.3.4 Microbiology of the processes

Once the heat treatment has been completed, the milk is cooled either to:

- 30°C prior to inoculation with a culture composed of one or more of the following species: *Lc. lactis* subsp. *lactis*, *Lc. lactis* subsp. *lactis* biovar *diacetylactis*, *Lc. lactis* subsp. *cremoris*, *Leu. mesenteroides* subsp. *cremoris*, for the production of one of the mesophilic milks (Scandinavian buttermilk or Laban).
- 22–25°C for the manufacture of traditional Kefir or Koumiss (Koroleva, 1991).
- 42°C for inoculation with equal numbers of *S. thermophilus* and *Lb. delbrueckii* subsp. *bulgaricus* for the manufacture of yoghurt. The cultures for the 'health-promoting' products that contain, for example, *Lb. acidophilus* and/or *Bifidobacterium* spp. are handled in the manner of the yoghurt culture.

The technology for handling these cultures can vary and is not reviewed here (Tamime, 1990; Tamime and Robinson, 1999) but, currently, there are two popular routes. The first is to produce a liquid 'bulk starter' using 'sterile milk' made from reconstituted skimmed milk powder and a freeze-dried or frozen culture purchased from a supplier. The inoculated milk is incubated until it just coagulates—at a pH of around 5.0 (Tamime, 1990)—and this pre-fermented bulk starter is added to the processed milk at a rate of 1–2 ml $100\,ml^{-1}$. The second route is to purchase a concentrated frozen or freeze-dried culture that can be added direct to the processed milk. For sheer convenience, this second route is becoming more popular; it is an essential process for the production of 'bio-yoghurts' where the total numbers of, and balance between, the various species is critical.

For normal yoghurt, the use of *S. thermophilus* and *Lb. delbrueckii* subsp. *bulgaricus* is historical in origin, in that they have frequently been isolated from natural yoghurt made by the indigenous tribes of the Middle East, but there are, nevertheless, good reasons for continuing with the tradition. Thus, when growing in milk, the two organisms interact synergistically, and this protoco-operation is based upon the facts that:

- *S. thermophilus* grows more rapidly than *Lb. delbrueckii* subsp. *bulgaricus* and releases lactic acid, carbon dioxide and formic acid, all of which stimulate the growth and metabolism of the *Lactobacillus*
- proteolytic action of the *Lactobacillus* on the milk proteins releases peptides, and the peptidase activity originating from *S. thermophilus*

makes available amino acids that are essential for further development of both species.

The end result of this mutual stimulation is threefold. Firstly, both species actively metabolise lactose to lactic acid, so that the fermentation is complete within 3–4 h. Secondly, the metabolites liberated by the two species give yoghurt a distinctive flavour, with acetaldehyde at levels up to 40 mg kg^{-1} as the major component; other compounds, such as free fatty acids, amino acids, diacetyl and keto or hydroxy acids also contribute to the final flavour. Finally, some strains of the two species can produce appreciable levels of exo-polysaccharides (EPSs), such as glucans, and the presence of these metabolites considerably enhances the viscosity and consumer appeal of the end-product. Thus, some of the polysaccharide forms a layer over the individual bacterial cells, and the remainder forms a network that binds the cells and the casein together into a viscous mass; commercial starter manufacturers have available a range of cultures that differ with respect to the types and levels of polysaccharide synthesised.

In contrast, *Lb. acidophilus* and/or *Bifidobacterium* spp. produce principally lactic acid along, in the latter case, with acetic acid, so that their influence on the flavour or physical properties of fermented milks should be minimal. It is also important that, compared with *S. thermophilus* and *Lb. delbrueckii* subsp. *bulgaricus*, these same organisms grow very poorly in milk, and thus their numbers approximately 1.0×10^7 cfu ml^{-1} of product (cfu, coloning forming units) are important only with respect to their alleged therapeutic properties.

6.3.5 Fermentation

Once the milk has been inoculated, it will follow one of the two routes shown in Figure 6.1 to give either a set or a stirred product. For mesophilic products, a fermentation time of 12–16 h may be necessary to achieve a final acidity of approximately 1.0 g 100 ml^{-1} lactic acid in the retail product but, with a yoghurt or 'bio-yoghurt' culture, the fermentation will be complete after 4–5 h; that is, the acidity of the milk will have risen to 1.2–1.4 g 100 ml^{-1} lactic acid (a pH around 4.2–4.3).

As the acidity exceeds 1.0 g 100 ml^{-1} lactic acid, the caseins in the milk become unstable and coagulate to form a firm gel composed of strands of casein micelles that interlock via hydrogen bonds to form a matrix of protein. Whey is entrapped within the matrix of strands and, in general, the higher the level of protein in the milk base, the stronger the resultant gel. For set products, the coagulated milk must be cooled to avoid over-acidification for, if this control is not exercised, then the product may develop an excessively sharp, sour taste; the protein gel may begin to shrink and cause whey to separate—on the surface of set products, the presence of excess whey is regarded as a fault.

When a fermented milk is produced in bulk, the coagulum will be gently stirred when the pH has reached 4.6 or less but, in this instance, immediate cooling is not so critical because the sugar and fruit or flavours present in most stirred products will mask any excess lactic acid. What is important is that the stirring action breaks the gel down into small pieces; the smaller the pieces, the greater the reduction in viscosity of the fermented milk. To some extent, the loss in viscosity that accompanies stirring can be nullified by the presence of stabilisers in the whey phase but, even so, excessive damage to the coagulum should be avoided if the retail product is to retain an attractive 'mouth-feel'

6.3.6 Final processing

For set products coagulated in the retail cartons, cooling can be achieved by blowing cold air through the incubation room or by carefully transferring groups of cartons in their retail trays to a chill room at a temperature of 2–4°C. In-tank cooling of the fermented milk base is often practised for stirred products, and this stage requires the circulation of chilled water (2°C) through the jacket or in-tank cooling system of the vessel. Alternatively, the product may be stirred and pumped through a plate or tubular cooler, but this process can lead to a loss in viscosity; as most stirred fermented milks are further processed, the initial cooling is usually to 15–25°C. In-tank mixing of the fruit or other flavouring mixture ($10–15\,g\,100\,g^{-1}$) can be used, but large factories tend to feed the fermented milk base and the fruit through a blending tube and then directly into cartons. The most popular size of carton is the individual portion (125–150 g), and the necessary cartons are either purchased preformed, filled and then heat sealed with an aluminium foil lid, or the polystyrene or similar material may come as a roll which is formed into cartons on the filling machine [the form–fill–seal (FFS) process]. Family packs (500 g) are available as well, usually with press-on lids to allow for consumption over several days.

6.3.7 Retail products

The chemical composition of the retail products will depend on the degree of fortification of the original milk, and some typical examples are shown in Table 6.2. In addition, all fermented milks should have high counts of bacteria of starter origin, and this presence is essential for any product for which therapeutic or prophylactic properties are claimed. In yoghurt, for example, the total bacterial population of starter origin may well exceed $2.0 \times 10^9\,\text{cfu}\,\text{ml}^{-1}$, and similar figures can be anticipated in 'bio-yoghurts' where the so-called 'therapeutic' minimum for *Lb. acidophilus* or *Bifidobacterium* spp. will be $10^6\,\text{cfu}\,\text{ml}^{-1}$. In Kefir, the total counts for lactic acid bacteria will be in the region

Table 6.2 Some typical values of the major constituents of an unfortified drinking yoghurt and some retail yoghurts (all units 100 g^{-1})

Constituent	Type of yoghurt			
	drinking	full-fat	low-fat	Greek-style
Water (g)	87.7	81.9	84.9	77.0
Energy value (kcal)	66.0	79.0	56.0	115.0
Protein (g)	3.2	5.7	5.1	6.4
Fat (g)	3.9	3.0	0.8	9.1
Carbohydrate (g)	4.8	7.8	7.5[a]	n.d.
Calcium (mg)	115.0	200.0	190.0	150.0
Phosphorus (mg)	92.0	170.0	160.0	130.0
Sodium (mg)	55.0	80.0	83.0	n.d.
Potassium (mg)	140.00	280.0	250.0	n.d.
Zinc (mg)	0.4	0.7	0.6	0.5

n.d., not determined. [a]The nutrient levels in fruit-flavoured yoghurt will vary with the type of fruit and stabiliser, but a typical value for carbohydrate might be 18 g 100 g^{-1}.

of 10^7–10^9 cfu ml^{-1}, along with 10^4–10^5 cfu ml^{-1} of acetic acid bacteria, and it may be that the alleged therapeutic properties of traditional Kefir result, at least in part, from this extensive bacterial microflora (Koroleva, 1991).

In contrast, the total colony counts for non-starter bacteria should be less than 10^3 cfu ml^{-1} and the yeast and mould count should be less than 10 cfu ml^{-1}. This latter count is especially important, for although the low pH of a well-made fermented milk should inhibit the growth of bacterial contaminants, yeasts or moulds are tolerant of acidity and can spoil retail products well within an anticipated shelf-life of 21 days. This restriction does not apply, of course, to products such as Kefir, and total yeast counts of 10^4–10^5 cfu ml^{-1} are to be expected (Koroleva, 1991).

6.4 Options for automation and mechanisation

6.4.1 Introduction

It has been mentioned already that millions of cartons of yoghurt and other fermented milks are produced each year and, if each batch of products is to have the characteristics outlined above, then it follows that the different stages of manufacture must be under strict control. Furthermore, this control must be exercised on a daily basis over extended periods of time so that automation of the separate stages allied to computer control has become essential.

Several approaches are employed for the preparation of the milk base during the manufacture of fermented milks, including yoghurt (see Figure 6.1). However, each manufacturer has its own specific requirements and, in effect, each plant supplied is tailor-made. Nevertheless, the primary objectives of automation of any dairy processing plant, including the application of computers, have been detailed by the IDF (1985, 1991, 1995), Auer (1995) and by Ferret and Trystram (1991), and can be summarised under the following headings:

- reduce waste of raw materials and product
- improve the quality and safety of the product
- increase operator safety and achieve better environmental control
- maximise site and plant utilisation and productivity
- provide management control information

Thus, taking these aspects into consideration when designing a dairy process line, Bylund (1995) reported that the project engineer has to compromise between different requirements, as follows:

- product requirements—these relate to the raw materials used, their preparation (i.e. fortification of milk SNF and/or standardisation of the fat content), processing and quality of the end-product
- process requirements—these relate to processing plant capacity, selection of components and their compatability, degree of process control and level of automation, availability of services (e.g. hot and chilled water) and cleaning of the process plant
- economic requirements—these correlate to the total cost of production and the stipulated quality standards
- legal and statutory requirements—these concern legislation that stipulates process parameters, as well as the choice of components and system solutions

In general, there are three different levels to which a dairy plant can be automated—manual, semi-automatic and fully automatic—and it is the fully automatic type which is detailed in this section. As fermented milk plants become large and complex, manufacturing a wide range of products every day, management, operators and marketeers may have great difficulty in ensuring that the process is secure and the business is manageable (Tamime and Robinson, 1999). Hence, the hierarchical approach to dairy factory automation is dependent on different levels of responsibility, from the operator on the production line to the director of the business, and the problems to which solutions are required are different. Ferret and Trystram (1991) have suggested four points at which a computer could be introduced in a fermented-milk factory:

- level 0 to regulate preparation of the milk base and processing
- level 1, cleaning-in-place (CIP)
- level 2, to coordinate between fermentation tanks and packaging station, to initiate and monitor operations, quality control and management of

stocks (i.e. raw materials and finished products) and to create dialogue with level 3
- level 3, production planning

As a consequence, such coordination can be realised by the use of process support and supplemented by some form of management information system (for details, see Tamime and Robinson, 1999). However, to optimise the productivity of a fully-automated fermented milks factory, the processing plant is divided into areas or departments. These divisions are inter-linked via a central data processing unit (CPU), but the following breakdown would be quite feasible:

- milk reception and storage
- process plants
- product storage and/or packaging
- product and CIP pipework system
- CIP system
- services supplies
- effluent system

Since some of these areas share common ground with other dairy products detailed in this book, only the section relating to fermented-milk process lines is discussed here.

6.4.2 Reception of milk

Milk collection from farms in industrialised countries is carried out in bulk with use of a road tanker. The facilities available for milk reception at a dairy, including automation, are detailed in Chapter 3 (see also Bylund, 1995; Tamime and Robinson, 1999). The general practice for handling milk after acceptance may include: measuring the milk by volume (i.e. the milk passes through an air eliminator, pump filter and metering device) or by weighing (i.e. weighing the tanker before and after unloading or by using special tanks fitted with load cells in the feet); transfer to an intermediate storage tank; thermisation and cooling (optional) or cooling only; and storage in silo tanks.

6.4.3 Processing plants

The technological aspects of fermented milks have much in common, and it is appropriate, therefore, to discuss in this chapter the process plants under one heading owing to the universal popularity of yoghurt. The stages of manufacture consist of preliminary treatment of the milk base (fat standardisation and fortification of the SNF), followed by homogenisation, heat treatment, cooling, fermentation, cooling, addition of fruit, packaging and cold storage. The automation of the partial stages is detailed below; however, variations in

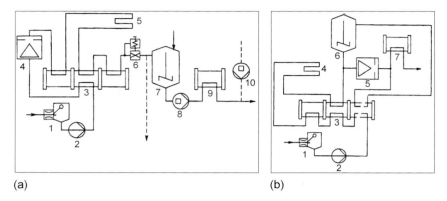

Figure 6.2 Simplified flow chart for the manufacture of (a) yoghurt, Kefir, Ymer and Ylette and (b) buttermilk. Part (a): 1, balance tank; 2, centrifugal pump; 3, plate heat exchanger (PHE); 4, homogeniser; 5, holding tube; 6, constant pressure valve; 7, fermentation tank; 8, positive pump; 9, plate cooler; 10, positive pump for fruit addition (dotted line for set-type product). Part (b): 1–3, as in part (a); 4, holding tube; 5, homogeniser; 6, fermentation tank; 7, plate cooler. Source: Bøejgaard [personal communication, Rannie Product Information, reference number 1.3-1(c) to 1.3-5(c)]. Reproduced by courtesy of APV Nordic A/S, Aarhus, Denmark.

plant design are tailored to suit the type of fermented milk being produced. For example, Figure 6.2a illustrates an APV Nordic simplified flow chart for the manufacture of yoghurt (set-type or stirred-type), Kefir, Ymer and Ylette; the production of buttermilk from skimmed milk is shown in Figure 6.2b.

Nevertheless, the processing conditions of each type of fermented milk vary in relation to the desired characteristics of the product. Table 6.3 shows these minor processing variables. For example, cultured buttermilk is made from skimmed milk (ca. 9.5 g 100 g^{-1} SNF and 0.6 g fat 100 g^{-1}) which is heated to 93°C for 15–20 s, cooled to 60°C, homogenised at 18–20 MPa, cooled and fermented at 20°C. After acidification, the gel is thoroughly stirred, homogenised at 5–10 MPa pressure and 20°C, cooled to 4°C and packaged. The purpose of postfermentation homogenisation is to make the product smooth without lumps (Bøejgaard, personal communication); a similar approach is used in Ymer making (see Table 6.3). However, the SNF and fat contents in cultured buttermilk may vary from one country to another, and in California, for example, they may range between 7.4–11.4 and 0.25–1.9 g 100 g^{-1}, respectively. If the skimmed milk is fortified with skimmed milk powder and milk fat is added, it is recommended by some researchers to use two-stage homogenisation (e.g. first stage at 13.8 MPa and second stage at 3.5 MPa) (Tamime and Marshall, 1997).

Fortification of the milk SNF and standardisation of the fat level of the milk base is highly recommended since the rheological and organoleptic properties of yoghurt are improved. The milk base is pre-heated to 60–70°C, de-aerated

Table 6.3 Process variables of some fermented milk products

Processing stage	Yoghurt	Kefir	Ymer and Ylette
Homogenisation			
Inlet pressure (MPa)	0.2–0.6	0.2–0.6	0.2–0.6
Pressure (MPa)	20–25	10–20	18–23
Temperature (°C)	65–70	ca. 70	65–70
Heating (°C)	95	–[a]	95
Holding	n.r.	n.r.	n.r.
Cooling (°C)	40–45	23	22
Fermentation (h)	2–3	24	18
Cooling and packaging[b]	15	4–6	n.r.

n.r., Not reported. [a] A heating temperature which is sufficient to denature the whey proteins. [b] Addition of fruit is optional. Note: during Ymer and Ylette production the milk is ultrafiltered 1.8-fold; after (optional) acidification the product is warmed to 40°C, homogenised (7.5–10 MPa), cooled and packaged; this second homogenisation improves the water-holding capacity and stability of the gel. A similar approach is used in buttermilk making (Section 6.4.3). Source: Bøejgaard [personal communication, APV Nordic A/S, Rannie Product Information, reference number 1.3-1 (c) to 1.3-5 (c)].

in a vacuum vessel (this stage is optional and is only used when powders are added to the milk), homogenised, heated to 95°C and held at this temperature for 5–10 min, cooled to 40–45°C, inoculated with starter culture and fermented to the desired pH. Therefore, the differences in the processing parameters of the milk base during the manufacture of different types of fermented milks are: homogenisation pressure and duration and temperature of fermentation (see Table 6.3).

In general technical terms, a highly automated fermented-milk process line is governed by two considerations: first, the availability of sensors to measure certain process parameters, such as fat and SNF contents, temperature and pH, accurately; second, the availability of devices that can be operated from a distance, and some examples are detailed in Chapter 3 (Section 3.4.3.1) regarding the control of heating and cooling of milk in a heat exchanger, including parameters to maintain pressure levels in a processing plant. Taking these factors into account, the automated manufacturing stages of fermented milk are discussed below.

6.4.3.1 Standardisation of milk fat

There are many systems available on the market for the fully automatic in-line standardisation of the fat content in liquid milk. Some examples are: the APV Compomaster type KCM or KCC, either model being fitted with an extra standardising loop (Hansen, 1996; Tamime and Robinson, 1999); Tetra Pak automatic direct standardisation systems (ADS; Bird, 1993; Bylund, 1995; Tamime and Robinson, 1999), and On-Line Instrumentation or Foss Electric. In each of these systems, the standardisation relies on measuring the density of the skimmed milk and cream phases after separation, and therefore on controlling the flow of each component to standardise the milk to the desired fat content.

However, the desired fat content of the milk base in yoghurt making can be achieved in different ways depending on the method of fortification used to increase the level of SNF. For example, if the fat content of the concentrated skimmed milk is standardised with cream, then the automatic standardisation system shown in Figure 6.3 could be used. The principles of operating the OL-7000 standardisation system are detailed in Chapter 3, Section 3.4.3.2.

Alternatively, if on-line fortification of the milk base with use of membrane processing or vacuum evaporation is used, the fat content is standardised before concentrating the milk, taking into account the factor of concentration. For example, to raise the protein content in skimmed milk from 3.2 to 5.0 g 100 g^{-1}, the concentration factor is ca. $\times 1.6$. Similarly, the fat is adjusted to 1 g 100 g^{-1} during the manufacture of semi-skimmed yoghurt containing 1.5 g fat 100 g^{-1} in the final product. Incidentally, when using membrane processing it is usual practice to concentrate the skimmed milk followed by fat standardisation, as shown in Figure 6.3. In the APV Nordic system of on-line evaporation of milk prior to the homogenisation and heat-treatment stages the fat content is standardised before the milk is concentrated (see Figure 6.4). Description of the Tetra Pak processing plant (i.e. evaporation, homogenisation and heat treatment of the milk base) is detailed elsewhere (Bylund, 1995; Tamime and Robinson, 1999).

6.4.3.2 *Fortification of milk solids–not-fat (SNF)*

Different automatic in-line approaches can be used to increase the protein content from 3.2 g 100 g^{-1} in milk to ca. 5 g 100 g^{-1} in the yoghurt milk base.

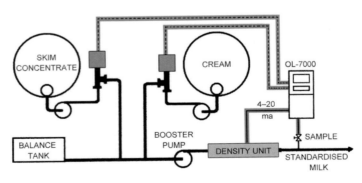

Figure 6.3 An illustration of a fully-automatic on-line standardising system. The OL-7000 system is described in Chapter 3, Section 3.4.3.2. Reproduced by courtesy of On-Line Instrumentation Inc., New York, USA.

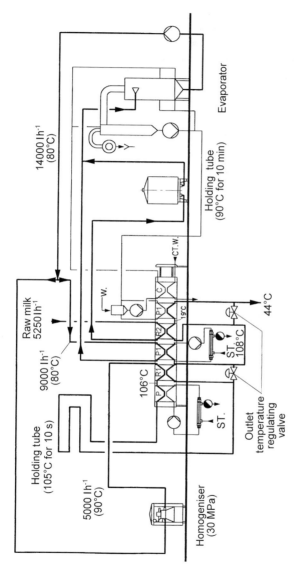

Figure 6.4 Large-scale milk base evaporation and heat processing. Regeneration section: P, heats the milk to 120–130°C; R1, cools the milk to 108°C; P1 warms the processed milk from 19–43°C where the energy is used from the condensate of the evaporator; R2, cools the processed milk from 108–19°C; C, cooler; ST, steam; CT.W., chilled/cold temperature water; W., water. Source: Bøjgaard (personal communication). Reproduced by courtesy of APV Nordic A/S, Aarhus, Denmark.

Some of these methods are:

- membrane filtration
- addition of powder(s)
- single-effect vacuum evaporation

Membrane filtration, such as ultrafiltration (UF) and/or reverse osmosis (RO), has been used in the industry to concentrate the total solids content of the milk base. The efficacy of automation of UF and RO processes is detailed in Chapter 11; however, the membrane filtration of milk for yoghurt production is shown in Figure 6.5. On-line UF or RO concentration of milk are as follows: the raw milk is pre-heated, separated, concentrated at 30–50°C, standardised (e.g. the fat), pasteurised at 72°C, cooled to 30–40°C and finally processed for yoghurt making. A typical composition of UF or RO retentate for plain yoghurt production is shown in Table 6.4.

The principles of in-line standardisation of the fat content in the concentrated milk are similar to those discussed in Section 6.4.3.1, and the concentration factor (CF) in the retentate could, hypothetically, be monitored by measuring the volume of permeate or retentate by using an in-line flow meter. Other methods could be used for more accurate and automatic measurement of the CF of the retentate; for further information refer to Chapter 11.

Addition of powder can be monitored automatically by measuring exactly the weight of powder prior to injection into the milk/water stream. The accuracy of such an approach is governed by the flowability of the powder. As a consequence,

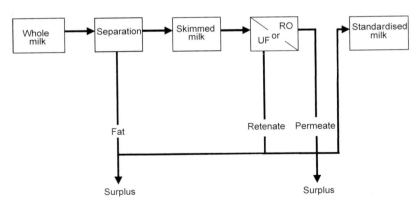

Figure 6.5 Continuous membrane filtration of yoghurt milk. RO, reverse osmosis; UF, ultrafiltration. Note: flow meters are used to measure flow in and out of the processing equipment; for example, the amounts of water or permeate removed are used to control the concentration level in the retentate; this method is also used as a standardisation control where, for instance, a UF plant may remove $1000 \, l \, h^{-1}$ of permeate from milk, and hence the UF plant is designed with a higher capacity than required and the surplus permeate is then mixed back with the milk (Bøejgaard, personal communication). Reproduced by courtesy of APV Nordic A/S, Aarhus, Denmark.

Table 6.4 Comparison of proximate chemical composition (g 100 g^{-1}) of milk, retentate and permeate in membrane filtration for yoghurt production

Constituents	Ultrafiltration			Reverse osmosis		
	milk	retentate	permeate	milk	retentate	permeate
Total solids	11.55	13.50	5.35	11.97	13.50	0.05
Total protein	3.32	4.31	0.16	3.31	3.73	0.03
Lactose and acid	4.83	4.85	4.78	4.81	5.43	0.01
Ash	0.75	0.86	0.41	0.75	0.84	0.01
Fat	2.65	3.48	0.00	3.10	3.50	0.00
Yield (kg)	1313.00	1000.00	313.00	1128.00	1000.00	128.00

Source: APV Nordic A/S, Aarhus, Denmark.

the powder dosage rate may fluctuate and ultimately affect the level of SNF in the milk base in a continuous process line.

Alternatively, a batch system for the recombination of powders can be used, where pre-heated skimmed milk/water is metered into one of the storage tanks. The circulation pump drives the liquid through a high-speed powder blending unit for rehydration. The weight of powder per batch can be weighed accurately, and, by providing more than one tank, the recombination process can be run continuously. The fat [anhydrous milk fat (AMF) or cream] can be metered into each tank for standardisation, or it may be injected in-line at 60–70°C before the milk is homogenised. Any milk fortified with powder must pass through a duplex filter to remove any undissolved or scorched particles before entering the plate heat exchanger (PHE), and then de-aerated before the homogenisation stage. The technology of recombination for the manufacture of a wide range of dairy products (e.g. market milk, fermented milks and cheeses) has been detailed elsewhere (Bylund, 1995; Nielsen et al., 1999; Tamime and Robinson, 1999); these processes are therefore not reviewed in this publication.

A single-effect vacuum evaporator may be used in-line to concentrate the milk base. Around 10–20 g 100 g^{-1} of the water is removed from the milk base by evaporation, and the milk solids content will be increased by 1.5–3.0 g 100 g^{-1}. The sequence of operation of the Tetra Pak A/B system is as follows (Bylund, 1995). The fat standardised milk base (i.e. taking into account the concentration factor) is pre-heated to 70°C in the regeneration section of the PHE by using the condensate from the evaporator. Subsequently, the milk is heated to 85–90°C in the heating section of the PHE and the pre-heated milk enters the vacuum chamber tangentially at a high velocity; the inlet is shaped as an expansion tube to prevent burning of the milk. In the evaporator, the milk forms a thin film rotating along the wall surface, where some of the water is evaporated, and the vapour is drawn off to the condenser. During evaporation, the temperature of the milk drops from 90 to 70°C; air and any other noncondensable gases are extracted from the condenser by the vacuum pump. However, the condensate from the milk is used to pre-heat the incoming cold milk, and this improves the thermal economy of the plant. Eventually, partially concentrated milk loses velocity in the vacuum chamber and falls to the inwardly curved bottom, where it is discharged. Part of the pre-concentrated milk base is recirculated by a centrifugal pump to the PHE for temperature adjustment and is then sent back to the vacuum chamber for further evaporation. Large amounts of the milk (Section 6.4.3.4) are recirculated four to five times until the desired degree of concentration is achieved. The recirculation cyle is controlled by the capacity of the vacuum chamber, evacuation pump and the float controller; during each recycle, about 3–4 g 100 g^{-1} of water is removed. The flow rate through such evaporators is up to five times the inlet flow of the processing plant. In some instances, the volume of condensate is measured to confirm the rate of concentration, and it is circulated in the PHE to pre-heat the incoming milk.

6.4.3.3 Homogenisation
The homogeniser is a high-pressure piston pump that reduces the size of the milk fat globules. The operating pressures used during the processing of the milk base range 15–20 MPA; much lower pressures are used to smooth the product after the fermentation stage (see Table 6.3).

In general, the homogeniser is placed upstream during the production of fermented milks (Bylund, 1995; Tamime and Robinson, 1999). A similar set-up is observed in the production of pasteurised milk, ultra high temperature (UHT) treated milk (indirect system) and cream. Some researchers have advocated post-heat-treatment homogenisation (i.e. downstream) of the milk base to improve the quality of yoghurt (Tamime and Robinson, 1999); such positioning of the homogeniser in a processing plant is similar to the direct system of UHT of milk. In this case, the homogeniser is of aseptic design to prevent microbial contamination of the product. However, in the APV Nordic yoghurt plant (see Figure 6.4) the milk base is heated twice and the homogeniser is positioned between the two heat-treatment stages so that an aseptic model is not required (see Section 6.4.3.4).

6.4.3.4 Heat treatment
Plate or tubular heat exchangers are widely used to process all dairy products, including fermented milks. These heating units are highly automated and are easily controlled from a remote station. For energy conservation, these units consist of regeneration and heating sections, and, for pasteurised liquid milk, up to 94% of the energy required to heat the cold milk (ca. 5°C) to pasteurisation temperature and to cool the heated milk (72°C) to 5°C is recovered. In fermented milk production plants processing 4000 l h^{-1}, possible energy recoveries of 47% and 33% for a milk base fortified with skimmed milk powder (SMP) or concentrated by evaporation, respectively, have been reported (Tamime and Robinson, 1999).

A wide range of time–temperature combinations has been used to process the milk base (Tamime and Robinson, 1999). One of the primary objectives of heating the milk base is to denature the whey proteins (β-lactoglobulin and α-lactalbumin) and encourage their interaction with κ-casein, which improves the gel firmness after the fermentation stage. Thus, the preparation of the milk base by evaporation (minimum 5%) and heating twice in an APV Nordic processing plant (Figure 6.4) is as follows. The fat standardised milk base enters the PHE at a rate of $52501\,h^{-1}$, where it is heated to 90°C and held for 10 min before it continues to the evaporator or de-aerator unit. The flow rate of the partially concentrated milk base is $140001\,h^{-1}$. Hence, a recirculation loop is fitted in the process line from the evaporator to the heating section of the PHE for temperature adjustment from 80°C to 90°C, it is then recirculated at a rate of $90001\,h^{-1}$, which is nearly twice the flow rate of the incoming raw milk base (see Figure 6.4). The recirculated milk, at a temperature of 90°C (i.e. partially concentrated), is mixed with heated raw milk at 90°C for 10 min and then recirculated through the evaporator until it reaches the desired concentration of total solids in the milk base. The concentrated milk (ca. 90°C) is homogenised at 30 MPa by using a specially designed multi-stage homogenising head LW or Microgab before it is heated in the second regeneration section (P) to 120–130°C (see Figure 6.4). The residence time of the heated milk is 10 s before it is partially cooled to 108°C in the regeneration section (R1). Afterwards, the milk is cooled to 19°C in the regeneration section (R2), which heats the incoming raw milk, and is re-warmed to 43°C in the PHE by using the energy from the condensate of the evaporator (P1; see Figure 6.4).

In this system of concentrating the milk base, the control of the concentration may be done by using flow meters on the condensate or by measuring the temperature drop and multiplying by the recirculated volume of milk, which corresponds to a specific amount of evaporated water (Bøejgaard, personal communication).

6.4.3.5 Fermentation

The equipment required for continuous production of large volumes and in a highly automated system depends on the type of fermented milks produced, that is set-type or stirred-type. Irrespective of what type of product is manufactured, the control parameters include accurate control of the incubation temperature, continuous measurement of the level of acidification and rapid cooling of the gel to control the metabolic activity of the starter cultures.

For the production of set yoghurt, the processed milk base is cooled to 37–45°C, inoculated with starter culture, flavoured (optional), packed and palletised, incubated, cooled, stored and dispatched. Bulk starter culture (i.e. liquid) can be easily and accurately metered in-line into the stream of milk as it is pumped from the buffer tank to the hopper of the filling machine. Alternatively, two buffer tanks are used in parallel, and to each tank the required amount of direct-to-vat inoculation (DVI) starter culture (e.g. freeze-dried or frozen) is added and mixed. Cabinets or tunnels can be used to incubate and cool the milk, and proper circulation of the hot air (moist or dry, depending on the sensitivity of the container) is essential in order to minimise temperature fluctuations. In some instances the cabinets are used as incubators only, and the product must be moved to another chamber for cooling. While the warm coagulum is in motion, some structural damage may occur. The damage may be minimised by using smooth rollers or conveyor belts similar to those adopted in a tunnel system that consists of a combined stationary incubation section and a continuous cooling tunnel (Bylund, 1995; Tamime and Robinson, 1999).

In contrast, stirred yoghurt is produced in bulk in insulated tanks and the gel is broken during in-tank cooling or prior to being pumped to plate or tubular coolers. The tanks are bottom filled, and most commercial tanks available on the market have a conical shaped design for easy discharge of the product from the base (see the review by Tamime and Robinson, 1999). Continuous production is achieved if fermentation tanks are provided in series. In general, the tanks are fitted with level sensors, pH meters and temperature indicators, which can be monitored and controlled from a distance. However, in the aseptic design of fermentation tank, the following extra specifications are provided: provision of filtered air capable of trapping particles of diameter greater than 0.3 μm; air velocity sufficient to create a positive pressure in the tank, corresponding to ca. 10 kPa; and an agitator with a double-shaft seal with a steam barrier to minimise contamination (Bylund, 1995).

6.4.3.6 Cooling, fruit mixing and packaging
After the fermentation stage, the stirred yoghurt is cooled to 15–25°C, mixed with fruit and packaged (Figure 6.6). The process parameters of the cooler are similar to those of heat exchangers. Contrary to the assumption that the gaps between the plates in the heat exchanger and cooler may be different, for yoghurt the gaps are the same in an APV Nordic unit. However, two different types are used: in a PHE, longer plates are installed to process the milk base; shorter plates are installed in a plate cooler. Furthermore, the cooling capacity of such a device to cool yoghurt from 43°C to the desired level is 2–3 times the size of the fermentation tank (Bøejgaard, personal communication).

Two possible options can be used to blend the yoghurt base with the fruit automatically. Option 1 (Figure 6.6, bold line) consists of two metering devices for dosing the correct amount of fruit and yoghurt, respectively, and a static in-line mixer. A typical example is the Tetra Pak static mixer (Bylund, 1995; Tamime and Robinson, 1999). The yoghurt–fruit blend passes through the blades in the mixer, ensuring a uniform distribution of fruit particles throughout the coagulm. The specifications of such a mixer are as follows:

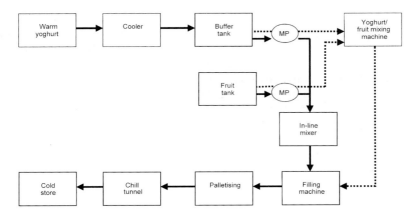

Figure 6.6 Production line of stirred yoghurt. MP, metering pump.

- length: 75–115 cm
- flow-through rate: up to 10 000 l h^{-1}
- number of twisted blades: up to 10
- pipe diameter: up to 6.35 cm

It is recommended that such a unit should be dismantled and rinsed before starting the CIP programme.

Option 2 (Figure 6.6, dotted line) consists of a blending unit fitted with two feeding pumps that draw yoghurt base and fruit from separate tanks into a mixing chamber. One such example is the Gast DOGAmix 60 (Benz & Hilgers GmbH, Germany; see Figure 6.7). Some of the specifications of such a machine are:

- maximum discharge rate of the yoghurt pump: 60 l min^{-1}
- discharge rate of the fruit pump: adjustable to a mixing ratio of fruit: yoghurt of between 1:5 and 1:20
- accuracy of blending: ±0.5%

The mixing chamber is fitted with a dynamic agitator. Microbial contamination of the product in the mixing chamber is avoided by isolating the moving parts (e.g. the rods of the plunger pump and the mixer drive of the dynamic agitator) from the surrounding atmosphere by a sterile air chamber. This unit is capable of being cleaned by CIP and of being sterilised by steam.

A wide range of filling machines is available on the market (Tamime and Robinson, 1999) and such machines are designed to fill yoghurt in rigid or preformed containers (e.g. in glass jars or plastic tubs) or in form–fill–seal (FFS) containers (e.g. plastic or laminated paperboard cartons). The ultimate choice of a packaging machine and type of container may be influenced by the following considerations (Tamime and Robinson, 1999):

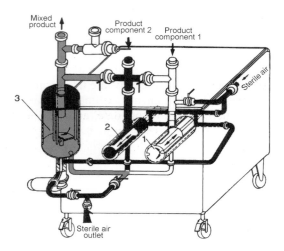

Figure 6.7 Schematic illustration of the Gasti DOGAmix 60. 1, Pump for first product component; 2, pump for second product component; 3, mixing tank with dynamic agitator. Reproduced by courtesy of Jagenberg (London) Ltd, Purley, UK.

- cost per unit container, speed of filling
- cost of packaging machine
- nature of the yoghurt products (e.g. liquid, viscous or concentrated)
- provision of product protection during storage, distribution and retailing
- volume of the unit container
- use of returnable or non-returnable packaging and, in the case of the former type, whether the container can be cleaned and sanitised
- requirements for a specific duration of shelf-life, including barrier properties (e.g. oxygen and light permeability) of the material
- marketing concepts and consumer acceptability
- application of modified atmosphere packaging
- proposed method of filling and sealing
- type of unit container being used
- desirability of filling under a controlled atmosphere
- degree of automation being sought
- need for a high standard of hygiene (e.g. all contact surfaces must be stainless steel and accessible for sterilisation or sanitisation)
- time required to change from one flavour to another or from one volume of carton to another
- versatility and reliability of the machine
- accuracy of filling and the elimination of drip between individual fills
- power and labour requirements of the machine
- other specifications, such as availability of date marking, method of dispensing the cups and safety measures (e.g. no cup, no fill)

The outputs of highly automated packaging machines may range from 15 000 to 100 000 cups h^{-1} (Tamime and Robinson, 1999). It is safe to assume that the hoppers of most filling machines are fitted with level probes (minimum and maximum) to ensure continuous availability of yoghurt to the filling head. Also, positive displacement or piston pumps are universal and the measures of the product are volumetric. In addition, most filling machines are equipped with marking attachments, such as 'best before date' and/or label attachment.

It would be impractical to discuss the automated systems of all the different types of yoghurt machines in detail. It has, therefore, been decided to illustrate a highly automated packaging line at Molkerei Alois Müller GmbH & Co. (UK Production), which is one of the largest yoghurt producers in the UK.

Two types of packaging machines are used at Müller Dairy [preformed plastic pots (PFPPs) and FFS containers]. The former machine is an Ampack that has a throughput of 28 000 pots h^{-1}; FFS types are Hassia and Erca and have throughputs of 15 000 pots h^{-1} and 30 000 pots h^{-1}, respectively. The filled yoghurt containers are nested in cardboard trays (see Figure 6.8a) before being palletised (Figure 6.8b).

6.4.4 Quick chilling, cold storage and retrieval of products

The packed products (i.e. the full trays) are transported to the palletiser, where they are pre-collated into a complete layer before they are stacked onto the pallet. In the case of FFS machines, the layer pattern is rotated 180° to help improve the stack stability. For the PFPP system, the trays are stacked into two or three units before they are pre-collated so as to allow the palletiser to achieve the required throughput.

When a full pallet is discharged from the palletiser to the pallet transport system, the warehouse management system (WMS) takes strategic control of destiny of the pallet. The pallet is given a unique identification number; linked to this is information about the line on which it was produced and the time and date of production. At this stage, samples of the finished product are taken; this is a manual operation. Several sampling regimes are followed for various quality tests. Afterwards, the pallets are driven along chain-and-roller conveyors out of the filling hall to one of three blast tunnels. The pallets of product are cooled from 20°C to less than 4°C and then transported to one of the stretch wrappers (see Figure 6.12, Section 6.4.6).

The pallets are stretch-wrapped automatically with a minimum of wrap, as many of the pallets will be destacked for order picking. They are then profile checked for any overhang from the pallet (e.g. 50 mm permissable), height and weight. This is done to check that the pallet is fit to be transported by an automatic storage and retrieval (ASR) machine. If the pallet fails any one of these checks, it is rejected from the storage system and returned to the production area to be reworked. The pallets conforming to standard are then driven into the

Figure 6.8 On-site automated handling and packaging of yoghurt at the Müller factory in the UK: (a) nesting of yoghurt cups in cardboard trays; (b) palletising of the cardboard boxes. Reproduced by courtesy of Molkerei Alois Müller GmbH & Co. (UK Production), Market Drayton, UK.

high-bay warehouse (HBW), where they are stored at 4°C (Figure 6.9). The ASR machines position the pallets of finished product at specific locations within the HBW. The pallets of 'fast-moving' products are stored in an area closest to the ASR in–out points; 'slower moving' products are stored further away. Similar products are stored in different ASR machine aisles to decrease the chances of not being able to retrieve a product should there be an ASR machine failure.

Full pallets are called to replenish the 'pick face' by the WMS from the HBW to enable orders to be put together (Figure 6.10). The WMS selects pallets by

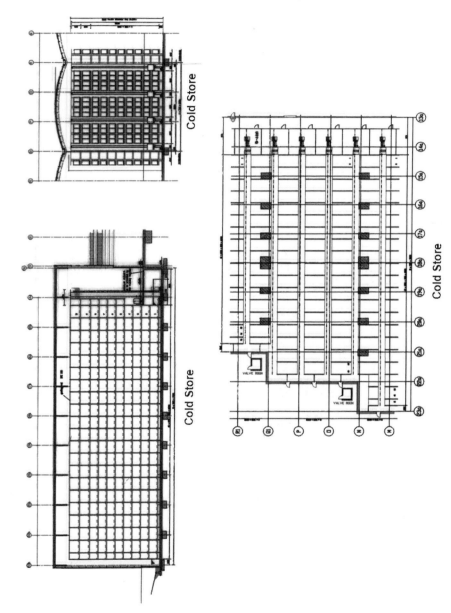

Figure 6.9 The highly automated design and construction of the cold storage area. Reproduced by courtesy of Molkerei Alois Müller GmbH & Co. (UK Production), Market Drayton, UK.

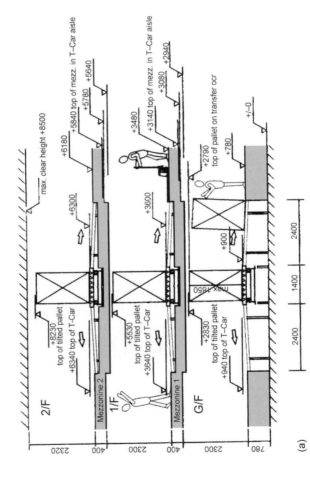

Figure 6.10 (a) Detailed cross-section and (b) top-view schematic drawing of the order picking area. Note: the scale is 1:100; all measurements are in mm. Reproduced by courtesy of Molkerei Alois Müller GmbH & Co. (UK Production), Market Drayton, UK.

YOGHURT AND OTHER FERMENTED MILKS 181

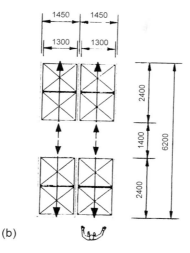

(b)

Figure 6.10 (*Continued*).

shortest 'best before date' to ensure effective stock rotation. Orders are picked manually; operators respond to instructions from a hand-held terminal and pick trays from the pick face accordingly. Orders are then palletised for delivery. A proportion of orders is dispatched as full pallets; these are not picked via the pick face but are taken from the HBW directly to the live storage racking. A new pallet label is printed and applied to the pallet before it is sorted. This label identifies the new pallet and contains data regarding the order.

All pallets that are picked for delivery are temporarily stored in the live storage racking, where they are held until the complete delivery is ready to be loaded onto a road vehicle. The drivers of the forklift trucks have a truck-mounted terminal to tell them from which location to pick up the pallet. The pallet identification label is verified at this point by the driver. The terminal will tell the driver through which loading bay to load the pallet onto the road vehicle; again, verification is done by scanning a barcode at the loading bay to ensure the pallet is loaded onto the correct vehicle. Finally, the picked pallets are transported in refrigerated (4°C) vehicles until delivery to the customer.

6.4.5 Product recovery

Recovery and re-use (optional) of fermented milks enhances production yields and reduces the cost of effluent treatment. Low-viscosity dairy products can be recovered by purging water through the installation, but this may lead to a dilution of certain products, such as yoghurt. Also, wastage of product may occur during start-ups, shut-downs and change-overs in any processing line. However, a purging scraper or a projectile known as a 'pig' can be used to displace product from pipes; a typical example is the APV Nordic pigging system (Figure 6.11).

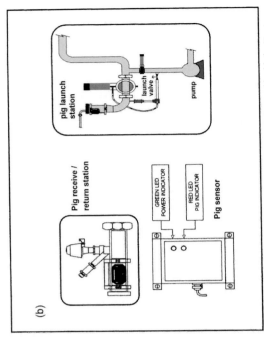

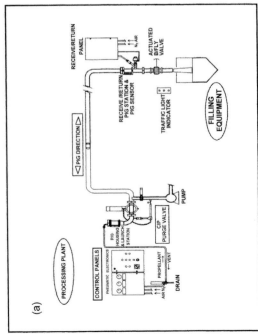

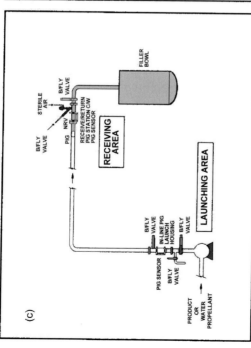

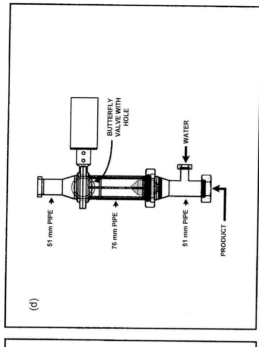

Figure 6.11 The 'pigging' system. This system reduces wastage during start-ups, shut-downs and product change-overs. CIP, cleaning-in-place; LED, light-emitting diode. Reproduced by courtesy of APV UK, Co. Ltd, Crawley, UK.

Bird (1996) and Bøejgaard (personal communication) reported the following advantages to the pigging system:

- the system reduces product losses during daily production
- the pig consists of flexible material (food-grade silicone) and can negotiate bends of 1.5 cm in radius
- the system enables fast change-over of products
- the system is designed and tailored to meet food hygiene requirements
- the system can be fully automated and can operate bi-directionally
- the design of the pig reduces propulsion pressures and is therefore safer
- the unit is fully cleaned with use of CIP, without need for removal from the plant installation
- the propellant is either compressed air, nitrogen, carbon dioxide or water
- the system reduces effluent and CIP costs

According to Bøejgaard (personal communication), in 1999 a factory in France producing fromage frais (a quarg-type product and possibly similar to concentrated yoghurt) with use of three separators, three process lines between the separators and buffer tanks, and eight filling lines, left an estimated 1.1 tonnes of product in the pipework. Recovery of the product by purging with water was 40%; with a pigging system the recovery rate was 98%. Taking into account the improvement of yield of fromage frais per annum and the increased profitability, the cost of investment of eleven pigging systems in the factory was achieved within one year.

According to Jones (personal communication), the automation of the pigging system requires the following components: (a) pig launch station; (b) pig receive–return station; (c) pig sensors (modified hall effect continuous-wave latching circuit); and (d) control panels linked to a programmable logic computer (PLC) (see Figure 6.11a). The modified hall effect sensor detects the magnetic core of the pig, and the sensor has a latching circuit which enables it to be fitted at suitable positions along the pipeline to monitor the progress of the pig and to provide a signal to be detected by the PLC. If the latching device was not included, the pig's signal could be missed by the PLC because the detection pulse could be shorter than that of the scan time of the PLC. The launch station (see Figure 6.11a and b) depicts the hygienic design of the system and also shows the pig, the receive–return station, pig sensor and return propellant valve.

At the end of production (e.g. for change over of fruit-flavour yoghurt or for using a different formulation of milk base), it is important to clean and empty the line of the first product to maximise the efficiency of the plant, and to eliminate the possibility of cross-product contamination in terms of flavour and hygiene. Hence, the feed pump to the destination (filler or buffer tank) is turned off by the PLC. The pig launch valve is, in this case, actuated through 90°. The pig is located in the launch housing and its presence is indicated to the PLC by the pig sensor positioned on the outside of the launch housing. Afterwards, the pig is launched by opening the propellant valve, allowing a pressurised medium (compressed gas or water) in behind the pig (see Figure 6.11); the bi-directional pigging system is also controlled by PLC. This forces the pig down through the launch valve and along the process pipeline. As the pig is a seal-fit in the pipe, it travels towards the destination point forcing the remaining product in the pipework out into the destination tank or filler.

Should the destination vessel not be large enough to accommodate the large amount of product that is contained within the pipeline, the high-level sensors positioned on the vessel will feed back to the PLC, which will stop the propellant driving the pig and halt its progress. The pig remains stationary until a signal is received by the PLC that the level in the vessel has dropped to a suitable level to restart the pigging operation. This process will continue until such a time as the pig reaches the sensor located adjacent to the receive–return station, which signals the completion of the pigging operation (see Figure 6.11d). The valve, which is located on the receive–return station for the return propellant, is then opened, allowing the pig to be driven back through the now empty line to the launch housing, where the sensor detects the arrival of the pig.

The product recovery phase of the pigging operation is now complete; the line can be either cleaned-in-place or is ready for the next product. If the line is to be cleaned, the cleaning-in-place (CIP) fluids can be pumped along the process pipeline, and some fluid is diverted up into the launch housing and cavity of the ball to ensure that both the valve and the pig are cleaned to the same level as the processing line. At each stage of the CIP cycle, the pig can be used to purge the line of chemicals to reduce the amount of rinsing required. The in-line launch housing (patent applied for) is an alternative method of launching the pig (see Figure 6.11c and d). It has been submitted to and passed a European Hygienic Equipment Design Group (EHEDG) test to prove its suitability for hygiene applications. The system operates by the gate valve holding the pig within the internal cage, and with the product flowing around the outside of the pig past the cage and out of the outlet of the launch chamber. At the commencement of the pigging operation, the product pump is turned off. The pig gate valve opens; water or other fluid propellant entering the launch chamber pushes the pig from the cage (i.e. past the now open gate value) and then travels along the pipeline to recover the product. On reaching the receive–return station, the pig is detected and the propellant is turned off; compressed gas then returns the pig to the in-line launch station.

6.4.6 *Automation in handling systems for finished product*

As can be observed in Section 6.4.4, highly automated systems are used in factories producing short-shelf-life dairy products. However, the final trail from factory to consumer is as follows:

1. processing
2. storage

3. order selection
4. consolidation
5. loading
6. retailers
7. consumer

Appropriate material (pallets, baskets and/or trolleys) may be re-used at stages 2 and 6.

This process is suitable for complete or partial automation. Information on the requirements of the customer form the initial premise upon which the processes are based. Production is geared to provide packaged material in the appropriate quantities and 'just in time' to avoid excessive stock in the buffer, refrigerated store. Options for the subsequent handling of materials in warehouses are discussed in an IDF (1991) bulletin, and detailed guidance is presented on the necessary steps for planning an efficient and appropriate system. The reader is also referred to this publication for illustrations of several practical systems ranging from the partially to the fully automated.

It is evident that present-day processors of fermented-milk products employ highly automated systems; a typical illustration of a layout of a yoghurt factory is illustrated in Figure 6.12.

6.5 Recent developments in some fermented-milk products

Many recent developments of a wide range of fermented-milk products have their origin in traditional processes, and some of the products have been industrialised to provide consumers with a wider choice. Some of these products are yoghurt-based, and the scientific and technological aspects have been extensively reviewed elsewhere (Tamime and Robinson, 1999). Milk from different mammals (sheep, goat, buffalo and/or camel) is used for the manufacture of yoghurt, and the processing plants are similar to those discussed in Section 6.4.3. However, only minor processing conditions, such as degree of milk solids fortification, no fat standardisation and processing temperature, are used to obtain products of the desirable physical properties and sensory profiles. In the present context, the automation of some yoghurt-related products is discussed below.

6.5.1 Long-life yoghurt

A wide range of time–temperature combinations has been reported in the literature as extending the shelf-life of drinking and fruit-flavoured yoghurt (Tamime and Robinson, 1999). Although a pasteurisation temperature of ca. 75°C and aseptic packaging may extend the shelf-life of the product to 2 months

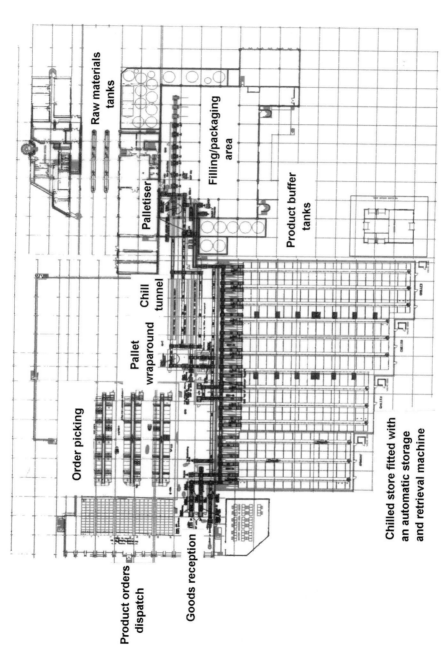

Figure 6.12 Layout of a highly automated yoghurt plant: product filling, storage and dispatch. Reproduced by courtesy of Molkerei Alois Müller GmbH & Co. (UK Production), Market Drayton, UK.

at 5°C, UHT ensures stability of the product for several months at room temperature.

Two possible options may be used for the production of long-life stirred yoghurt. According to APV Nordic (Bøejgaard, personal communication), the stabilised yoghurt base is heated in a PHE or tubular heat exchanger (THE) to 75°C for 15 s, cooled to 25°C with use of two-stage cooling (i.e. first in the regeneration section and second by cold water at ca. 15°C), and then forwarded to a sterile buffer tank. However, if the fruit is not sterile it is heated separately by using a THE, then cooled to 25°C, blended with the heated yoghurt and packaged with use of an aseptic-type filling machine. The alternative option is shown in Figure 6.13; the stabilised yoghurt base is pumped from the fermentation tank(s) to a plate cooler to be cooled to 20°C, blended with fruit with use of a dosage pump and mixed in one of the two tanks in parallel so that the process becomes continuous. Then the fruit/yoghurt mix is forwarded to a THE to be pasteurised at ca. 75°C for 15 s, cooled to ca. 25°C, stored in a sterile buffer tank and finally packaged with use of an aseptic filler. However, the process controls of such a processing plant (e.g. heating or cooling) are more or less similar to those applied to liquid milk processing (see Chapter 3).

6.5.2 Strained or concentrated yoghurt

Two mechanical systems—separators and ultrafiltration—are available to manufacture strained yoghurt, and they are similar to those used for the production of quarg. Mechanical separators (also known as quarg or nozzle separators) are used for the production of strained yoghurt from skimmed milk (Figure 6.14). Two different options are available to produce strained yoghurt, and in either process the fermented skimmed milk is stirred vigorously, thermised at 57°C for 1–2 min, filtered in a duplex filter to remove any large clots, cooled to ca. 40°C, concentrated to 15–18 g 100 g^{-1} solids blended with cream or fruit (optional) and, finally, packaged (Bylund, 1995; Tamime and Robinson, 1999). However, the extended shelf-life of concentrated yoghurt (i.e. second option; see Figure 6.14b) is pasteurised at 70°C, cooled and packaged. The heat-treatment conditions and process controls of the PHE are similar to those discussed in Chapter 3.

Process controls of the nozzle separator are governed by model type and size of separator, number and diameter of the nozzles, temperature of separation, concentration factors, and the solids content in the whey. In quarg production, for example, the solids content in the skimmed milk, product and whey are 9, 18 and 6 g 100g^{-1}, respectively, and thus the concentration factor is 4. In Table 6.5, nozzle diameters needed to provide particular flow rates in the separator at 37°C are listed. If the total number of nozzles in a separator is 12, 8 of diameter 0.8 mm

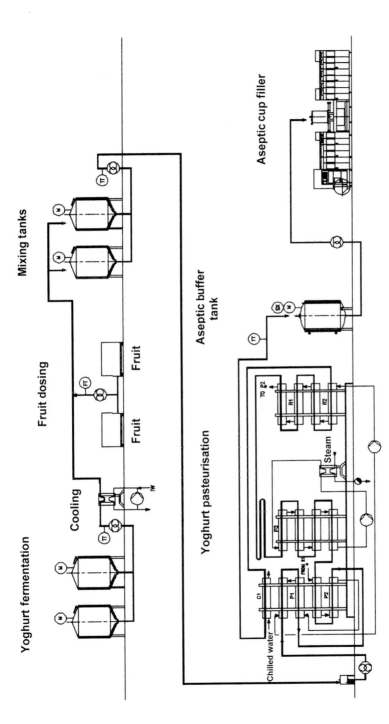

Figure 6.13 Illustration of a processing plant for the manufacture of pasteurised yoghurt. Source: Bøejgaard (personal communication). Reproduced by courtesy of APV Nordic A/S, Aarhus, Denmark.

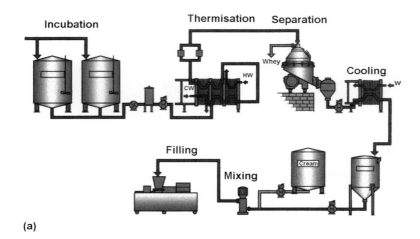

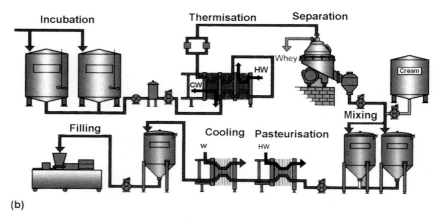

Figure 6.14 Flow charts for the manufacture of strained yoghurt by mechanical separation: (a) thermisation method; (b) combined thermisation and pasteurisation of the end-product. CW, chilled water; HW, hot water; W, mains water. Reproduced by courtesy of Tetra Pak (Processing System Division) A/B, Lund, Sweden.

Table 6.5 Nozzle separator specifications: flow rates achieved with specified nozzle diameters at an operation temperature of 37°C

Nozzle diameter (mm)	0.6	0.7	0.8	0.9	1.0
Product throughput per nozzle (kg h^{-1})	106	144	188	238	293

Source: Nilsson (personal communication).

and 4 of diameter 0.9 mm, the overall throughput is 2456 kg h^{-1} [i.e. (8 × 188 kg h^{-1})+(4×238 kg h^{-1})]. The feed rate is equal to the concentration factor (4) multiplied by the product throughput and is therefore equal to 9824 kg h^{-1} (i.e. 4 × 2456 kg h^{-1}). The volume of whey produced is 7368 kg h^{-1} [i.e. (9844–2456) kg h^{-1}]. The yield, Y, of product can be calculated by using the following formula:

$$Y = \frac{TS^{skim} - TS^{whey}}{TS^{prod} - TS^{whey}} \times 100 \tag{1}$$

where, TS^{skim}, TS^{whey} and TS^{prod} are the total solids content of the skimmed milk, whey and product, respectively.

By increasing the number of nozzles, by using larger diameter sizes and increasing the temperature of separation, the flow rate can be increased, but as a consquence the solids content of the whey may increase. The operation of the separator may vary from one dairy to another, and for strained yoghurt the following conditions may be recommended:

- block some nozzles so that the separator bowl is quickly filled in order to reduce losses in small production lines
- when, owing to wear and tear, the nozzles are replaced, alternate by changing every second nozzle with a new one—replacement of nozzles should be every 6 months or more or when the colour of the whey becomes 'milky', which indicates high losses of milk solids in the whey stream
- the flow rate of the separator should be reduced when there is a high protein content in the skimmed milk

The ultrafiltration (UF) system has been used for the production of strained yoghurt (see Figure 6.15; Tamime and Robinson, 1999). The fat standardised milk base is processed in a similar way as for the production of yoghurt (i.e. prewarmed to 60–70°C, homogenised at 17.3 MPa and heated to 95°C for 3–5 min), cooled to 43°C and inoculated with a thermophilic of a starter culture which does not produce exo-polysaccharides (EPSs). At pH 4.4–4.6, the fermentate is stirred, cooled in a PHE and pumped to a buffer tank. Thereafter, the cold yoghurt is thermised at 58°C or more for 3 min, cooled to 45°C and ultrafiltered in a multi-stage UF plant (see Figure 6.15). The degree of concentration factor of yoghurt in a two-stage UF plant, for example, could be 2 at the first stage and 3.3 at the second stage; however, in large industrial operations up to a four-stage UF plant is used. The strained yoghurt (i.e. retentate) is cooled to ca. 20°C in a plate cooler, stored in a quarg silo and finally packed. However, the permeate is

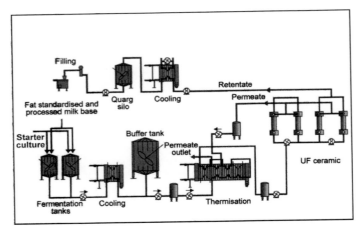

Figure 6.15 Production line of strained yoghurt by ultrafiltration (UF) of warm fermented milk. Reproduced by courtesy of Tetra Pak Filtration Systems A/S, Aarhus, Sweden.

cooled in the refrigeration section of the same PHE that thermised the yoghurt before the UF stage (Harbo, personal communication).

Membrane fouling during the concentration of fermented milk can affect the flux rate, and some flux enhancement strategies that are used commercially may include one of the following (Cheryan, 1998). At start-up of UF, some of the permeate is diverted back to the balance tank or inlet product feed to the membrane. This concept, sometimes known as 'slow-start', results in a high long-term flux and a minimum long-term fouling. Alternatively, periodic reversal of the permeate back into the feed channel by maintaining a high permeate back pressure (i.e. at start-up and throughout the filtration cycle) or co-current permeate flow (also known as uniform transmembrane pressure) helps to dislodge accumulated particles from the membrane surface. As a consequence, the fouling rate is reduced and a higher flux rate is maintained for longer operation before membrane cleaning is required.

6.5.3 Dried fermented milks

Skimmed yoghurt and buttermilk have been dried for use in food formulations as replacements of whole or skimmed milk powders. The drying technology is similar to that detailed in Chapter 4. The difficulty in drying these products is the evaporation stage, where the acidity restricts the concentration to a maximum level of $30\,g\,100\,g^{-1}$. Also, drying is difficult and, under normal practice, up to three drying stages are used (e.g. one spray and two fluid-bed dryers). The optimum air inlet temperature is 160°C or less in order to avoid excessive deposits in the dryer or scorching of the product (for further details, refer to the reviews by Masters, 1991; Caric, 1994; Pisecky, 1997; Tamime and Robinson, 1999).

6.5.4 Frozen yoghurt

Frozen yoghurt products are classified into three main categories: soft, hard or mousse (Tamime and Robinson, 1999). The stabilised yoghurt base containing the desirable sugar level is processed as follows: (a) the yoghurt base is blended with flavours and frozen by using a continuous ice-cream freezer; the product is then packed (lolly stick, bar, cup or bulk) and hardened to produce hard frozen yoghurt; and/or (b) the yoghurt mix is either packed or sterilised before packaging aseptically to be used for the production of soft-serve frozen yoghurt in retail shops.

Technical data and process parameters for the manufacture of ice cream have been discussed in detail by Marshall and Arbuckle (1996).

6.5.5 Drinking yoghurt

This product is categorised as stirred yoghurt with low viscosity, but it is consumed as a drink. The processing aspects are somewhat similar to those detailed in Section 6.5.1; however, the fruit is replaced with a flavouring agent.

6.6 Process control systems

6.6.1 Introduction

As with other fermentation processes, the production of fermented milk is widely automated, but a specific step of the process which is not well controlled is the fermentation stage. During this stage, the complexity of the microbial and biochemical reactions cannot be followed owing to the lack of suitable sensors for in-line monitoring. For example, the simple in-line measurement of the evolution of milk pH is generally impossible for safety reasons because most pH probes are made of glass and cannot be placed directly inside the tank. Moreover, the calibration procedures are time-consuming, and the lifetime of the probes is reduced owing to fouling of the sensitive elements by proteins in the milk. However, Watanabe *et al.* (1994) have reported that different sensors [e.g. pH measurement, yoghurt measuring apparatus FALA-1, (developed by the same authors) gel monitoring development and a quick measuring method for lactic acid development by the starter cultures in synthetic medium] are being developed. Indirect in-line measurement of pH during yoghurt fermentation is possible with use of cheap and convenient sensors, such as the relative pH probe, electrical conductivity probe, coagulometer probe and temperature probes. Finally, the trends in process automation include statistical process control and prediction of final process time.

6.6.2 Controlled variables

The most widely controlled parameters in any processing plant are pressure, temperature, level, agitation speed and flow. For the manufacture of set yoghurt, the temperature of the cabinets or the temperature and speed of the conveyor belt in tunnels are regulated to a fixed value. Generally, the processing cycle is worked on a time basis and not to a pH target, despite the heterogeneity between yoghurt containers and the daily variation in acidification rates.

For the manufacture of stirred yoghurt, the level in the fermentation tank is controlled during the filling of the milk base and the pumping of the coagulum. The temperature of the milk base is controlled to a pre-set level from outside the tanks by the heat exchanger and is monitored and maintained at 40–45°C or 30°C during the incubation period with no real possibility of control. In fact, during this period, heat transfers are very limited because the stirring of the milk and of the coagulum could cause syneresis. If partial cooling takes place in the tank, the agitator speed is controlled when a pH target value (close to 4.5) is reached. The pH of the coagulum is not measured in-line with conventional pH probes, and in many cases the measurements are performed manually on samples. However, the collection of representative samples necessitates attention because of pH heterogeneity of the coagulum near the tank walls.

6.6.3 New reliable sensors for fermentation monitoring

6.6.3.1 In-line measurement of pH with relative pH probes

Recently, a new kind of ceramic steel pH probe has been developed by Pfaudler-Werke GmbH (Schwetzingen, Germany; Figure 6.16). Such a probe consists of three electrodes: the first electrode is sensitive to cationic potential and reduction–oxidation potential; the second electrode is sensitive to pH and reduction–oxidation potential only; and the third electrode is a reference electrode. The potential difference between the first two electrodes gives a signal proportional to the pH and cationic concentration of the medium. In a medium such as milk the concentration of cationic ions does not change significantly. Consequently, the probe is able to measure a relative pH. When the probe is calibrated directly in the milk, the accuracy of the measurements is ±0.1 pH over the entire acidification range. Figure 6.17 shows the good agreement between ceramic steel and glass pH probes (Ingold-Mettler, France) for monitoring the acidity level during the manufacture of yoghurt.

Experience with this pH probe in yoghurt production has shown that a calibration is required for each type of milk base used (Corrieu *et al.*, 1998).

Figure 6.16 An illustration showing a ceramic steel pH probe.

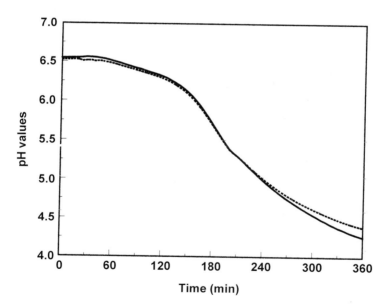

Figure 6.17 Comparative accuracy between ceramic steel (- - -) and glass (—) pH probes for monitoring acidity during the manufacture of yoghurt.

Calibration is carried out by comparison with pH values determined with a classical glass probe; however, no new calibration is required afterwards for up to two months. Table 6.6a summarises the main advantages and disadvantages of ceramic steel pH probes.

Table 6.6 Advantages and disadvantages of using different sensors and probes for milk fermentation monitoring: (a) ceramic steel pH probe; (b) electrical conductivity probe; (c) coagulation-time detector; (d) thermal method for coagulation time detection

Advantages	Disadvantages
(a) Ceramic steel pH probe 1. Only one calibration is required per milk base 2. It is suitable for steam sterilisation 3. It can be cleaned-in-place 4. It is constructed from a nonglass element 5. It has a short response time (i.e. less than a few seconds) 6. There is no shift in accuracy over a few months	1. It requires calibration for each type of milk base used 2. Accuracy is within ± 0.1 pH units 3. It requires a correction law when the temperature of fermentation changes 4. Specific heat treatment (e.g. steam for 5 min or hot water at 80°C for 30 min) is required after cleaning the tank with caustic soda 5. It is corroded by caustic soda, which decreases the lifetime of the probe 6. It requires a specific electronic device for signal amplification 7. It is expensive
(b) Electrical conductivity probe 1. No calibration is required for a direct electrical conductivity measurement 2. Accuracy is within ± 0.08 pH units and does not shift over the years 3. Excellent accuracy is achieved at the beginning of the fermentation stage 4. It has similar advantages to 2–5 in part (a) above 5. Not expensive and has a long lifetime	1. It requires complex calibration for indirect measurement of pH and when using different types of milk bases or starter culture blends 2. It requires a correction law when temperature changes 3. There is reduced accuracy at the end of the fermentation stage (i.e. at pH < 4.6)
(c) Coagulation-time detector 1. Has similar advantages to 2–4 in part (a) above 2. No calibration is required 3. Accuracy for the determination of the pH at coagulation is ± 0.04 4. There is no contact with product for the outer tank wall probe type 5. It has a very long lifetime, it is cheap and there is no shift in accuracy for years	1. It determines the coagulation time only 2. It requires specific signal processing for the calculation of the first derivative with respect to time 3. It requires information on the state of the tank whether in use or not
(d) Thermal method for indirect measurement of pH 1. It has similar advantages to 2–5 of part (a) and 2 and 5 of part (b) above	1. Complex calibration is required for each type of milk base used 2. It requires a specific signal for the determination of the coagulation time 3. It requires special information on the state of the tank whether in use or not

6.6.3.2 Indirect measurement of pH with an electrical conductivity sensor

During the fermentation stage, the electrical conductivity of the milk base increases significantly owing to the metabolism of lactic acid bacteria that produce ionic compounds as lactate and small amounts of ammonium. Thus, it is possible to replace pH probes by electrical conductivity probes (Figure 6.18) that are more reliable and less expensive in the long term. However, no linear theoretical relationship is known between these two variables because of the influence of the composition of the medium. Moreover, this relationship changes with fermentation temperature and the growth medium of the different species of lactic acid bacteria (Latrille et al., 1992). In order to identify the nonlinear relationship between pH and electrical conductivity measurements, a neural network model with two inputs (conductivity and temperature) and one output (pH) has been used (Corrieu et al., 1996). Experimental values of pH and conductivity were required to establish this relationship. The indirect measurement of pH from electrical conductivity measurements is obtained with an accuracy of ± 0.08 pH for different types of growth media and starters. Figure 6.19 shows the evolution of the electrical conductivity of the milk during the acidification stage; the indirect measurement of pH compared well with that using a glass pH probe.

Table 6.6b summarises the main advantages and disadvantages of an electrical conductivity probe for indirect measurement of pH. One interesting advantage is that a significant evolution in pH value appears at a very early phase of the yoghurt fermentation and therefore measurement of pH improves the early detection rate of growth failure of starter cultures.

6.6.3.3 Detection of the coagulation time

The use of hot-wire sensors for in-line detection of coagulation time is widely used in cheesemaking processes (Hori, 1985; Corrieu and Picque, 2000), but few examples have been reported in monitoring fermented milk processes at

Figure 6.18 A typical electrical conductivity probe.

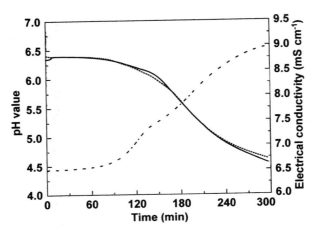

Figure 6.19 Evolution of the electrical conductivity (- - -) of milk during the fermentation period; note the good accuracy obtained with the indirect measurement of pH (· · ·) compared with a glass pH probe (—).

pilot-scale (Corrieu et al., 1996). However, the in-line Coagulometer MSC 1000 (Ysebaert, Frepillon, France) has been used successfully to detect the coagulation time of milk during the manufacture of cheese and fermented milks. The probe looks like a typical temperature sensor, but it has an additional small electrical heater inside the stainless steel tube. When coagulation of the milk occurs, the heat transfer is modified and the temperature of the probe is increased by a few degrees (Figure 6.20). Continuous monitoring of the temperature and the calculation of its first derivative with respect to time allows automatic detection of the coagulation that occurs at a defined pH value between 5.5 and 5.3 (depending on the composition of the milk base); the accuracy of the measured value is within ±0.04 pH units (Table 6.6c).

Another type of sensor based on measuring changes in thermal properties of the milk base gives similar results to the hot-wire sensor. It is a flat probe and is stuck on the outside wall of the fermentation tank to monitor the temperature of the wall surface continuously. During the fermentation period, the temperature of the tank remains the same. However, the moment the milk base starts to coagulate, the temperature of the wall surface decreases drastically as a result of: the great temperature difference between the inside and outside surfaces of the tank; a decrease in the effective thermal conductivity of the coagulated milk base. Monitoring of the temperature and the calculation of its first derivative with respect to time allows the automatic detection of coagulation time, but with a delay of 3 min in comparison with detection with use of the hot-wire sensor. This leads to an underestimation of the coagulation pH by 0.04 units. Figure 6.20 shows a comparison of the detection of coagulation time with use

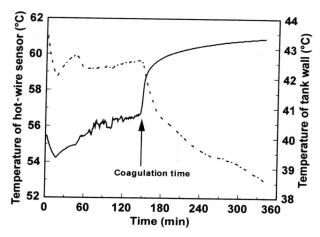

Figure 6.20 Comparison in the detection of coagulation time during the fermentation period in a large tank with use of a hot-wire sensor (—) and a tank wall surface temperature probe (- - -).

of these two types of probes in a large vat during the fermentation period. Table 6.6c lists the main advantages and disadvantages in using coagulation-time detectors for fermented-milk production. The main advantages of the tank wall suface temperature probe compared with the hot-wire sensor are: low cost and the fact that there is no contact between the product and the wall probe.

6.6.3.4 Indirect measurement of pH through measurement of temperature increase

The acidification of milk by lactic acid bacteria produces heat owing to the hydrolysis of lactose to lactic acid. As mentioned earlier, the temperature of the processed milk base cannot be regulated accurately in the fermentation tank. As a consequence, a very low level of heat loss occurs during the fermentation period, which is compensated for by a slight increase in temperature 1–2°C as a result of the metabolic activity of the starter microflora which is directly linked to lactic acid production. Also, the shift in temperature is indirectly related to a drop in pH and the buffering properties of the milk base (Corrieu *et al.*, 1997). The two phases in temperature change are: a pre-coagulation period of the milk base, characterised by low, but not negligible, heat loss; and a period after coagulation which does not show any heat loss.

Therefore, the indirect measurement of pH requires two devices—a coagulation probe (as described above) and a temperature sensor. The temperature sensor consists of a sensitive element which is fitted inside the fermentation tank at a distance of at least 30 cm from the wall. The in-line Coagulometer MSC 1000 measures the coagulation time, and the temperature sensor monitors the increase of temperature, which is then converted to pH values by using a calibration law. The calibration is established by pH sampling from previous batches of fermented milk (i.e. by a reference method) with use of different starter cultures and the same level of solids in the milk base. Figure 6.21 shows one such example of the temperature increase in a large fermentation tank and the accuracy (± 0.08 pH) of the indirect pH measurement during incubation compared with the reference method. The main drawback in using this thermal method is the need for a meticulous calibration procedue, which is time-consuming. Table 6.6d lists the main advantages and disadvantages of using the thermal method for indirect measurement of pH.

6.6.4 Advanced monitoring: prediction of the final process time

In a fermented-milk production line, the control of the rate of acid development in a succession of fermentation tanks is a useful tool for production. Taking into account the desired target pH in the final product, such process control can be achieved by the prediction of pH drop over the fermentation period. Two methods are available to predict the evolution of pH in fermented milks (Latrille et al., 1994).

First, for isothermal microfloras a reference curve is calculated for each type of starter culture used by averaging data on pH evolution of previous cultures. This reference curve, combined with a sliding geometrical method, is used to perform a comparison with the actual fermentation. The final fermentation

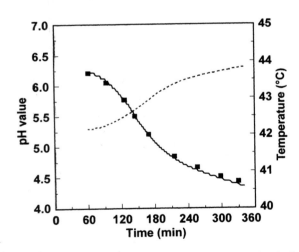

Figure 6.21 An illustration showing temperature increase ($\cdots$) of milk during the fermentation period in a large, industrial-sized tank; comparison between pH measurements with use of a laboratory glass electrode (■) and the indirect (thermal) method (—).

time, which occurs at a predetermined pH, is predicted with an accuracy of less than 10%.

Second, as starter cultures grow at different temperatures a dynamic model of pH evolution is required. It can be obtained as a recurrent neural network that is able to simulate any dynamic system. Because the growth temperature is an important parameter for explaining different types of dynamic behaviour in the process, this temperature is explicitly taken into account. The final fermentation time is predicted with an accuracy of less than 10%.

In the near future, the use of neural network modelling or modern statistical tools will be more common as the computing power in production lines increases.

6.6.5 Statistical process control and future trends

Statistical process control employs statistical means as a part of the total quality management in order to ensure that the process is under control and that the manufactured products are within defined specifications. In the case of statistical process control of fermented-milk production, statistics of properties such as evolution curves of the pH or electrical conductivity may be used (Corrieu *et al.*, 1996). Confidence intervals of reference curves are also useful for fault detection of unexpected delays in fermentation time resulting from bacteriophage infection or unbalanced bacterial populations.

In the future, new on-line sensors for key process variables and quality attributes will be required for efficient control. For example, the on-line measurement of the viscosity of fermented milk is a very complex but essential challenge for total quality management. Also, the personal computer has finally arrived on the plant floor, although doubts and criticisms have been expressed. Modern control systems can easily incorporate on-line and off-line data into their knowledge pool, but one of the major problems is communication between the different sensors and control systems. The principle of independent 'agents' and 'applets' capable of communicating with each other could be visualised as a future solution to communication problems. Moreover, the development of smart sensors including microprocessor software to overcome problems caused by non-linearity and calibration would benefit the food industry (Linko and Linko, 1998).

Acknowledgement

The Scottish Agricultural College (SAC) receives financial support from the Scottish Executive Rural Affairs Department (SERAD).

References

Auer, A. (1995) An integrated computer-aided quality assurance system using processing and laboratory data of dairy plants. *Bull. Int. Dairy Fed.*, **308** 61-7.
Bird, J. (1993) Milk standardization. *J. Soc. Dairy Technol.*, **46** 35-7.
Bird, B. (1996) APV pigging system. *Milk Industry Int.*, **98**(3) 36-7.
Bylund, G. (1995) *Dairy Processing Handbook*, Tetra Pak Processing Systems A/B, Lund.
Caric, M. (1994) *Concentrated and Dried Dairy Products*, VCH Publishers, New York.
Cheryan, M. (1998) *Ultrafiltration and Microfiltration Handbook*, Technomic, Lancaster.
Corrieu, G. and Picque, D. (2000) Operation programming and automation, in *Cheesemaking: From the Science to Quality Assurance* (Eds. A. Eck and J.C. Gillis), 2nd edn, Lavoisier, Paris, pp. 624-36.
Corrieu, G., Perret, B., Picque, D. and Latrille, E. (1996) Computer-based fermentation process control, in *Computerized Control Systems in the Food Industry* (Ed. G.S. Mittal), Marcel Dekker, New York, pp. 260-5.
Corrieu, G., Latrille, E. and Béal, C. (1997) *Recent improvements in fermentation processes*. Lactic Acid Bacteria 97. Congrès LACTIC 97, France, 10–12 September. Adria Normandic and Presses Universitaires de Caen, pp. 185-94.
Corrieu, G., Perret, B., Quemener, P., Latrille, E. and Mana, J. (1998) La fermentation lactique en temps réel. *Revue Laitière Française*, **579** 34-5.
Ferret, R. and Trystram, G. (1991) The role of the computer in production and manufacturing processes in the dairy industry. *Bull. Int. Dairy Fed.*, **259** 12-23.
Hansen, S.D. (1996) Flexible standardising units. *Scandinavian Dairy Information*, **10**(4) 24-5.
Hori, T. (1985) Objective measurement of the process of curd formation during rennet treatment of milks by the hot wire method. *J. Food Sci.*, **50** 911-17.
IDF (International Dairy Federation) (1985) How to automate. *Bull. Int. Dairy Fed.*, **185** 10-24.
IDF (International Dairy Federation) (1991) Application of computers in the dairy industry: guidelines for solving in-plant materials handling systems in dairy plants. *Bull. Int. Dairy Fed.*, **259** 164.
IDF (International Dairy Federation) (1995) Applications of computers in the dairy industry. *Bull. Int. Dairy Fed.*, **308** 164.
Koroleva, N.S. (1991) Products prepared with lactic acid bacteria and yeast, in *Therapeutic Properties of Fermented Milks* (Ed. R.K. Robinson), Elsevier Applied Science, London, pp. 159-79.
Kurmann, J.A., Rasic, J.L. and Kroger, M. (1992) *Encyclopedia of Fermented Fresh Milk Products*, Van Nostrand Reinhold, New York.
Latrille, E., Picque, D., Perret, B. and Corrieu, G. (1992) Characterizing acidification kinetics by measuring pH and electrical conductivity in batch thermophilic lactic fermentations. *J. Fermentation and Bioengineering*, **74** 32-8.
Latrille, E., Corrieu, G. and Thibault, J. (1994) Neural network models for final process time determination in fermented milk production. *Computers and Chemical Engineering*, **18** 1171-81.
Linko, S. and Linko, P. (1998) Developments in monitoring and control of food processes. *Trans Icheme*, **76**(c) 127-37.
Marshall, R.T. and Arbuckle, W.S. (1996) *Ice Cream*, 5th edn, Chapman & Hall, London.
Marshall, V.M.E. and Tamime, A.Y. (1997) Physiology and Biochemistry of fermented milks, in *Microbiology and Biochemistry of Cheese and Fermented Milk* (Ed. B.A. Law), 2nd edn, Blackie Academic & Professional, London, pp. 153-92.
Masters, K. (1991) *Spray Drying Handbook*, 5th edn, Longman Scientific and Technical, Harlow.
Metchnikoff, E. (1910) *The Prolongation of Life*, revised edn, translated by C. Mitchell, Heinemann, London.
Nielsen, W.K., Bøejgaard, S.K. and Vesterby, P. (1999) *Technology Update: Recombined Dairy Products*, technical bulletin 0499, APV Nordic A/S, Aarhus, Denmark.
Pisecky, J. (1997) *Handbook of Milk Powder Manufacture*, Niro A/S, Soeborg, Denmark.

Rao, J.F. and Steffe, M.A. (Eds) (1992) *Physical Properties of Foods*, Elsevier Applied Science, London.
Robinson, R.K. (Ed.) (1995) *Colour Guide to Cheese and Fermented Milks*, Chapman & Hall, London.
Robinson, R.K. and Tamime, A.Y. (1990) Microbiology of fermented milks, in *Dairy Microbiology*, Volume 2 (Ed. R.K. Robinson), 2nd edn, Elsevier Applied Science, London, pp. 291-343.
Robinson, R.K. and Tamime, A.Y. (Eds) (1991) *Feta and Related Cheeses*, Ellis Horwood, Chichester, Sussex.
Robinson, R.K. and Tamime, A.Y. (1993) Manufacture of yoghurt and other fermented milks, in *Modern Dairy Technology*, Volume 2 (Ed. R.K. Robinson), 2nd edn, Elsevier Applied Science, London, pp. 1-48.
Salji, J.P. (1993) Acidophilus milk, in *Encyclopaedia of Food Science, Food Technology and Nutrition*, Volume 1 (Eds. R. Macrae, R.K. Robinson and M. Sadler), Academic Press, London, pp. 3-7.
Sellars, R.L. (1991) Acidophilus products, in *Therapeutic Properties of Fermented Milks* (Ed. R.K. Robinson), Elsevier Applied Science, London, pp. 81-116.
Tamime, A.Y. (1990) Microbiology of starter cultures, in *Dairy Microbiology*, Volume 2 (Ed. R.K. Robinson), 2nd edn, Elsevier Applied Science, London, pp. 131-201.
Tamime, A.Y. and Marshall, V.M.E. (1997) Microbiology and technology of fermented milks, in *Microbiology and Biochemistry of Cheese and Fermented Milk* (Ed. B.A. Law), 2nd edn, Blackie Academic & Professional, London, pp. 57-152.
Tamime, A.Y. and Robinson, R.K. (1999) *Yoghurt: Science and Technology*, 2nd edn, Woodhead Publishing, Cambridge, pp. 129-388.
Tunail, N. (2000) Biology of *Enterococcus* spp., in *Encyclopaedia of Food Microbiology*, Volume 3 (Eds. R.K. Robinson, C. Batt and P.D. Patel), Academic Press, pp. 1365-73.
Watanabe, H., Fujioka, S., Motohashi, R. and Imai, E. (1994) Development of sensors for lactic acid fermentation process. *Japanese Technol. Reviews (Section E: Biotechnol.)*, **4**(2) 44-52.

7 Cheddar cheese production
B.A. Law

7.1 Introduction

There are degrees of mechanisation and automation at practically any level of Cheddar cheese production, beginning with simple mechanical temperature programmes for scalding curds in whey, mechanical stirring paddles, cutting knives and chipping mills. This chapter deals with levels of mechanisation of curd formation and handling appropriate to cheese plants whose daily throughput is measured in tonnes rather than kilograms, necessitating high levels of reproducibility (in cheese yield and quality), low levels of maintenance and ease of cleaning to maintain production hygiene.

Milk preparation and standardisation for cheesemaking are dealt with generically in Chapter 3, and starter culture technology likewise forms a special category of technology described in detail elsewhere (Hoeier *et al.* 1999). Additional information on culture production and choice in cheese plants can be obtained from the culture suppliers, but it is worth mentioning here a particular new development in the automation of culture inoculation resulting from collaboration between a culture company and a dairy equipment supplier. Many large cheese plants use the direct-to-vat starter system whereby very concentrated pre-grown cultures are bought from the culture supplier without the need for bulk starter preparation equipment. One of the disadvantages of this culture technology is that quite bulky packs of frozen or freeze-dried starter have to be opened up in the cheese room, thawed or hydrated manually, then added manually to the vats.

Tetra Pak Dairy & Beverage Systems (Sweden) and Chr. Hansen A/S (Denmark) have recently introduced the patented automatic inoculation system (AISY; Figure 7.1) to remove this series of manual operations for frozen direct-to-vat concentrated starters by integrating and automating the thawing, mixing and dosing operation from the culture pack (left-hand side of Figure 7.1) right through to the cheese vats (on the right-hand side of Figure 7.1).

Frozen culture is emptied from the pack directly into a buffer tank, diluted with water and stirred. The culture is then inoculated in-line to the milk stream or pumped at the correct dose rate straight into the cheese vat (or milk fermentation vessel). If kept below 10°C, the starter in the buffer tank can be used for multiple sequential inoculations over a period of 24 h. The whole system is designed to be coupled to automatic cleaning-in-place (CIP) systems.

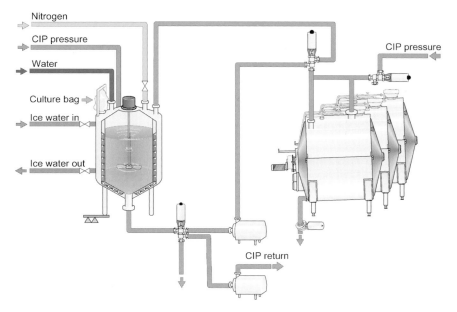

Figure 7.1 Diagrammatic representation of the automatic inoculation system (AISY) for Chr. Hansen frozen direct-to-vat system starter cultures. The culture pack opener and the mixing and holding buffer tank are to the left of the diagram. The buffer tank contents are dosed via the pump and lines either into the milk feed line or directly to the cheese vats, shown to the right of the diagram. Reproduced with permission of Tetra Pak Dairy & Beverage Systems AB, Lund, Sweden and Chr. Hansen A/S, Hoersholm, Denmark.

In this chapter I will concentrate on the coagulation and acidification stages, whey separation, curd texturisation and salting, and finally, pressing and packing for storage and maturation. This will involve not only a review of current mechanical and control systems, but also a critical look at the applied research literature for insights into the next generation of mechanisation and control technology.

7.2 Cheesemaking as process engineering

Small-scale cheesemaking is a craft, but converting hundreds of thousand of litres of milk into cheese every day with predictable outcomes (yield, composition, grading quality, maturation potential) is process engineering. The cheesemakers' craft is still needed in setting the boundaries of the recipe in terms of the starter culture, rennet gel strength for cutting, the time for separating curds and whey, milling acidity and other such experience-learned parameters, but the shear volume of large-scale production, and the economic consequences of critical variations within the process, necessitate help from machines and computers. Although continuous cheesemaking remains a goal of the engineers, continuous cheddar manufacturing is not yet a commercial reality, even though the APV/Sirocurd process came very close some 10–15 years ago. Despite this limitation, process engineers have achieved very significant labour-saving innovations in cheese vat design, whey separation and cheddaring, and they have smoothed the transitions between the boundaries of the unit operations that make up the whole process (Figure 7.2).

In practice, the vat filling and dosing operations can be automated via programmable logic controllers (PLCs) for valves and pumps, but this remains a unit operation, as does the next stage in the vat, involving the formation of the coagulum, or rennet gel. Cutting and stirring operations are amenable to automation through new sensors and research-based understanding of optimum cutting and stirring speeds, all taking advantage of modern cheese vat design, whereby mechanical tools are built in and are interchangeable between in-vat operations. The remaining unit operations of traditional cheddar manufacture (right through to the wrapped cheese blocks ready for the maturation store) still exist in the sense that they require individual control, but the equipment designers and manufacturers have more or less removed the practical boundaries between them. Given sufficient capital, cheese plants can be equipped with mechanised and automated curd drainage, cheddaring, milling, salting, pressing and block formation, and bagging. Applied research into sources of process variation and their effect on cheese yield and quality has gone hand in hand with advances in computer-based data handling to further improve the smooth and consistent running of Cheddar plants. This is achieved through supervisory control and data acquisition (SCADA) systems (see Chapter 1), either connected directly to sensors and the PLC network or by feeding data to an expert system to guide the plant operators. Let us now consider these developments in detail.

7.3 Coagulation of milk and curd formation

The critical recipe parameters are milk composition, rennet dosage, starter type and dose, and 'ripening time' (to allow the starter to begin growing and producing lactic acid). Automation in this area is a matter of meters, pipes, valves and PLCs, dealt with elsewhere in Chapter 1. Here we are concerned with automatically starting the cutting process at the right curd gel strength every time, minimising damage to the delicate curd particles during cutting and stirring, and optimising the timing and mechanics of whey separation when the curds have reached optimum acidity in the whey. The science and technology of curd formation, including the initial coagulation stage, were described and reviewed most recently by Lomholt and Qvist (1999).

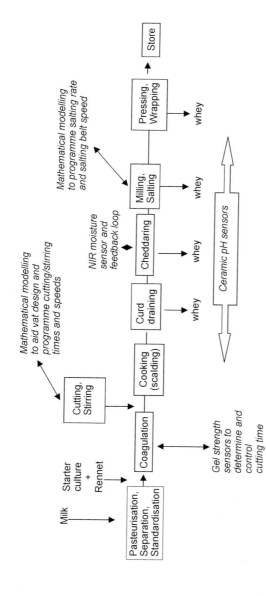

Figure 7.2 Schematic representation of Cheddar cheese manufacture through a succession of constituent unit operations, indicating the inputs which are involved in mechanising and automating the flow between them. NIR, near infrared.

7.3.1 Vat design

Cheddar cheese has traditionally been made in open vats, even on a large scale (i.e. vats of up to 10 000 l capacity) to facilitate access to the curd during cheddaring, allowing the matted curd to be manually cut, turned and stacked to achieve the required 'flow' for texturisation and whey drainage. Although a degree of mechanisation was possible in cutting and stirring within this configuration, modern hygiene standards were difficult to apply, especially when complex drive mechanisms have to be situated immediately above the vat, when plant personnel lean into it to handle the process and when a large surface area of milk and curd is open to the plant room environment. Almost all modern Cheddar plants now use various forms of enclosed cheese vats of the vertical double-O (e.g. APV, Damrow, Wincanton) or horizontal (Tetra Tebel OST) design, with operating capacities of up to 30 000 l. Two well-developed commercial examples of such vats are illustrated in Figures 7.3 and 7.4. The APV and Tetra-Tebel equipment is described here to illustrate the design principles and advantages of enclosed automated cheese vats.

Cheesemakers need accurate, reproducible temperature programmes for maintaining the setting temperature and then increasing it to maximum scald. Both types of enclosed vat provide this through good jacketing insulation, combined with well-distributed indirect heating by steam or hot water circulating in heat exchangers and through sprinklers within the vat jacket. The manufacturers use their own proprietary designs to optimise vat shape in relation to the geometry of the stirrer blades to ensure that the heat in the jacket is picked up evenly and quickly so that there are no static regions in the vat which could

Figure 7.3 The APV Curdmaster double-O design cheese vat, with ancillary control unit, temperature control pipework and valves, filling and emptying pipes and integral cutting and stirring gear. Reproduced with the permission and cooperation of APV Ltd, UK.

CHEDDAR CHEESE PRODUCTION

Figure 7.4 A pair of horizontal design Tetra Tebel OST vats installed for Cheddar cheese manufacture, with control unit, all pipework and valves for filling and emptying, and integral cutting and stirrer gear. Reproduced with the permission and cooperation of Tebel-MKT B.V., Holland.

create hot or cool spots. Such design differences are best discussed directly with the equipment manufacturers in relation to specific installations.

7.3.2 Cutting and stirring

The design of the cutting and stirring tools in these enclosed vats is not only for clean, shatter-free cutting and stirring, which minimises damage to the curd particles, but also to deliver a controlled, protected environment at this critical stage of the cheesemaking process. They can be cleaned-in-place with minimal need to make and break joints. Indeed, the OST vat can be cleaned in place entirely, with no pipe joints having to be broken and remade even when the vat is in continuous use. It is a feature of the horizontal vat design, of which the OST is an example, that the whole cycle of filling, curd making, emptying, cleaning and refilling is done with no need to expose the pipework or the vat interior to outside contamination. The cutting and stirring tools of enclosed vats are always in place; indeed, the same hardware does both jobs in the double-O design. If the paddles are rotated one way, they present a cutting edge to the curd. When they are later reversed, they act as stirrer blades for most of their working area and as surface scrapers for the rest to ensure that no curd lumps stick to the sides or bottom of the vat and become overacidified or overheated. This is all very helpful in the task of mixing into the milk the coagulant, starter culture and any other processing aids, for taking the work out of cutting the coagulum and for stirring it well but gently during scalding, but next we have to consider the process of cutting the curd as a critical step in successful cheesemaking.

Without instrumentation as a guide, cheesemakers traditionally judge the firmness or strength of the milk gel formed by the action of rennet by feeling the pressure necessary for a finger or a knife to break the gel and part the broken surfaces. This is a proven, reliable method, but there are fewer and fewer working cheesemakers in the industry with sufficient experience to make this fine judgement consistently. In any case, modern enclosed cheese vat design makes access for such a test increasingly difficult, where the priority is on mixing dynamics, efficiency of filling and emptying, and ease of cleaning-in-place. It is therefore a logical step, both in vat design and automation, to introduce instrumental methods into the coagulation process and, although no single machine or operating principle has yet emerged as the obvious way forward for commercial instruments, there are some promising systems under intensive trial and investigation. It is not the purpose of this chapter to review the history of coagulation parameters, except to note that many promising designs have been proposed, based on changes in the viscosity and resistance to torsional forces of the milk as the rennet gel forms, but none of these is in widespread use as proven technology (Laporte *et al.*, 1998). The biggest problem for such instruments is that they measure the same phenomenon that the experienced cheesemaker measures, but they do not provide any more objective a value or parameter which can be used to decide when to cut the curd at the optimal gel strength. This is the key to successful instrumentation-based automation here. It does not matter whether or not the precise moment when the gel is cut is absolutely right to balance curd shattering against a good fat-retaining, smoothly syneresing gel. What does matter is that the instrument should consistently, faithfully and accurately tell the cheesemaker or the automation controller (PLC) when the gel strength is the same as that used by the cheesemaker to make the decision to cut. Once that is done, the PLC can be programmed to respond to the critical instrument value whatever the seasonality or composition of the milk or whatever type of coagulant is in use. Indeed, the PLC could be adjusted to respond slightly differently if, based on applied research, the coagulant supplier advises that a particular coagulant displays different coagulation kinetics (e.g. a microbial coagulant, compared with chymosin, may show such variation whereby the gel firms up at a different rate after the initial gelation occurs).

The instruments currently available for factory trials are based both on thermal and on optical techniques. The thermal instruments are typified by the Coagulometer, which is described in a French patent application (Noel *et al.*, 1988). This instrument 'detects' milk coagulation through a heated thermal probe (sometimes referred to as a 'hot wire') whose temperature is measured as the milk coagulates. The viscosity of the milk increases as a function of the aggregation of casein micelles under the influence of the coagulant, and this reduces heat loss by convection from the probe, so it heats up from an initial steady state. The rate of temperature increase reaches a maximum as the milk gels and convection ceases, but then its temperature reaches a new steady state as heat loss by conductance matches heat input. Fluctuations in background temperature are compensated by a second control probe, and the corrected temperature change data are processed to measure percentage coagulation, which is presumably calibrated against an experienced cheesemaker to pinpoint the ideal percentage coagulation for cutting the curd. The Coagulometer is a robust instrument which has had limited uptake in automated systems. Its chief drawback seems to be its insensitivity to changes in the gel structure following the gelation phenomenon itself. In practice, such changes are as important as gelation time in deciding the right time to cut and, in this regard, light-measuring instruments seem to have the advantage.

The Gelograph, a Swiss-made instrument, has proven successful in UK commercial pilot plants and may find application in the factory. Its mode of operation is similar to that of the Coagulite instrument (Payne *et al.*, 1993) used in the new Food Science Australia (FSA) cheese pilot plant at Werribee, Victoria, and described by Roupas (1998). Both instruments use light scattering by casein micelles to measure changes in micelle size during the enzymic-induced (coagulant-induced) swelling phase, during the aggregation phase when the milk gels and finally during the gel-firming phase as cross-linking within the gel increases, making it firmer. It is this final stage that is not easily measured by thermal probes, yet this is the most critical stage in governing the extent to which the curd will shatter and produce fines when cut (if it is too hard) but will still have enough cross-linking to retain fat globules efficiently (i.e. is not too soft). The degree of cross-linking at cutting is also a significant factor in the vital process of syneresis (water expulsion) from curd during cheddaring.

The use of visible light and fibre optics to measure and record the light scattering in coagulating and gelling milk allows for very compact, nondestructive, safe and easily cleaned instrumentation. The application in the FSA pilot plant is integral to each cheese vat and feeds information into a SCADA system. To my knowledge, none of these instruments is in routine use in the tough day-to-day cheese factory environment, but their apparent success in commercial pilot plants may lead to designs which will cope with prolonged use and exposure to repeated manufacturing and CIP cycles.

Laser light scattering is suggested by ten Grotenhuis (1999) as a further option for the design of a predictive instrument, because it not only measures changes in particle (aggregate) size during gelation by the coagulant but also measures the mobility of the aggregated particles within the cross-linked gel structure. The detection and measurement of multiple laser scattering produced by particles in concentrated systems is called diffuse wave spectroscopy (DWS) and, like normal light-scattering techniques, makes use of compact and robust fibre-optics hardware to transmit the collimated laser beam into the milk and collect the scattered light for amplification and analysis by the correlator board in a personal computer (PC). The correlation function between scattering and particle size or mobility is automatically calculated and shown as a curve in which inflection points represent critical times when aggregation, gelling and gel-firming events happen. It is from this curve that the cutting time is calibrated and fed into the reporting system, either to trigger the cutting knives in the vat or to alert the operator to turn on the cutters. This technology is at the trial stage, but appears to give a particularly faithful cutting time in comparison with the subjective judgement of experienced cheesemakers. NIZO food research (The Netherlands) has developed a trial commercial instrument based on this applied research.

Near infrared (NIR) spectroscopy has also been suggested as the basis for instruments to predict optimum curd-cutting time (Laporte et al., 1998) and this certainly looks promising as an accurate measure of post-gelation reactions which firm up the gel. Foss-NIR Systems, UK is developing such an instrument, but I have no knowledge of its widespread industrial adoption.

7.3.3 Theoretical aids to the optimisation of the cutting and scalding stages

As indicated above, cheesemakers must judge the time or gel firmness at which the coagulated milk (rennet gel) is cut so that it is sufficiently well structured to retain globular milk fat and to develop texture during whey expulsion (syneresis). At the same time, the coagulum must not be too firm and shatter on cutting so that fine curd particles (fines) are formed then lost later on whey separation. However, even after this stage has been negotiated well, there is no room for complacency. The final textural quality of the curd and the maturation potential of the cheese is still dependent on how well the curds are treated in the whey during the next phase, in which the temperature in the vat is raised progressively from 30°C to between 39°C and 40°C (the 'scalding' stage). The scald programme itself is, of course, controlled by a PLC connected to the steam or hot water supply valves and pumps, but the speed and duration of cutting and the mechanical vigour of stirring in the heated whey are also critical, chiefly in terms of fat loss, further shattering (loss of fines) and final curd moisture. All of these parameters are vital in optimising the efficiency of the plant, expressed as cheese yield, and also in determining whether the resultant cheese will be suitable for short-hold high-moisture cheese or long-hold low-moisture cheese destined for the mature market. I do not intend to review the factors involved in optimising yield in this chapter as they have been expertly set out and discussed already, and I commend the works of the International Dairy Federation (IDF, 1991), Lucey and Kelly (1994) and Morison (1997).

Here I will consider the factors in mechanisation and automation that are important to control and optimise curd formation. Generalisation in this area of cheesemaking technology is not a useful exercise because the final output of individual plants is very much the result of the skill and experience of the manager and key operational staff. However, there are in the technical research literature some valuable data on the operation of vertical double-O (Damrow) and horizontal (OST-type) cheese vats which can be used as generic guidelines in process optimisation (Johnston et al., 1991, 1998).

In essence, Johnston and co-workers have demonstrated a relationship between curd particle size [CPS; expressed as a cumulative percentage of curd particle size less than 7.5 mm diameter], percentage fat in whey and the number of revolutions of the cutting tools in a Damrow (double-O) cheese vat. The term 'cumulative' refers to the particle size after the cumulative effects of cutting and stirring (see below). They have subsequently worked out equivalent relationships which apply during the operation of a typical OST horizontal vat (Figure 7.5). The hypothesis developed from this relationship allows, not only for optimisation of curd yield, but also for adjustments to the plant to achieve predetermined moisture targets in the cheese. Thus, Johnston and his group suggest that although the cutting sequence in mechanised cheese vats is a major determinant of the curd particle size and the level of fat 'lost' in the whey, the subsequent stirring program, prior to scalding, has to be determined in tandem with the cutting program in order to best control these parameters.

In practice, this means that a cutting program that has an insufficient time–speed combination leaves curd particles which are too large to be stirred without the stirring causing shattering. Particle shattering in turn causes high fat losses. Therefore, once the cheesemaker has decided on a cutting speed which minimises initial curd shattering when applied to the firmness of curd which suits the cheese recipe, the best strategy for minimising shattering and fat loss at the stirring stage is to prolong cutting to reduce curd particle size according to the model data, to a percentage CPS value which will not decreased (by stirring) between cutting and whey draining. At the time when Johnston and co-workers were developing their hypothesis and models there was a clear difference in cutting knife density between the Damrow double-O vats (145 mm spacing) and

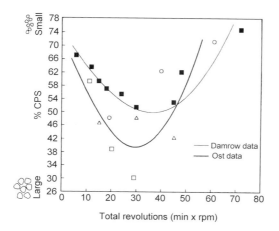

Figure 7.5 The effect of speed and duration of cutting in double-O vertical (Damrow) and Tetra Tebel OST cheese vats on the cumulative percentage of curd particles with diameter less than 7.5 mm (%CPS) during Cheddar cheese manufacture. ■, Damrow data; □, OST, 3 rpm; △, OST, 5 rpm; ○, OST, 7 rpm. Reprinted from Johnston et al. (1998), Copyright 1998, with permission from Elsevier Science.

the Tetra Tebel OST design (40 mm spacing), which meant that curd particle size was reduced in OST vats more quickly than in double-O vats, so the total speed–time combination (cutting revolutions) that is optimal in OSTs is less than that in double-O vats.

Figure 7.6 summarises the model developed for Tetra Tebel OST vats, superimposed on the Damrow double-O model. It shows how variation in cutting speed and duration of cutting followed by a constant stirring speed determines curd particle size distribution in the vats. Dotted lines represent an estimated trend, whereas solid lines are taken from curd particle size measurements. The models show how shorter cutting times produce large particles, which shatter during stirring, so that fat losses are relatively high and final curd particle sizes are small at draining. On the other hand, longer cutting times also produce smaller curd particles, but they do not shatter on stirring, reducing fat losses. The model, therefore, predicts the best combination of cutting speed and duration to give a combination of low shattering overall and optimum particle size at draining to control loss of fines and cheese moisture.

The comparative curves in Figure 7.6 for double-O and OST designs show that much less cutting (in speed and time) is needed in OST vats to avoid shattering, suggesting a more efficient design, though the difference in knife

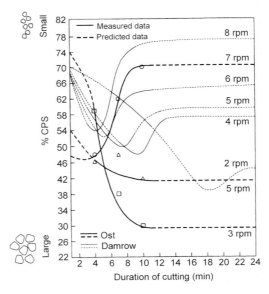

Figure 7.6 Model of the change in the cumulative percentage curd particle size (%CPS) less than 7.7 mm diameter with increasing duration of cutting in double-O vats (Damrow) and Tetra Tebel OST vats used for Cheddar cheese manufacture. ■, Damrow data; □, OST, 3 rpm; △, OST, 5 rpm; ○, OST, 7 rpm. Reprinted from Johnston *et al.* (1998), Copyright 1998, with permission from Elsevier Science.

'density' between these particular vats may explain most of the difference in 'efficiency' measured by this criterion. Also, the tendency of curd particles to shatter more in Damrow double-O vats after cutting durations similar to those in Tetra Tebel OST vats can be compensated in the former by faster cutting followed by slower stirring before scalding. This reduces fat losses and results in larger curd particles at draining.

7.4 Curd draining, cheddaring, milling and salting

The next stage in Cheddar manufacture involves separating the curds and whey after scalding (traditionally called 'drawing') and allowing the curd particles to coalesce ('mat') and stretch to a limited extent ('flow') in the stage called 'cheddaring'. Through this stage, the curd loses more moisture, becomes more acidic and acquires a 'chicken-breast' texture. Traditional cheesemakers do this largely by hand, draining the whey from the vat then using the vat itself as the cheddaring vessel, cutting, turning and stacking the curd mass until it reaches the required acidity to be milled (shredded) and hand-salted prior to packing in moulds ('hoops') and pressing (Johnson and Law, 1999).

In large throughput factories, the draining and cheddaring stages are mechanised within one machine, which receives curds and whey directly from the cheese vats and delivers cheddared curd to the shredding mill. The configuration of the vat is important in this stage of mechanisation in the sense that the mechanised enclosed vats are designed to empty 'quietly', causing little turbulence, which might otherwise damage the curd. The vats must be stirred during emptying to ensure that the curds remain suspended in the whey, but this must be done with minimum shear, and no curds and whey must be left in the bottom of the vat. This is ensured by soft, angled scrapers on the ends of the stirrer paddles and by smooth, corner-free, vat surfaces. Also, the pumps which empty the vats have a frequency control so that their speed can be varied according to the amount of curds in the whey, as this varies during emptying. This is vitally important to ensure an even layer of curd particles on the draining belt and a consequent even moisture content in the curd mat, delivered to the first cheddaring belt.

Although cheddaring machines are made and supplied as a module, in practice they are installed together with a salting machine and belt placed at the end of the machine; also they are usually supplied and commissioned by the manufacturer of the cheddaring machine. There are two main types of automated mechanical cheddaring machines. Both have in common a curd-draining belt preceded by a screen to hold back fines from curd shattering, preventing their loss into the whey. The essential difference between these types of machine is in the method of cheddaring after whey drainage. One type uses a series of slowly moving belts placed one above the other to allow the curd to drain and then to mat together. The matted curd is then inverted onto the next belt and allowed to flow and stretch so that texture develops. This type of cheddaring machine is typified commercially by the AlfOmatic made by Tetra Tebel (Figure 7.7) and the Cheddarmaster, made by APV (Figure 7.8).

The Cheddarmaster can be used in several different configurations, depending on the amount and size of the space allocated within the cheese plant (Figure 7.9; this figure also shows the position of the salting belt). Thus, the Cheddarmaster can be set up in a configuration with only belts for matting and turning (cheddaring) the curd, with a large cheddaring tower after the draining belt, or

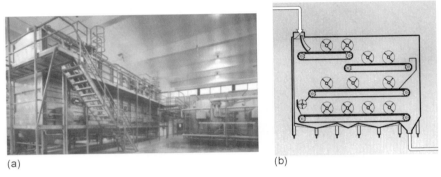

(a)   (b)

Figure 7.7 Automated mechanical manufacture of Cheddar cheese curd: (a) fully installed AlfOmatic curdmaking machine; (b) diagrammatic cross-section of the AlfOmatic machine showing the succession from top to bottom of draining belts, curd matting and cheddaring belt, and inverting mechanism to second matting and cheddaring belt; the bottom belt is for salting the milled curd. Reproduced with permission and cooperation of Tebel-MKT B.V., Holland.

Figure 7.8 Automated mechanical manufacture of Cheddar cheese curd: fully installed 'belt-only' Cheddarmaster system (type OCS), with salting belt, made by APV. Reproduced with permission and cooperation of APV Ltd, UK.

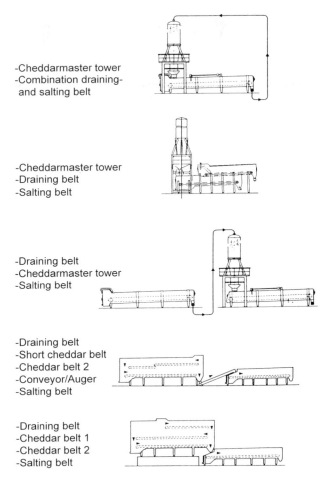

Figure 7.9 Scale drawings (1:125) of typical Cheddarmaster plant configurations ranging from belt-only systems to those feeding drained curd to a cheddaring tower for texturisation and milling. Reproduced with permission and cooperation of APV Ltd, UK.

with a tower and a combined draining and salting belt. This tower configuration is the alternative to the cheddaring belt system mentioned above. The drained, unmatted curd is blown by compressed air to the top of the tower, and the compression effect, together with movement down the tower as cheddared curd is removed from the bottom, provides the flow and drainage needed to texturise it. A typical cheddaring tower is shown in Figure 7.10.

At the top of the tower is a conical inlet which deflects curd evenly into the interior of the tower and directs exuding whey down a central draining tube. As the curd particles matt and are compressed in the tower, movement for the

Figure 7.10 A typical cheddaring tower with guillotine and curd milling gear at its base. Reproduced with permission and cooperation of APV Ltd, UK.

cheddaring effect is created by a gradual reduction of the internal diameter from top to bottom. The acidity of the exuded whey (and, more recently, the moisture of the curd) is monitored as the curd progresses down the tower, so that it reaches the correct recipe value by the time it reaches the knife at the outlet gate.

In belt-only configuration, the Cheddarmaster and AlfOmatic are similar in operation, delivering a cheddared curd mat to the shredding and chipping mill prior to transfer to the salting belt. However, the Cheddarmaster configurations which include a cheddaring tower deliver a guillotined block of curd to the chipping mill at the base of the tower (not to be confused with the cheese blocks delivered by blockformers to bagging machines later in the process).

Not all Cheddar producers use cheddaring machines or cheddaring towers during acidification and syneresis of the curd in the dry (postdrawing) stage. The alternative is dry stirring, by which the curd, after whey draining, is continuously stirred to prevent matting while acidity develops. This method is less labour and equipment intensive than the above mechanisation methods and is often used to make short-hold Cheddar because microbiological control in the cheese room is more difficult owing to the high level of exposure of curd surfaces during the 'dry stage'.

Although not yet in general use, some cheese plants are experimenting with moisture sensors, based on multiple-reflectance NIR spectra in curd as it forms texture during the cheddaring stage. These data are important in both feedback mode and reporting mode. In the feedback mode, the moisture data over time give a measure of the vigour of the starter culture, and can be fed back to help decisions on the next starter recipe (acidification characteristics and/or phage relatedness) and to influence the scald time–temperature and whey drawing times if the curd moisture loss is deviating too much from the standard limits. Infrared Engineering (UK) make a commercial NIR sapphire probe (InfralabTM 5000) that the NZ Dairy Group (Lichfield plant) uses off-line to automatically and rapidly determine curd fat and moisture (replacing Gerber tests and moisture determination by oven method). The data are fed into the SCADA system for the plant as part of the production management network (Anon., 1996). The next stage would be to put a series of sapphire 'window' probes into the walls of a cheddaring tower to monitor moisture loss on-line and control the passage of the curd through the tower to produce uniform, pre-set moisture values in the curd. Alternatively, the NIR moisture data could control the salting rate downstream from the curd mill, to maintain constant salt-in-moisture values in the cheese blocks downstream from the salting belt.

The Food Science Australia pilot plant also uses a TM 5000 probe as the basis of its rapid compositional analysis system for curd and cheese. Fat, protein and moisture data are generated in 40 s and fed into the management information system (Roupas, 1998).

Ideally, cheese plants should have an automated pH or acidification monitoring system on-line, but no commercially available electrode sensors have yet proven reliable or robust enough for the job.

Milled curd at a pre-determined acidity, in the form of 'chips' several inches long, is next delivered to a belt similar to the draining and cheddaring belts of the preceding equipment (usually perforated polypropylene). However, this conveyor has horizontal rotary stirrers to prevent the curd from matting, and vacuum dried salt is delivered onto the curd from multiple nozzles above the conveyor belt as it moves slowly along towards the cheese moulds (boxes, or 'hoops'). Stirrers also mix in the salt after it has been dropped into the curd chip layer, and the final stage of the belt is quiescent, to allow the salt to penetrate the curd (called 'mellowing') before it is pressed onto blocks either in hoops or an automatic blockformer. Even distribution of salt onto the salting belt is a very important factor in determining the quality of the cheese, and various mechanical and model systems are in use to try to achieve this.

Mechanical devices are based on three main principles. First, the outlet of the chipping mill is designed to distribute the curd chips evenly across the width of the salting belt, and the belt speed is coordinated with the rate at which curd chips are delivered from the mill. Second, the stirrer paddles 'tidy up' the layer of chips by their even-speed action on the curd chip layer. Last, the valves which control the amount of salt emerging from the nozzles above the moving belt are directly connected to pivoted sensing forks which rest on the surface of the curd chip layer immediately before the position of the salt nozzles. Their vertical movement, which 'follows' the depth of the curd layer passing below, changes the size of the apertures of the salt nozzles and therefore the amount of salt delivered, directly in relation to the depth of the layer and the amount of curd.

This type of mechanical sensing and dosing system for salting the curd is reasonably efficient, but it is by no means perfect and can result in variations of salt-in-moisture in cheese blocks of up to 50% when blockages occur in the salt nozzles or when the flow of curd from the mill is interrupted. Salt-in-moisture is one of the most critical quality parameters in Cheddar cheese, and once it is made uneven within a cheese block by uneven delivery it is not redistributed by diffusion or equilibration during maturation (Baldwin and Wiles, 1996; Wiles and Baldwin, 1996). These authors have published models relating the (sinusoidal) distribution of salt across cheese blocks to both the rate and lateral distribution of salt from the nozzles and to linear distribution of salt, governed by the speed of the conveyor belt. They have shown that variability in salt in cheese blocks is twice as sensitive to changes in linear distribution along the salting belt as to lateral distribution through the multiple nozzles. The published model shows how to coordinate the speed of the belt, the amount of curd on it and the rate of salt delivery to keep salt variation within ± 0.75 g salt $100\,\mathrm{g}^{-1}$ cheese within 15 g sample areas.

7.5 Production of pressed cheese blocks ready for maturation

In traditional Cheddar cheese manufacture, the salted and mellowed curd is packed into cylindrical moulds, or hoops lined with cheesecloth, then pressed in vertical mechanical presses by using mechanical force via a screw-thread and spring action. Modern mechanised, semi-automated cheese plants use either horizontal gang presses (pressing tunnels) in which pressure is applied evenly through a diaphragm to large numbers of curd blocks at a time (e.g. the APV SaniPress system; see Figure 7.11) or blockformer towers, which use gravity and a vacuum to create a continuous vertical cheese block which can be cut to standard 20 kg blocks for bagging.

This equipment can be installed as part of a module to include automatic mould filling, lid placement, transfer to the pressing tunnel, timed pressing at stepwise increasing pressure according to the recipe, removal from moulds by compressed air, weighing and trimming, metal detection and vacuum bagging for transfer to the store (optionally via a blast cooler to rapid equilibration with the maturation store temperature, commonly 8°C). The entire module is designed to clean the moulds automatically, wash them, re-line them and recycle them to the filling point. SaniPress also has a full CIP system built into its design.

Cheddar plants are making increasing use of vertical blockformers to save space and to smooth the flow from curd-forming plant to the cheese bagging stage. These machines are produced by many of the major equipment suppliers

Figure 7.11 A fully installed SaniPress multiple autofeed–autoempty Cheddar cheese pressing system for the production of 20 kg blocks from milled, salted cheese curd. Reproduced with permission and cooperation of APV Ltd, UK.

such as Tetra Tebel and Wincanton Engineering. They all work on the basic principle of a filling, compressing, block-cutting and bagging cycle. A typical multiple blockformer tower installation is shown in Figure 7.12.

Each tower has four main working units; the curd tower itself, a barrier plate between the tower and its base unit, the base unit with a guillotine and movable elevator plate, and a vacuum pump. Curd from the salting and mellowing belt is sucked up to the top of the tower by the vacuum pump and, when the tower

Figure 7.12 A fully installed multiple cheese blockformer plant with control units, curd transport and feed gear, block guillotines and conveyors to curd bagging equipment. Reproduced with permission and cooperation of Tebel-MKT B.V., Holland.

is full, a vacuum is applied to aid gravitational forces in consolidating the curd mass, squeezing out cracks and air trapped between curd boundaries. At a pre-determined time, the base of the tower opens and the curd column drops onto an elevator plate under gravity and under the pressure difference between the tower and the base unit. The elevator plates move automatically up or down in relation to the guillotine above it so that, after a short settling period, and application of a pre-press force, the guillotine cuts a block of a size weighing 20 kg. While this is all going on the tower is topped up with curd and a vacuum is applied again to consolidate it. As soon as the block in the base unit has been ejected into the bagging machine, the lowering, block-sizing and cutting cycle can begin again. Blockformer towers usually operate at about 680 kg h^{-1} each, and produce a block every 1.5 min (Bylund, 1995). Residence time in the tower is about 30 min for a 5 m tower height (overall plant height is 8 m).

Some blockformer installations use manual bagging in the sense that an operator places the 'plastic bag' over the block exit from the base unit each time a block is cut and ejected and ensures that the bagged block is directed towards the vacuum-sealing machine. However, automatic bag loading and vacuum sealing of cheddar blocks is available commercially (e.g. Tetra Tebel cheese block packer; Figure 7.13), presumably to operate on the output of pressing or block forming systems. This system can be also be adapted to 227 kg barrel-filling equipment.

7.6 Storage and maturation of cheese

This phase of Cheddar cheese production is mechanised only in the sense that transport and stacking systems are mechanical, because of the sheer size and weight of palletted cheese blocks stacked in large temperature-controlled

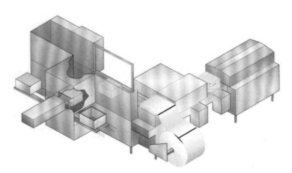

Figure 7.13 Cheese block (20 kg cheddar blocks) packing equipment showing automatic bag loading and vacuum sealing of the blocks ready for boxing, palleting and storage in maturation rooms. Reproduced with permission and cooperation of Tebel-MKT B.V., Holland.

stores. The inventory is controlled through computerised records of production, chemical composition and recipe, backed up by in-house grading and sensory evaluation (McBride and Muir, 1999) to regulate the sale of cheese to retail outlets and/or processing companies according to the maturation potential of the cheeses. The introduction of automated grading is not an immediate prospect, although compositional analysis as an aid to grading could be with us within a few years if developments in NIR technology continue at their present pace (Sørensen and Jepsen, 1998). Much has also been written about the so-called 'electronic nose' (Shiers, 1995) as an objective quality control instrument, but such equipment is not yet sophisticated enough to cope with the complexities of cheese flavour and aroma.

For the present, therefore, a typical mechanised semiautomated cheese plant has automated culture and rennet feed into enclosed, fully mechanised cheese vats under the control of a PLC, with programmable cutting and stirring and automated vat emptying to a curd draining and cheddaring machine. The cheddaring machine delivers texturised, milled curd at the correct (known) acidity and moisture to a salting machine, linked either to an autofeed multiple pressing system or to a blockformer system and bagging and labelling machine.

Acknowledgements

The author is grateful to Tetra Pak A/B, Sweden and Tetra Pak B.V., Holland, for permission to use illustrations from the *Dairy Processing Handbook* and *Sales Information* respectively. He is also grateful to APV Ltd for arranging and supplying material from their technical bulletins.

References

Anon. (1996) Testing times down under. *Dairy Inds. Int.*, **61**(4) 21.
Baldwin, A.J. and Wiles, P.G. (1996) Dry salting of cheese, 2: Variability. *Food Bioprod. Processing*, **74** 133-9.
Bylund, G. (1995) *Dairy Processing Handbook*, Tetra Pak Processing Systems A/B, Lund, Sweden.
ten Grotenhuis, E. (1999) Prediction of cutting time during cheese production. *Eur. Dairy Magazine* (February) 40-1.
Hoeier, E., Janzen, T., Henriksen, C.M., Rattray, F., Brockmann, E. and Johansen, E. (1999) The production, application and action of lactic cheese starter cultures, in *Technology of Cheesemaking* (Ed. B.A. Law), Sheffield Academic Press, Sheffield, UK, pp. 99-131.
IDF (International Dairy Federation) (1991) Factors affecting the yield of cheese. *Special Issue, Int. Dairy Fed.*, **9301**.
Johnson, M. and Law, B.A. (1999) The origins and development of cheesemaking technology, in *Technology of Cheesemaking* (Ed. B.A. Law), Sheffield Academic Press, Sheffield, UK, pp. 1-32.
Johnston, K.A., Dunlop, F.P. and Lawson, M.F. (1991) Effects of speed and duration of cutting in mechanised cheddar cheesemaking on curd particle size and yield. *J. Dairy Res.*, **58** 345-54.

Johnston, K.A., Luckman, M.S., Lilley, H.G. and Smale, B.M. (1998) Effect of various cutting and stirring conditions on curd particle size and losses of fat to the whey during cheddar cheese manufacture in OST vats. *Int. Dairy J.*, **8** 281-8.

Laporte, M.-R., Marbel, R. and Paquin, P. (1998) The near-infrared optic probe for monitoring rennet coagulation in cows' milk. *Int. Dairy J.*, **8** 659-66.

Lomholt, S.B. and Qvist, K.B. (1999) The formation of cheese curd, in *Technology of Cheesemaking* (Ed. B.A. Law), Sheffield Academic Press, Sheffield, UK, pp. 66-98.

Lucey, J. and Kelly, J. (1994) Cheese yield. *J. Soc. Dairy Technol.*, **47** 1-14.

McBride, R. and Muir, D.D. (1999) The grading and sensory profiling of cheese, in *Technology of Cheesemaking* (Ed. B.A. Law), Sheffield Academic Press, Sheffield, UK, pp. 281-314.

Morison, K.R. (1997) Cheese manufacture as a separation and reaction process. *J. Food Eng.*, **32** 179-98.

Nöel, Y., Bellon, J.L., Herry, J.M., Cere, O., Pain, J.P. and Antonini, G. (1988) Methode d'étude et controllé des changements d'état d'un milieu liquide ou gelifié par mesure différentielle de characteristiques thermiques dudit milieu et disposif capteur pour la mise en oevre de cette methode. *Demande de brevet francais* no. 8800803.

Payne, F.A., Hicks, C.L. and Shen, P. (1993) Predicting optimal cutting time of coagulating milk using diffuse reflectance. *J. Dairy Sci.*, **76** 48-61.

Roupas, P. (1998) Cheesemaking: from art to precise science. *Australian Dairy Focus* (February) 37-8.

Shiers, V. (1995) A study of the functions and advantages of the electronic nose. *Eur. Food & Drink Rev.* (Winter) **61** 63-5.

Sørensen, L.K. and Jepsen, R. (1998) Assessment of sensory properties of cheese by near-infrared spectroscopy. *Int. Dairy J.*, **8** 863-71.

Wiles, P.G. and Baldwin, A.J. (1996) Dry salting of cheese, 1: diffusion. *Food Bioprod. Processing*, **74** 127-32.

8 Semi-hard cheeses
G. van den Berg

8.1 Introduction

8.1.1 Cheese varieties involved

This chapter deals with the mechanisation and automation of the manufacturing process of various semi-hard cheese types such as

- Gouda, Edam and Maasdam (Dutch)
- Danbo, Fynbo, Havarti and Samsø (Danish)
- Mimolette and St Paulin (French)
- Norvegia and Jarlsberg (Norwegian)
- Fontal and Fontina (Italian)
- Drabantost, Herrgårdsost and Sveciaost (Swedish)
- Tilsit and Loaf Edam (German)
- Jaroslavskij and Hollandskij Brushkovyi (former USSR)
- Manchego (Spanish)

This list mentions many countries, but most of the cheese types are made all over the world and the process technology is rather similar. They are made with various fat contents [majority 40–50 g $100g^{-1}$ fat in dry matter (FDM)] and the water content in fat-free cheese (WFFC) ranges from ca. 55–62 g $100g^{-1}$. Weights can be between about 0.4 kg and 20 kg, and the shape can be a flat cylinder, a loaf, a rectangular block or a sphere. They are matured from 1 month to 1 year in a natural way (covered with a plastic coating layer or sometimes with a smear layer on the surface in the case of Tilsiter, Havarti or Danbo) or in foil or wax. For information about the different characteristics of these cheese varieties, the reader is referred to Fox (1993).

8.1.2 General technology

The different varieties of semi-hard cheese are manufactured according to procedures that have most stages in common. Coagulation is by the action of 'renneting' enzymes, chymosin (calf rennet) being most commonly used. Cultures containing mixed strains of mesophilic lactic acid bacteria (LAB) serve as the starters, consisting of combinations of acid-producing *Lactococcus lactis* subsp. *lactis* and *cremoris* strains and citrate-fermenting and carbon-dioxide-producing

Leuconostoc lactis and/or *Leuconostoc mesenteroides* subsp. *cremoris* (L-starters) or *Lactococcus lactis* subsp. *lactis* biovar *diacetylactis* and *Leuconostoc* strains (DL-starters).

DL-starters ferment citrate quicker than do the L-starters and are used for stronger 'eye' formation. In some cases eye formation must be very limited or even absent, and a starter without citrate-fermenting bacteria is applied (O-starter). For types such as Jarlsberg and Maasdam cheese, propionibacteria are added as an extra starter in order to achieve propionic acid fermentation to obtain the typical big holes. Nowadays, there are also cheese types that are made with an extra, but attenuated, starter culture of thermophilic nature to enhance flavour development (van den Berg, 1990). The production and applications technology of cheese starters is detailed elsewhere by Hoeier *et al.* (1999), and the use of special equipment for bacteriophage-free preparation of cultures (ready for inoculation into the cheese vat) is advised by CSK Food Enrichment, Leeuwarden, The Netherlands, supplying starters under the name NIZOSTAR.

After renneting, the gel is cut so that after curd preparation the average curd size will be approximately 5 mm for full-fat cheese. Low-fat cheese needs coarser cutting to larger curd particles, to control moisture content. The curd scalding temperature is moderate to obtain the desired moisture content and is usually controlled by adding an amount of warm water after part of the whey (released after cutting) has been removed (the so-called 'first whey'). This water is called curd-washing water because it also serves to decrease the lactose content of the curd. For moulding, the curd is collected under the whey and the whey is drained off, except for cheeses from granular curd such as Tilsit, Havarti, Maribo and Prästost. After pressing, the cheese will often be held for 1–2 h for further acidification. Subsequently, the cheese is usually brined to take up the salt and then put in ripening rooms to mature.

In mechanised factories, these manufacturing stages can still be recognised in the type of equipment installed. This is illustrated in Figure 8.9, (page 243) which shows diagramatically a large cheese factory for manufacturing rectangular block cheeses. (Figure 8.9 will be discussed in more detail in Section 8.6.2 in relation to mould handling.)

8.1.3 General historical background

Since the 1960s, there has been an almost complete change from hand-made to fully mechanised and automated production. The first successful attempts to mechanise the production of semi-hard cheese started in 1957. This was an initiative of the association of cooperative dairies in Friesland, Leeuwarden, The Netherlands (Roosenschoon, 1972). Ideas about the first pre-pressing vat for curd drainage were combined with a German idea on the curdmaking tank (the Käsefertiger, made by the firm Schwarte, Ahlen). In a few years this development, together with the factory of Bijlenga in Leeuwarden, The Netherlands (now Tetra Pak Tebel BV), appeared to be crucial for the mechanisation of factories for semi-hard cheese. Equipment based on these ideas was installed during the 1960s in most Dutch cheese factories and also quickly outside The Netherlands. However, for curdmaking a (still open) oval deep curd vat of 10 000–12 000 l was used.

Around 1970, the first closed cheese vats were introduced and several machine manufacturers in various countries started the production of such cheesemaking equipment. More process stages were mechanised, and remote control was applied in most modern cheese factories at the end of the 1960s. In the beginning, the transport of curd and whey was done by gravity to avoid curd damage by pumping; the cheese vats were installed above the pre-pressing vats. This made the investment in buildings more expensive. However, later the cheese vats and the pre-pressing vats were installed on the same floor because curd damage could be reduced by using better pumps and pipeline systems.

In the 1970s and 1980s, the pre-pressing vats were replaced to a great extent by more continuous drainage and moulding systems such as the Casomatic® (Hansen, 1979). The mechanisation and automation of all parts of the factory, such as milk treatment, brining and ripening, were also further developed. In addition, a well-integrated process for the whole factory demanded a sophisticated transport system. The basis for these expensive investments was also increasing by further amalgamation of individual factories into big companies with a limited number of production plants. A similar tendency can be noted concerning the supply industry. As a result, modern factories manufacture 30 000 tonnes or more of cheese per year; production continues 24 hours a day at a capacity of 50 000 l h^{-1} of milk, stopping only twice a day, or sometimes less, to clean the plant.

In addition, higher demands are made on process control to ensure quality and safety of the final product. This has a strong influence on the whole layout of the factory, needing not only a good quality raw material but also good hygienic control of the process and design of the equipment used. Moreover, other important demands exist concerning yield, cost price, logistics, environment, water supply and energy. It is obvious that the modern factory can work only with an adequate system of total process control. Nowadays, this is also part of the whole system of certification.

8.2 Basic technology

The process for making semi-hard cheese varieties has been comprehensively reviewed and documented (Walstra *et al.*, 1993, 1999a, 1999b; van den Berg *et al.*, 1996) and will not be detailed here. The focus here is on process, but only from the point of view of mechanisation and automation.

8.3 Milk handling and processing

The bulk handling of milk in a large dairy factory is detailed in Chapter 3. In The Netherlands, milk for semi-hard cheese is treated by thermisation at 66°C for 10 s, cooled and stored overnight at less than 8°C before processing as cheese milk. The primary objective of thermisation is to inactivate and control the growth of psychrotrophic bacteria, many of which produce lipolytic and proteolytic enzymes as they grow in cooled raw milk. Although the bacteria are killed by pasteurisation, the enzymes are not, and they can reduce the flavour quality of the cheese (Stadhouders, 1982; van den Berg, 1984). Raw milk in bulk silos has to be stirred to prevent separation. Most modern semi-hard cheese factories use mechanical stirring rather than air agitation to aviod foaming, which later might interfere with eye formation in the cheese.

8.3.1 Milk fat standardisation

At many cheese factories a batch of 300 000 l of milk will be prepared by mixing (at low temperature) calculated amounts of thermised whole milk, thermised skimmed milk and whey cream, which has been heated (95°C for 1 min) in order to avoid bacteriophage contamination, because it is dangerous to the activity of the starter bacteria in the cheese. After mixing, the fat content is adjusted (by adding small amounts of skimmed milk or whey cream) in proportion to the protein content of the milk or, more correctly, the casein fraction. The operator will measure the quantities of the different flows, based on analytical results received from the factory laboratory. Most modern laboratories use rapid methods based on near infrared (NIR) techniques. In effect the aim is to control the amount of fat in dry matter (FDM) of the final cheese, taking account of the uptake of salt at brining and the fat and protein losses during production (Lolkema, 1991).

Automatic in-line standardisation of milk to a desired fat: casein ratio has also been advised to cheese factories, but this is still not popular, especially at factories that produce only one cheese type. However, processing 24 hours per day and 6 days a week is now quite usual, and the demand to minimise variation and maximise yield may push most plants to adopt this measure in the future. Milk standardisation equipment is dicussed in Chapter 3.

8.3.2 Control of sporeformers by bactofugation and microfiltration

In Northern and Western Europe, in particular, milk is unavoidably contaminated with spores of *Clostridium tyrobutyricum*, the most troublesome type of butyric acid bacteria (BAB). The number of spores of these anaerobic bacteria increases during winter when the cows are fed mainly with silage, but they should not be disregarded in summer (Stadhouders *et al.*, 1983). They survive pasteurisation and may cause a defect called 'late blowing' through the butyric acid fermentation (BAF). This defect may show up after some weeks or months in semi-hard cheeses, in which the conditions are rather favourable because they have, besides a low redox potential and sufficient lactate, a very low salt content in the centre of the cheese during the first weeks because of the brining process. During ripening, the final salt concentration is important for controlling BAF in the more matured cheese. Further, the pH in the cheese is not very low (5.2–5.3 for Gouda, and 5.4–5.5 for Maasdam cheese) and the ripening temperature is usually 12–15°C, except for foil-ripened cheese. The higher ripening temperature for cheese with propionic acid fermentation increases the risk of BAF considerably. Traditionally, cheesemakers have added potassium or sodium nitrate to the milk to inhibit BAB and avoid this defect. However, modern market demands increasingly limit the use of nitrate and the salt content of the cheese. The use of lysozyme as an alternative to nitrate is also possible but the mechanical methods (bactofugation and microfiltration) discussed below are cheaper for the larger factories (van den Berg, 1990).

During the past two decades bactofugation technology has been improved to increase nominal capacity of (for example) the Tetra Pak machine to $25\,000\,l\,h^{-1}$. Application of the new hermetic separator principles also minimises air inclusion in the bactofugate, a big problem for makers of semi-hard cheeses. The bacteria-removing separator of Westfalia Separator, Oelde, Germany (now GEA Process Technology Division) has a similar capacity and efficiency (van den Berg et al., 1988). The design of the bowl of this manufacturer is shown in Figure 8.1.

The feed (item 2) is at the top of the bowl, coaxial with the outlet for the 'bacteria-free' milk (item 3). The milk flows from the bottom of the bowl into the distribution holes of the disk stack. The bowl is provided with a self-de-sludging system to clean the sediment holding space via the ejection ports (item 10). The continuous discharge of the heavy phase (item 1), called transport liquid, can also be removed via a separate outlet as during the test run of the Westfalia machine (van den Berg et al., 1988), or this fluid is mixed with the incoming raw milk for reprocessing. The spore-removing efficiency appears to be slightly lower but acceptable under the condition that the de-sludging frequency is sufficient. In this case, the concentrated sludge, representing approximately 0.1–$0.2\,ml$ $100\,ml^{-1}$ of the milk volume, contains all separated spores and can be removed in those countries where mixing with the cheesemilk is not allowed. Otherwise, sterilisation after mixing with some milk, to have sufficient volume for the steriliser, is possible before addition to the cheesemilk. This so-called one-phase type bactofuge is also made for higher capacities, accepting a somewhat lower efficiency (Tetra Pak Processing Systems, personal communication). However, it is worth noting that a decrease of the bactofugation efficiency from 98 to 96% doubles the spore level in the cheesemilk.

The current bactofugation process for cheesemilk is illustrated in Figure 8.2. The milk (item 1) is pre-heated in the regeneration section of the pasteuriser

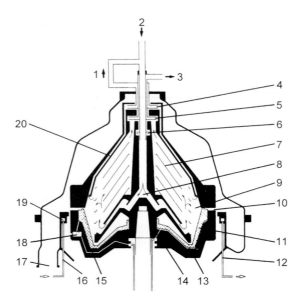

Figure 8.1 Cross-section of the bowl of a bactofuge, type CND 215, made by Westfalia Separator. 1, Discharge, transport liquid; 2, feed; 3, discharge, bacterially clarified product; 4, centripetal pump, transport liquid; 5, centripetal pump, bacterially clarified product; 6, hydropump; 7, disk stack; 8, soft-stream-inlet; 9, sediment holding space; 10, sediment ejection ports; 11, sliding piston; 12, cooling water discharge; 13, closing water chamber; 14, closing water duct; 15, opening water duct; 16, cooling water valve; 17, discharge, sediment; 18, piston valve; 19, cooling chamber; 20, separating disk. Reproduced by courtesy of Westfalia Separator/GEA Process Technology, Germany.

to bactofugation temperature. The continuously discharged bactofugate (item 2), (about 3 ml 100 ml^{-1} of the milk volume) is heated to 130°C in an infusion steriliser and mixed with half the volume of the bacteria-free milk (item 6) to cool it before being filtered to remove possible sediment from the steriliser and mixed with the remaining milk ready for pasteurisation and cooling to cheesemaking temperature. The intermittent discharge from the bowl is still often mixed with the continuously discharged bactofugate and sterilised so that no protein is lost to affect cheese yield (Daamen *et al.*, 1986; van den Berg *et al.*, 1988). The holding time at the sterilisation temperature depends on the liquid level in the steriliser unit, which is provided with a water-cooling jacket to prevent burnt sediment on the hot wall. In case of Maasdam cheese, and using no nitrate, double bactofugation is necessary. When two, two-phase machines are used in-line, the continuously discharged heavy phase from the second is fed back into the milk supply of the first, and the heavy phase from the first, together with the intermittent discharges from both machines, is sterilised. The spore-removing efficiency is then be greater than 99.5%.

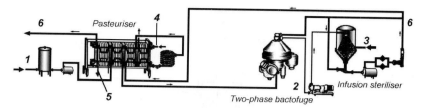

Figure 8.2 Operation line consisting of pasteuriser, bactofuge and infusion steriliser, as offered by Tetra Pak Processing Systems, Lund, Sweden. 1, Milk; 2, bactofugate; 3, steam; 4, heating medium; 5, cooling medium; 6, bactofugated milk. Reproduced by courtesy of Tetra Pak Processing Systems, Sweden.

The technique of microfiltration (MF) is an alternative to bactofugation and has good separation efficiency on an industrial scale. More details of this technique are given in Chapters 3 and 11. The efficiency is similar to double bactofugation, but the cream has to be separated first and then mixed with the spore-containing MF retentate for separate heat sterilisation. The total investment is higher than for double bactofugation, but all raw milk bacteria are removed more effectively than by bactofugation. As a consequence, the use of this equipment for raw milk cheese will cause not only the removal of the unwanted flora, but also those bacteria which might be, in part, responsible for the desired flavour notes (McSweeney *et al.*, 1993).

8.3.3 Pasteurisation

This is the final aspect of control of the raw material with respect to the manufacture of a safe product. Nowadays, not only must pathogenic micro-organisms causing tuberculosis be killed but also those which cause listeriosis and illness associated with *Mycobacterium arium* subsp. *paratuberculosis*. Traditionally, the demand was to heat the cheesemilk only up to the phosphatase negative level, but during the 1990s the heat load has generally been increased in order to kill all such pathogens. A pasteurisation temperature of 74–75°C with a holding time of 10–15 s is now more usual.

8.4 Cheese vats and curd production

A wide range of cheese tanks (vats) are manufactured by different equipment suppliers, and the technology involved in their design and use has been reviewed extensively (Tamime, 1993; Robinson and Wilbey, 1998). In this chapter, only two examples will be detailed, with special emphasis on design, automation and other special specifications required for their use in semi-hard cheese manufacture. Traditionally, the open-type cheese vat was used for all cheesemaking operations, but major developments in currently available commercial vats now include

- enclosed vat design (vertical and horizontal)
- integral cutting and stirring devices
- built-in filling, emptying and dosing pipework and metering devices
- control and automation modules attached

In terms of automation and mechanisation in cheesemaking, the current approach in technology identifies the separate cheese vat functions as fermentation, formation of the coagulum, cutting and stirring of the coagulum, and scalding (with washing) or cooking and whey separation. Let us now consider the two chief design types in detail.

8.4.1 Horizontal vats

The Tetra Tebel OST vat is basically designed as a horizontal cylinder (Figure 8.3); by increasing the length of the vat the capacity can be increased from 10 000 l to 27 000 l (a volume of more than 20 000 l is not used for semi-hard cheese). When installed, the vat is positioned with a 4% slope towards the outlet, situated at the bottom of the tank to facilitate emptying. The milk inlet (item 1) is at the top of the tank and is designed to minimise the inclusion of air during filling. In practice, it is becoming more usual to 'bottom-fill' the tank by mounting the milk pipe-line on the curd outlet, with an automatic switching valve.

The blades for cutting and stirring (item 2) are mounted on the horizontal shaft over the whole length of the tank for effective cutting. The power unit to

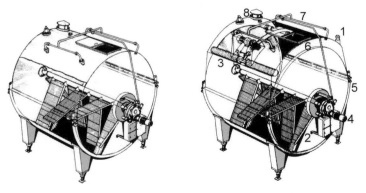

Figure 8.3 Cut-away diagram of the Tetra Tebel TM OST vat (formerly called OST IV curd-making tank). 1, Milk inlet; 2, cutting and stirring blades; 3, tubular whey strainer; 4, electric motor and transmission unit to drive the horizontal shaft; 5, jacket with distribution pipes for heating or cooling; 6, manhole (with safety grid); 7, cleaning-in-place system with four nozzles; 8, air vent. Reproduced by courtesy of Tetra Pak Tebel BV, The Netherlands.

drive the central shaft is mounted in the middle of the front of the tank (item 4). Whey removal is through a whey strainer (item 3) which is lowered into the curd–whey mix just below the surface and kept in that position during draining. The whey strainer consists of an outer perforated tube and an inner tube provided with a limited number of holes to create a quiet inflow. Curd washing water is introduced via the strainer and partially via the built-in cleaning-in-place (CIP) entry system nozzles (item 7). The tank has a jacket with hot and cold water feeds (item 5) for rinsing the inner wall, and an inlet for steam injection on the outer wall during curd cooking. Installation of an optional rennet dosing system is advised.

The Tetra Tebel OST vat has been tested independently (de Vries and van Ginkel, 1984) and performed well in terms of losses to the whey, control of curd clumping, efficiency of emptying and low curd shattering (low proportion of 'fines').

8.4.2 Vertical vats

A typical example of a vertical cheesemaking vat is the Damrow 'double-O' design, now supplied by GEA Damrow, Kolding, Denmark; a DB type vat is shown in Figure 8.4. Although the double-O was originally developed for

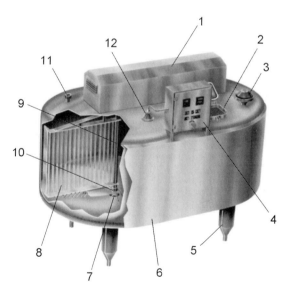

Figure 8.4 Cut-away diagram of the Damrow 'double-O' (DB type) curdmaking tank. 1, Electric motor with gear; 2, manhole with safety grid; 3, air vent; 4, control panel; 5, leg with adjustable ball feet; 6, insulated wall; 7, dual bottom outlet; 8, assembly for cutting and stirring; 9, Shaft of the cutting and stirring assembly; 10, bottom bearing; 11, cleaning-in-place head; 12, inspection lamp. Reproduced by courtesy of GEA Damrow A/S, Denmark.

Cheddar and related cheese in the USA (see Chapter 8), it was adapted at its introduction in Europe for semi-hard cheesemaking with use of a whey strainer and wash water dosage. The tank is fitted with two vertical frames of cutting knives. When they are rotated by the vertical drive shafts in the same direction, they cut the curd efficiently because it moves along the inside walls of the double circle to the point at which the two circular sections intersect, from where it is moved towards the centre of the tank and cut by both cutting and stirring assemblies. This motion also gives very even and nondestructive stirring of the fragile curd cubes during scalding. During stirring, the assembly is moving in the opposite direction with an activated stirring blade downside the frames. The cutting assemblies have thin sharp blades, but, for the FB type, near the axis of the drive shaft the spacing is wider to avoid squeezing the gel and shattering it. The good performance of the vat design in these respects was confirmed for the FB type by de Vries and van Ginkel (1980). Figure 8.4 shows a cut-away view of the newer DB type (with a smaller spacing at the drive shaft but cutting at lower speed). The various types differ in ancilliary equipment and bottom design, for example to facilitate emptying without tilting (Figure 8.5).

8.4.3 Preparation of the curd

It is up to the cheesemaker to decide at which gel strength (curd firmness) to start cutting. In order to achieve the desired curd particle size (related to final moisture content) the gel firmness at cutting should be chosen in relation to the effectiveness of the cutting programme, which depends in turn on the design of the cutting assembly and the shape of the tank. For example, the cutting time

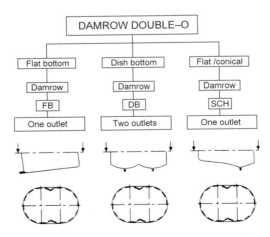

Figure 8.5 Schematic illustration of the bottom designs for three types (FB, DB and SCH) of Damrow double-O curd-making tanks. Reproduced by the courtesy of GEA Damrow A/S, Denmark.

should normally be about 15 min, with a maximum of 20 min, starting at low speed and progressively increasing speed. At the start of cutting in the horizontal vat, when the curd has low density, cutting is best done in intermittent mode (Tetra Tebel OST vats; de Vries, 1979; de Vries and van Ginkel, 1984). During the final minutes before sucking the first whey, when the desired curd size is already obtained, the mass can be stirred. The movement of the knives will then be switched in the opposite direction; this switch in direction also occurs in vertical curd-making tanks, with an extra stirring blade coming in action when stirring begins (de Vries and van Ginkel, 1980, 1983; van Ginkel *et al.*, 1987). The use of the knives as agitators provides an extra reason to give the knives a smooth back edge in order to control losses of fines. It is important to realise that drainage of the whey when stirring (e.g., with a stationary whey strainer in the wall) is not desirable because it will cause a very high loss of curd fines and possibly blockage of the strainer as a result of curd sticking into the perforations.

Within 20 min from the start of cutting, initial syneresis is complete, the knives are stopped and the curd is allowed to settle for 2 min. Then, 40–45 ml $100\,\text{ml}^{-1}$ of the original milk volume is drained off (the first whey), preferably within 5 min, to avoid fusion of the curd resulting in curd lumps that are difficult to disperse during the following stirring stage without excessive losses.

As soon as the whey has been drained off, the strainer is brought back into the upper position and the remaining curd and whey mixture is stirred to prevent lumping. Then curd washing water is added via the whey strainer and sometimes via the cleaning nozzles. The curd should not be allowed to settle further in-spite of increasing density by continuing syneresis. A contact time between curd and diluted whey of approximately 25 min is normal, by which time more than 90% of the theoretical maximum removal of lactose is achieved. After curd preparation in the curdmaking tank, the curd and whey mixture has to be pumped either to the buffer tank before continuous drainage and moulding or into the batch-wise operating drainage and pre-pressing vat. Therefore it is important that the curdmaking tank be emptied easily without leaving behind much curd. Special centrifugal pumps are generally used and the curd and whey mixture is usually transported at a speed of approximately $2\,\text{m}\,\text{s}^{-1}$ to a limited height with minimal restrictions in the pipeline.

8.4.4 *Instrumentation to control and automate curd cutting time*

The control of the renneting process is important, as it is the first step in cheese-making. Although it looks an inaccurate method, experienced cheesemakers use their visual judgement to decide to cut and actually manage the cut at fairly constant gel strength. However, this is not so easy with modern closed tanks and not very acceptable with respect to hygiene. Before 1990 there were no robust, practical instruments to replace the traditional 'manual' assessment (van Hooydonk and van den Berg, 1988). However, during the past decade some promising results have been obtained with techniques that use small fibre-optic probes mounted through the wall of the tank. The measuring ends of the fibres are mounted in a small round plate, extending a very small distance into the milk so that the knives can operate normally. This measuring head is automatically cleanable during the normal tank-cleaning procedure. Such methods use light scattering to measure decreasing mobility of the casein micelles (and their aggregates) during the renneting. The first method used was based on the diffuse reflectance technique (Payne *et al.*, 1993) and was marketed under the brand name CoAgulite by Reflectronics, Lexington Ky, USA. The first derivative of the reflectance profile measured gives a typical maximum (inflection point) at about the time when gel formation starts. This indicates that the renneting process is in progress, but the cheesemaker has still to decide when to cut (e.g. after 1.6 × the time that elapsed from rennet addition till the inflection point was reached). However, the time to cut depends on factors such as fat content, protein content, pH and temperature, so the process must be well standardised. Nevertheless, the system is already used in practice where subsequently the process in all cheese vats within a factory can be measured. A similar measuring system is applied in the new Gelograph, from Gel Instrumente AG (Thalwil, Switzerland) for test purposes and by Referenz Messtechnik (Duisburg, Germany) for practical cheesemaking.

The most recent approach was presented by NIZO food research (ten Grotenhuis, 1999). A laser fibre-optic device has been developed to use diffusing wave spectroscopy (DWS) to measure the complete course of coagulation and gel firming. Laser light is collimated into an optical fibre and conducted into the milk. The light is scattered by the particles in the milk (casein micelles, fat globules) and detected by one or more single-mode fibres. The light is detected by a photomultiplier tube–amplifier–discriminator train and fed to a computer-mounted correlator board, which produces the autocorrelation function. Half-decay time is defined as the autocorrelation time. This increases at the beginning of aggregation and continues to do so thereafter. When plotted as a function of real time, an upward curve indicates increasing firmness of the coagulum. A series of tests with practical cheesemaking gave good results and the production of commercially applicable instruments is expected.

8.5 Curd drainage and moulding

8.5.1 Buffer tanks

A buffer tank is required between the batchwise operating curdmaking tanks and the continuously operating whey drainage and curd moulding facilities (see Figure 8.9, page 243). It will contain the whole batch of curd and whey before it is pumped to the drainage unit, and the primary function is to supply a homogeneous curd suspension all the time to ensure the correct operation of the drainage unit. The buffer tank is specially designed with a conical bottom and a double jacket to cool the content a little. The six-blade continuously variable speed turbine agitator has a vertical axis with the motor on top of the tank.

Air entrained into the curd–whey mix during cyclic emptying and re-filling of a single buffer tank has led to the increasing use of two buffer tanks to give more time for air to escape and at the same time decrease residence time in the cheese vats. From the buffer tank the curd and whey mixture is pumped by positive dosing pumps to the curd drainage systems, which will be described next.

8.5.2 Casomatic® systems

The Casomatic® system is used in almost all mechanised production of semi-hard cheese. It has been developed over several decades (Hansen, 1979) and only the latest versions will be considered here. For example, a Casomatic® for rectangular cheese (e.g. Euroblock) was developed and later extended with a cutting device for smaller cheeses such as Loaf Edam (Hansen, 1987). By doing this, Tebel BV presented a number of new versions of the SC (single-column) type machine. The MC (multi-column) type is a different type.

8.5.2.1 Casomatic® SC

Very recently the newest version was launched; the Tetra Tebel Casomatic® SC MK6. The capacity has been doubled in comparison with the older versions because of the addition of another control system for whey drainage. Figure 8.6 shows that the top of the column has been widened to reduce the influence of the whey level, which varies by whey drainage, and discharge of the curd block. There are three drainage sections, but the lower two have a larger surface area. For each column section, the pressure difference from the whey level at the top is measured (items 2A–C). A certain difference per section, depending on the type of curd, is automatically maintained by regulating the whey discharge by means of the valve in the discharge tube. The difference is bigger in the lower sections because the curd is more compacted. The curd and whey mixture is pumped by one dosing pump per column in order to achieve a better controlled supply for each column. Whey that would raise the level too much is directly discharged from the top, which may diminish possible bacterial growth in that part of the machine. From the base of the column, the curd blocks are cut off at regular intervals. A 'dosing plate' is pushed up just under the knife plate, which is opened, and the curd column on the dosing plate is lowered to the desired level with respect to the final weight of the cheese. The block is cut off, surrounded by a movable perforated cylinder and pressed for a few seconds against the knife plate by pushing up the dosing plate again. The curd block then descends to the level of the stationary plate on which it slides by pushing

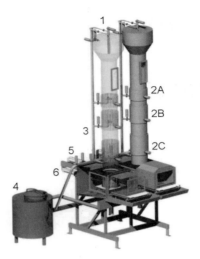

Figure 8.6 Diagramatic illustration of the latest generation of the Tetra Tebel Casomatic® SC MK6. 1, Curd and whey supply; 2, pressure indicators for the three sections; 3, whey collection tube; 4, whey collection vessel; 5, pneumatically operated cylinder for opening and closing the curd knife; 6, pneumatically operated cylinder for removing the curd block to fill the cheese mould. Reproduced by courtesy of Tetra Pak Tebel BV, The Netherlands.

the perforated cylinder to the mould filling position. The cylinder has to be back under the column before the next block can be cut off. The design of the knife, the dosing plate and the perforation of the moving cylinder has been improved in the newest version. It is expected to result in a more compact curd block, fewer losses during cutting and minimisation of the danger of 'air rim' in the cheese. It is obvious that the capacity of this drainage and moulding equipment is determined mainly by the number of columns installed.

8.5.2.2 Casomatic® MC

The Casomatic® SC is not a very flexible machine with respect to the shapes of cheeses that can be produced. Even the cutting of a rectangular curd block into smaller units, as can be done with this machine, remains complicated when a constant weight is demanded. To this end, about 15 years ago Stork Friesland introduced a drainage apparatus with exchangeable perforated drainage columns within a common jacket (Hansen, 1986). The diameter and/or shape of these perforated columns were adapted to the different cheese types to be made. Stork Friesland sold the rights on this apparatus to Tetra Pak Tebel BV. This firm improved it further and it is marketed now as the Casomatic® MC, shown schematically in Figure 8.7.

The unit consists of four curd columns, but the number may be more for smaller cheeses, or less (e.g. one large column) for Baby Gouda to 25 kg block

SEMI-HARD CHEESES

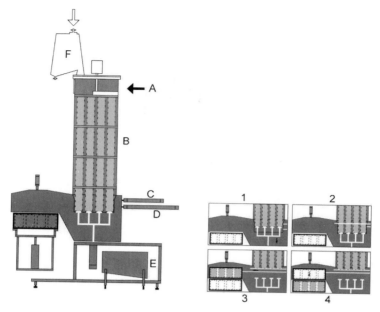

Figure 8.7 Sections through the Tetra Tebel Casomatic® MC continuous whey drainage and portioning machine with a flexible number of curd columns adapted to the size of the cheese. A, Curd and whey supply; B, drainage columns with three whey sections; C, pneumatically operated cylinder for opening and closing the curd knife; D, pneumatically operated cylinder for moving the curd block(s) to fill the cheese mould(s); E, whey collection vessel; F, whey strainer (optional). 1, Dosing the curd block(s); 2, cutting the curd block(s); 3, removing the curd block(s); 4, putting the curd block(s) into a single mould or a multimould. Reproduced by courtesy of Tetra Pak Tebel BV, The Netherlands.

size. The perforated columns are mounted in three plates just fitting inside the outer jacket. These plates separate the area outside the perforated columns into three whey chambers, each having their own pressure regulator. There is always a constant pressure difference between each chamber and the whey level at the top of the machine. This difference increases in the lower sections and is maintained in the same way as in the Casomatic® SC MK6 (Section 8.5.2.1). The discharge of the curd blocks also follows a similar cycle. The cutting height and the assembly for moving the blocks has to be adapted to the size and shape of the cheese when there is a change of cheese type. The curd blocks from one discharge should be placed in a number of moulds in one cassette, the so-called multi-mould. The drainage machine is fed in a similar way to the Casomatic® SC, via a reservoir at the top by a variable dosing pump with a whey and curd mixture from a buffer tank. The reservoir is equipped with level control and a slowly moving agitator to be sure that all columns are equally provided with curd. The capacity of such a drainage unit is approximately 2000 kg cheese per hour when making larger cheese types. All Casomatic® types can be adapted for granular curd cheese as indicated in Figure 8.7.

8.5.3 Pre-pressing vats

As batchwise operating alternatives to Casomatic® systems, pre-pressing vats consist of a flat rectangular reservoir containing a perforated conveyor belt above the base. At filling, a quantity of whey is pumped before the main flow of curd, so the curd will always arrive in whey to avoid air between the curd grains after drainage. Raking over the whole surface spreads the curd in a nearly uniform layer after drainage; the thickness is determined by the amount of curd and the adjustable vertical back-plate at the end of the perforated bottom plate. The whey is then drained off and the curd will mat. Perforated plates are laid on to compact the curd further. A single pneumatically operated plate just before cutting is often considered to be sufficient to achieve this effect. After taking off the plates the curd mat is moved forwards to be cut in blocks by a combination of knives. A number of vertical knives, equally divided over the breadth of the curd mat, cut the mat during moving, and a guillotine knife cuts off the desired length from the row of blocks. Thus, the pre-pressing vat produces only square or rectangular curd blocks to fill the moulds.

The newly designed Tetra Tebel Pressvatic® is shown in Figure 8.8. It has various improvements besides those for CIP. The curd is pumped via a distributor moving both ways over the whole vat (item 1). The system can be adapted for the manufacture of cheeses from granular curd. A large perforated pressing plate on top of the apparatus will press the whole curd mass for several minutes

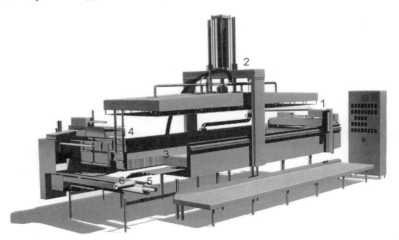

Figure 8.8 Illustration of the Tetra Tebel Pressvatic® for whey drainage, pre-pressing and cutting the cheese blocks. 1, Curd distributor; 2, pressing system; 3, curd layer; 4, cutting knives; 5, roller on the conveying belt; 6, Conveying belt for the curd blocks. Reproduced by courtesy of Tetra Pak Tebel BV, The Netherlands.

before cutting (item 2). After the pressing plate has been lifted, a computerised system detects the unevenness (the 'topography') of the curd surface, and will automatically adjust the length of the row that has to be cut off. The position of the back plate can also be automatically controlled. Under the draining belt there will often be some air, which can be automatically removed, otherwise it may become entrapped in the curd above. The curd blocks are forwarded after being cut off onto a conveyor belt in front of the pre-pressing vat and moved transversely to an automatic mould-filling station. APV and GEA Damrow also supply similar pre-pressing vats with slightly different details.

It is worth noting that long pressing times with more pressure in such a vat do create a large block of cheese. This can be cut into smaller blocks before brining, but the rind of the cheese does not exist around the whole cheese. Such blocks may be used for foil-ripened cheese.

The difference in time between filling the first and the last mould of one batch and the deformation of the curd block after cutting leads to variations in weight and moisture content. This variation is easier to control in Casomatic® systems, which are easier to automate and control than pre-pressing vats, and they have more means to compensate for variations.

8.6 Pressing

The cheese has to be in a mould to obtain the right shape and a closed rind. During the past half century much has been changed concerning moulds and presses. However, the aim is still the same: to obtain a well-modelled cheese with a closed rind. The development of cheese-mould design is beyond the scope of this chapter, but the reader can gain much information from the many technical product brochures of equipment companies such as Servi (Laugeais, France), Laude BV (TerApel, The Netherlands), Sonoco-Crellin BV (Rotterdam, The Netherlands) for Kadova, and APV (Them, Denmark) for Perfora moulds.

8.6.1 Cheese pressing

Cheese presses have been developed as long tables to hold moulds containing all cheeses of one batch (e.g. four moulds per row and 30 rows over the length). All suppliers of cheesemaking equipment deliver this type. The filled moulds are automatically conveyed to their position, each exactly under its own pneumatic pressing cylinder with pressing plate fixed to the piston rod in order to press the lid of the mould in a horizontal position. Formerly, this plate was often not fixed and more cheeses were pressed under one cylinder. As a consequence, many cheeses had a somewhat oblique upper side because all cheeses were not of equal height.

In spite of better control of cheese weight there are still some variations. In fact, this holds also for smaller cheeses in multi-moulds, which have one lid for the whole cassette. It is possible to avoid differences in the effective pressure by a Laude system, which has springs mounted for each lid of the individual moulds to level out the slight differences in height during pressing (Hansen, 1981).

The effective pressure on the cheese is expressed in kPa cheese (upper) surface. A pressing cylinder is provided with a much higher pressure to deliver the desired pressure on the cheese because the surface of the piston of the pressing cylinder is much smaller than the surface of the cheese. This is different from the cushion-type cheese press, as represented by the SaniPress of APV Nordic Cheese A/S (Them, Denmark). The top of this press tunnel is provided with a plastic cushion that can be inflated. When the tunnel is filled, the lids of the moulds are covered over their whole surface with this cushion, so the pressure in the cushion is directly converted to the cheese. As a result the pressure in the cushion at the end of pressing will be, for example, 30–40 kPa.

The effective pressure on the cheese to obtain a well-closed rind is the pressure provided by the pressing system minus the resistance of lowering the lid along the inner wall or net of the mould. This is different for the various mould types because of the different materials used and the way in which the lid fits into the mould. The final pressure applied on the cheese depends on this but is often in the range of 20–30 kPa applied in 3–4 gradually increasing steps.

8.6.2 Mould handling

Nowadays, a continuous mould transport system with different cheese handling stations is used in order to achieve smoothly running continuous production (Figure 8.9). The empty and cleaned moulds are filled at the drainage unit (item 3). The lid is automatically positioned (item 4) and the mould is automatically fed to the correct position in the cheese press (item 5). When the press has been filled with one batch of cheese, the next press will receive the moulds with curd of the next batch. After pressing, the press is emptied at the opposite side while filling takes place again. When making rectangular block cheeses, as in Figure 8.9, the moulds with cheese and the lid are moved to the two holding stations for acidification (item 6), where they are slowly moved and placed row by row on the next conveyor belt. Then the lid is lifted (item 7), the moulds are emptied upside down by compressed air and the cheeses are placed on the next conveyor belt to the weighing station (item 8) and the brining basin (item 11) successively. During transport and pressing, a substantial quantity of whey leaks from the moulds onto the conveying belts and therefrom into gutters. This demands specially designed equipment for efficient cleaning. The weighing of the cheeses before brining is also part of the moisture control system, as explained by Lolkema (1991). Moulded, pressed Gouda-type cheeses are held for about 1 h before brining so that their pH reaches 5.5–5.6.

SEMI-HARD CHEESES

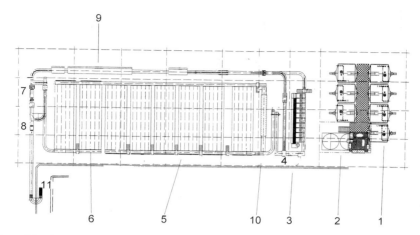

Figure 8.9 Plan of a typical production line for the manufacture of semi-hard cheese in rectangular blocks. 1, seven curd-making tanks (OST vats); 2, two curd and whey buffer tanks; 3, drainage and moulding station by ten Casomatic® single-column; 4, lid placing; 5, six cheese presses; 6, two holding tables for acidification; 7, lid removing, turning, blowing and mould lifting to be emptied; 8, cheese weighing; 9, mould and lid cleaning; 10, mould storage; 11, brining bath. Reproduced by courtesy of Tetra Pak Tebel BV, The Netherlands.

8.7 Brining

The modern salting systems for semi-hard cheese are characterised by the creation of a brine flow on which the cheesess float as a means of transport. This means that the brine must have a density sufficiently higher than that of the cheese, which is approximately 1.07 g ml^{-1} depending on composition. Sodium chloride is the salt used. For dietary purposes other salts may be used but the peculiarities of this are not within the scope of this book.

8.7.1 Brine composition

The pH of the brine should be 4.4–4.6 for Dutch-type cheese varieties. Brine composition should be checked at least every day and salt must be added on the basis of the brine surplus and the calculated amount of salt taken up by the cheese. Nowadays, salt is delivered in bulk and stored in a silo. A flow of brine circulates downside the silo to take up salt when needed.

8.7.2 Hygiene measures

There are various sufficiently salt-tolerant microorganisms that can survive (but not grow) in brine and thus threaten cheese quality. It has been found that salt-tolerant lactobacilli represent such a problem. Traditionally, brine 'maintenance' is a matter of cleaning everything above brine level (Stadhouders *et al.*, 1985) to keep microbiological counts low, but automatic membrane filtration units are now available to remove unwanted microorganisms and clean the brine physically (Ottesen and Konigsfeldt, 1999). The flow via the bypass used for brine cooling can be used for filtration. However, the application of membrane filtration is still very limited.

8.7.3 Brining systems

Nowadays, mostly deep-brining systems are installed. Such a mechanised brining plant is shown in Figure 8.10. A roller-conveyor (item 1) transports the cheese into the entrance ditch with floating brine. In this case the cheeses arrive first in the cages (item 4). After they are filled, they are emptied again and the cheeses are received in the cages (item 5) to be brined as usual. The cheeses then leave the cage and are taken up by a discharge conveyor (item 12) to the cheese store. During this transport the cheeses pass a blow drier (item 11) to remove excess brine. By this changing of cages, the plant uses a 'first in, first out' system for timing the brine exposure, which has to be very accurate for smaller cheeses.

8.7.4 Dry salting

Brining is the usual way to salt cheese. However, this has a number of consequences for the plant such as the need for large brining rooms and equipment,

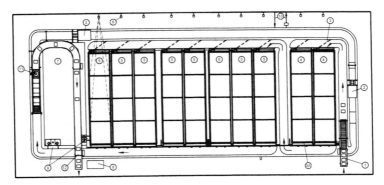

Figure 8.10 Layout of the DeepBrine® system designed by APV Nordic Cheese A/S, Denmark. 1, Cheese inlet conveyor; 2, two propellers for creating a flow; 3, in-let and 3A, outlet gates; 4, two turning cages each with ten racks; 5, eight brining cages each with ten racks; 6, a hydraulic system for lifting and lowering the cages; 7, buffer basin for level control; 8, circulation and brine cooling; 9, screens for filtering the brine; 10, automatic salt dosing from an exterior silo; 11, blow drier; 12, discharge conveyor. Reproduced by the courtesy of APV Nordic Cheese A/S, Denmark.

and the need to handle a brine surplus. Recently, APV presented the possibility of dry-salting the cheese to be used for foil ripening (Teilgård and Busk, 1995). The salt is spread on both sides when packing. Moisture will come out of the cheese shortly after salting but it is expected that the resulting 'brine' will be reabsorbed.

8.8 Treatment during natural ripening

In the ripening room, the rind of the cheese should dry to a certain extent to keep the rind in good condition. The cheese is further protected by the use of a cheese coating containing natamycin to inhibit yeast and mould growth. Weight loss of the cheese must also be well controlled because of the need to control the final moisture content of the cheese. Climatic conditions suitable for surface-ripened cheese types are more humid than they should be for coated-cheese types.

8.8.1 Cheese handling systems

In modern cheese factories, automated and more integrated systems are desired for the treatment of cheeses and shelves that carry the cheeses (the number depends on cheese size). In the meantime, satisfactory solutions are being found. Figure 8.11 shows a combination of machines and conveyors as delivered by Elten Systems (Barneveld, Holland) that can handle approximately 600 shelves per hour.

Boxes loaded with shelves carrying four cheeses each and coming from the ripening room are brought in (item 1) and the shelves with cheese are taken out of the box (item 2). They are individually transported to a machine where the shelves are turned 180° sideways (item 5). The cheeses are conveyed to the coating apparatus (item 6). In the meantime, the shelves are transported dirty side down to the soaking and washing machine (item 8) and are cleaned with brushes only on that side. The shelves are rinsed and wiped off (item 9) and transported, with the still clean and dry side up, to the cheese loading station (item 7). In the meantime, the cheeses coated at the upper side arrive there too and are automatically positioned onto the dry side of the shelf. The shelves containing the cheese are then loaded into the boxes (item 4) used earlier (item 2). The underside of the shelves will dry in the cheese ripening room. The coating machine (item 6) consists of two assemblies, each of two turning bars with flexible flaps to spread the plastic dispersion evenly over the cheese. A set dose of this dispersion is applied per cheese. The bars with flaps are mounted shortly above the cheese obliquely to the moving direction of the cheese to be sure that the whole upper side of the cheese will be covered. It is simply designed and can be opened easily in order to facilitate cleaning. Via the two-way conveying belt (item 10), brined cheese can easily enter or ripened cheese can be delivered.

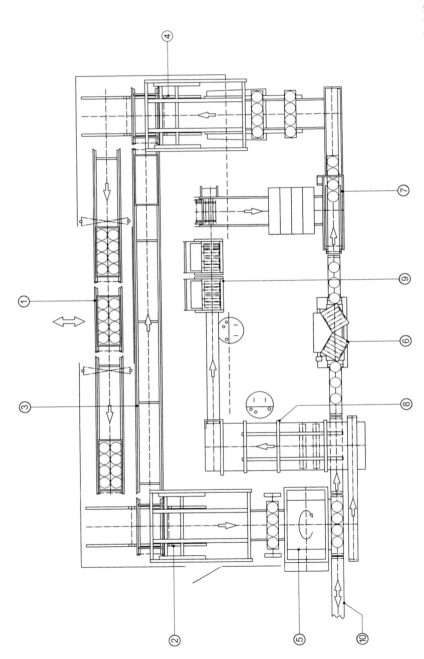

Figure 8.11 Schematic illustration of a handling system for cheeses and the shelves (racks) in a modern Gouda cheese store. 1, In-out and output shelves; 2, shelf unloading; 3, empty-box conveyor; 4, shelf loading; 5, cheese turning; 6, cheese coating; 7, shelf loading with cheese; 8, shelf presoaking; 9, shelf cleaning; 10, input and output cheeses. Reproduced by courtesy of Elten Systems, The Netherlands.

8.8.2 *Conditioning of the ripening room*

The climate in the ripening room is humid to avoid too harsh a drying of the cheese rind. The building should therefore be well insulated to avoid cold spots on the walls where water may condense and mould growth take place. Moreover, the humidity and the flow of the air must be well controlled. A typical climate-control system for a semi-hard cheese store is shown in Figure 8.12.

The air from the ripening room is conditioned (item 1) by cooling to condense water, which is then removed, whereupon the air is dried by heating. The conditioned air is transported by a centrifugal fan (item 2) through air supply ducts (item 3). A large number of nozzle tubes (item 4) in the ripening room are connected to these supply ducts to divide the air over the room. These vertical tubes are provided with nozzles. There is one nozzle for one or two shelves. A nozzle blows the air along the shelf over the cheeses. At the other end of the shelf there is a similar assembly of tubes but here they serve as suction tubes (item 5). In this way each shelf with cheeses is provided with conditioned air. The ripening room is preferably built inside the building. Some air from the ripening room is conducted through the air return channel (item 6) and serves as a conditioner for the wall of the ripening room. In this way, it is not possible for cold spots to form on the wall of the ripening room. The air returns to the air-conditioning unit either via the double ceiling between the roof and the ceiling of the ripening room or via ducts. The building itself has a double wall (item 7) and a double roof (item 8). It will be obvious that all these ducts and rooms to circulate air have to be accessible for inspection and cleaning.

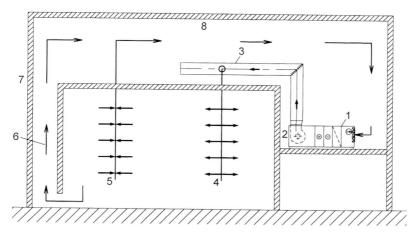

Figure 8.12 Schematic illustration of a climate-control system for a cheese store. 1, Air-conditioning unit; 2, ventilator; 3, air supply ducts; 4, special nozzle tubes for air induction; 5, suction tubes; 6, air return channel; 7, double wall; 8, double ceiling. Reproduced by courtesy of GTI-NAVEP, The Netherlands.

The conditioning of the air has to be done independent of the amount of cheese in the ripening room, so the difference between the water activity (a_w) at the surface of the cheese and the relative humidity of the air, the so-called moisture deficit, should be constant. The best means of controlling water evaporation from the cheese surface and ripening temperature is to regulate the absolute moisture content of the air that enters the ripening room according to the well-known Mollier Humidity to Temperature (HT)-diagram.

The supply nozzle blows the air over the first cheese on the shelf and this incoming flow induces even more air than the incoming airflow itself to flow around the cheeses. On its way to the suction tube it collects the water vapour from the cheese and the air actually becomes more humid. This small difference in conditions on the shelf is one of the reasons for changing the position of the cheeses when they are turned and/or coated, as indicated in Figure 8.11. The air velocity is more relevant for water evaporation during the first weeks after cheese manufacture. Later, the air flow is necessary, mainly for transport of water vapour and for air conditioning.

References

van den Berg, G. (1990) New technologies for hard and semi-hard cheese in dairying in a changing world, *Proceedings of the XXIII International Dairy Congress, Volume 3*, Agriculture Canada, Ottawa, pp. 1864-75.

van den Berg, G., Daamen, C.B.G., de Vries, E., van Ginkel, W. and Stadhouders, J. (1988) Test of the bacteria-removing separators, manufactured by Westfalia Separator AG, for the manufacture of Gouda cheese. Report R127, NIZO, Ede.

van den Berg, M.G. (1984) The thermization of milk. *Bull. Int. Dairy Fed.*, **182** 3-11.

van den Berg, M.G., van den Berg, G. and van Boekel, M.A.J.S. (1996) Mass transfer processes involved in Gouda cheese manufacture, in relation to casein and yield. *Neth. Milk Dairy J.*, 50 501-40.

Daamen, C.B.G., van den Berg, G. and Stadhouders, J. (1986) Test of the bactofugation efficiency of a self-cleaning hermetic bactofuge, type BMRPX 618 HGV, manufactured by Alfa-Laval A/B, Lund (Sweden). Report R124, NIZO, Ede.

Fox, P.F. (Ed.) (1993) *Cheese: Chemistry, Physics and Microbiology, Volume 2: Major Cheese Groups*, 2nd edn, Chapman & Hall, London.

van Ginkel, W.A., de Vries, E. and Daamen, C.B.G. (1987) Test of a curd-making tank, type 3113-15 000, code No. 3113-1049, with a capacity of 15 000 litres, manufactured by Pasilac Silkeborg. Report R125, NIZO, Ede.

ten Grotenhuis, E. (1999) Prediction of cutting time during cheese production. *European Dairy Magazine* (February) 40-41.

Hansen, R. (1979) The modified Casomatic. *North European Dairy* **45** 66-76.

Hansen, R. (1981) Multi-cheese moulds. *North European Dairy* **47** 208-11.

Hansen, R. (1986) Curd draining and moulding machines with multitubes. *North European Dairy* **52** 275-9.

Hansen, R. (1987) The new flexible Casomatic. *North European Dairy* **53** 126-34.

van Hooydonk, A.C.M. and van den Berg, G. (1988) Control and determination of the curd-setting during cheesemaking. *Bull. Int. Dairy Fed.*, **225** 2-10.

Hoeier, E., Jangen, T., Henriksen, C.M., Rattray, F., Brockmann, E. and Johansen, E. (1999). The production, application and action of lactic cheese starters in *Technology of Cheesemaking* (Ed. B.A. Law), Sheffield Academic press, Sheffield, UK, pp. 99-131.

Lolkema, H. (1991) Cheese yield used as an instrument for process control experience in Friesland, The Netherlands, in Factors affecting the Yield of Cheese, *IDF Special Issue* **9301**, 156-97.

McSweeney, P.L.H., Fox, P.F., Lucey, J.A., Jordan, K.N. and Cogan, T.M. (1993) Contribution of the indigenous microflora to the maturation of Cheddar cheese. *Int. Dairy J.*, **3** 613-34.

Ottesen, N. and Konigsfeldt, P. (1999) Microfiltration of cheese brine. *European Dairy Magazine*, (August) 22-4.

Payne, F.A., Hicks, C.L. and Shen, P. (1993) Predicting optimal cutting time of coagulating milk using diffuse reflectance. *J. Dairy Sci.*, **76** 48-61.

Roosenschoon, C.F. (1972) Bakens in de tijdstroom; een kenschets van de bond van cooperatieve zuivelfabrieken in friesland bij het 75-jarig bestaan (A 75 year history of the Association of Cooperative Dairies in Friesland). Bond van Coöperatieve Zuivelfabrieken in Friesland, Leeuwarden, (in Dutch).

Robinson, R.K and Wilbey, R.A. (1998) *R. Scott Cheesemaking Practise*. 3rd edn, Aspen, Rockville, MD, USA.

Stadhouders, J. (1982) Cooling and thermization as a means of extending the keeping quality of raw milk. *Kieler Milchwirtsch. Berichte*, **34** 19-28.

Stadhouders, J., Hup, G. and Nieuwenhof, F.F.J. (1983) Silage and cheese quality. *Communication* M19A, NIZO, Ede.

Stadhouders, J., Leenders, G.J.M. Smalbrink, L. and Klompmaker, T. (1985) Vermindering van de besmetting van slappe pakel met zoutresistente lactobacillen door een betere hygiene in het pekellokaal (Reduction of the contamination of weak brine with salt-tolerant lactobacilli by improving the hygienic conditions in the brining room). In Dutch, with English summary. *NIZO-Nieuws Zuivelzicht*, **77**(9) 892-4.

Tamime, A.Y. (1993) Modern cheesemaking: hard cheeses, in *Modern Dairy Technology Volume 2* (Ed. R.K. Robinson), 2nd edn, Elsevier Applied Science, London, Volume 2, pp. 49-220.

Teilgård, J. and Busk, P. (1995) A method of salting cheese. *International Patent Publ. nr*. WO 95/30335.

de Vries, E. (1979) Test of a curd-making tank, type OST III, with a capacity of 10 000 litres, manufactured by α-Tebel. Report R110, NIZO, Ede.

de Vries, E. and van Ginkel, W. (1980) Test of a curd-making tank, type Damrow 'double O', with a capacity of 16 000 litres, manufactured by DEC International. Report R113, NIZO, Ede.

de Vries, E. and van Ginkel, W. (1983) Beproeving van een gesloten wrongelbereider met een inhoud van 16 000 l, fabrikaat Alfons Schwarte GmbH. [Test of a curd-making tank, with a capacity of 16000 litres, manufactured by Alfons Schwarte GmbH, Ahlen (Westphalia, Federal German Republic)]. In Dutch, with English summary. Report R118, NIZO, Ede.

de Vries, E. and van Ginkel, W. (1984) Test of a curd-making tank, type Ost IV, with a capacity of 10,000 litres, manufactured by Tebel BV. Report R120, NIZO, Ede.

Walstra, P., Geurts, T.J., Noomen, A., Jellema, A. and van Boekel, M.A.J.S. (1999a) Principles of Cheese Making, in *Dairy Technology*, (Ed. P. Walstra), Marcel Dekker, New York, pp. 541-53.

Walstra, P., Geurts, T.J., Noomen, A., Jellema, A. and van Boekel, M.A.J.S. (1999b) Process steps, in *Dairy Technology*, (Ed. P. Walstra), Marcel Dekker, New York, pp. 555-600

Walstra, P., Noomen, A. and Geurts, T.J. (1993) Dutch-type varieties, in *Cheese: Chemistry, Physics and Microbiology, Volume 2: Major Cheese Groups*, (Ed. P.F. Fox), 2nd edn, Chapman & Hall, London, pp. 39-82.

9 Soft fresh cheese and soft ripened cheese

H. Pointurier and B.A. Law

9.1 Introduction

Soft ripened cheeses and fresh cheeses together represent almost two thirds of total cheese production in France and Germany. Their popularity is also increasing in Japan, as well in the USA, Canada, Australia and the United Kingdom, long associated with the consumption of Cheddar and similar varieties, rather than soft cheese. Although the best and most widely known soft ripened cheeses are Camembert and Brie, there are numerous other varieties such as Pont L'Evêque, Munster and Italian Crescencia, which are made with a bacterial smear culture rather than the more familiar white surface mould. Similarly, if the best-known fresh cheeses are now the smooth, homogeneous types sold in plastic pots, there are also firmer products such as le Suisse and le Quark Allemand, and the moulded, drained products, of which the best-known are goats' milk cheeses. Also, cottage cheese is very different from the smooth soft fresh cheeses, yet it remains an economically important product in most cheese-consuming countries.

Alongside the two main soft cheese categories are also the blue–green mould cheeses such as Roquefort, Danish Blue and Gorgonzola, whose technology is close to that of the soft ripened cheeses.

Here we present the main physico-chemical characteristics of soft ripened and soft fresh cheeses, together with their essential manufacturing parameters and flow charts. We then proceed to discuss the special issues involved in mechanising and automating their manufacture. The biochemistry and microbiology of mould-ripened soft cheeses have been reviewed recently by Gripon (1997), and the cultures used to ripen all of the blue, surface-mould and smear cheeses have been reviewed by Bockelmann (1999). For detailed descriptions of process recipes, the reader is referred to Scott's *Cheesemaking Practice* (Robinson and Wilbey, 1998).

9.2 Characteristics of ripened and fresh soft cheeses

9.2.1 Soft ripened cheeses (les fromages a pâte molle)

Physically, the soft ripened cheeses have a crumbly, brittle texture just before their ripening stage, necessitating curd transfer and handling practices which are much more subtle and less robust than those used in plants making hard and semi-hard pressed cheeses. Also, unlike film-wrapped Cheddar and plastic-coated Gouda, these cheeses have a distinct difference between the open, fractured surface zone and the more homogeneous, smooth interior. During ripening, various exchanges of material and chemical equilibria exist between surface and interior in progressive gradients, which change throughout ripening. They depend on the initial mineral composition of the curd, the surface microflora and the 'climatic' conditions of ripening (humidity, temperature, airflow). The regulation of these climatic conditions is an essential factor governing the quality of the final cheese and is amenable to automatic control.

This type of cheese is sold to the consumer as a whole cheese unit, as manufactured in the cheese plant, and the packaging is not moisture-proof. Therefore, another important area of mechanisation and automation lies in the need to weigh out the curd for each cheese unit with minimum error to lower the pack weight variation at the retail end of the chain; any such error cannot be corrected by trimming at the packaging stage.

Finally, these cheeses have a reduced margin of error in their ripening time compared with hard and semi-hard cheeses, as well as strict, shorter shelf-life limits. This means that production must run on a tightly controlled process flow at high productivity in order to maintain supply—another compelling reason to mechanise and automate.

Soft ripened cheeses have relatively high water contents; for example, Camembert, with $45\,g\,100\,g^{-1}$ fat in dry matter (FDM), has $46\,g\,100\,g^{-1}$ moisture. Low-fat soft cheese carries even more water: $65–70\,g\,100\,g^{-1}$ for Brie and $65–67\,g\,100\,g^{-1}$ for Camembert. Salt content is also higher in soft ripened cheese than in most other types: $2.5–3.5\,g\,100^{-1}$ in surface mould cheese, $4.0–4.7\,g\,100\,g^{-1}$ in smear cheese and $6.0\,g\,100\,g^{-1}$ in blue cheese. Salting is sometimes done with dry salting machines or with granular salt spread onto the outside 'crust' of the cheese.

The final manufacturing pH of the cheese should be 4.8–4.8 except in the so-called 'stabilised' types; this goes lower during the first 4 or 5 days following manufacture. One day after manufacture there is still $0.5–1.5\,g\,100\,g^{-1}$ lactose in the cheese (more in 'stabilised' types) and the art of maturing good soft ripened cheese is in managing the residual lactose (Mietton, 1986). The relatively acidic nature of soft cheese curd ensures a low proportion of calcium in dry matter (DM), in the order of $0.7\,g\,100\,g^{-1}$ in Brie, and $1.3\,g\,100\,g^{-1}$ in Camembert, signifying that calcium ions solubilised by the low pH are dissolved in the whey and expelled during syneresis. Similarly the proportion of calcium ions to phosphorus ions is quite low, at about 1.4%.

Bacteriologically, soft ripened cheeses are characterised by strong acidification by the lactic acid bacteria (LAB) used as starter cultures (initially mesophilic LAB, but later thermophiles) at the first stage of manufacture. Next, a diverse population of yeasts develops, dominated by *Kluyveromyces* spp., which ferment lactose. If the surface ripening culture is based on moulds, the dominant one is *Penicillium candidum* (*camemberti*), but 'modern' cultures also contain *Geotrichum candidum* to raise pH by utilising lactic acid and to boost aroma through fatty acid metabolism (mushroom aroma) and amino acid catabolism (pungent aroma) (Gripon, 1997). Smear bacteria also benefit from yeast and mould growth to raise surface cheese pH so that coryneforms and micrococci can colonise it and produce special aromas.

9.2.2 Fresh cheese (fromage frais)

Varieties within this category include cottage cheese, with a drained curd having a brittle structure, and the smooth fromage frais varieties conditioned in plastic pots. Some of the latter are firmer than others, for example le Suisse, le Quark Allemand, Mascarpone Italien and Labneh oriental.

The curd body is often 'aerated', containing air or carbon dioxide at up to 5–10%, so that the cheese is quite compressible during transfer and handling. The resulting mechanical stress can change its texture between initial curd manufacture and dispatch in packaged form from the plant, and this has to be taken into account in designing mechanical automated plant to speed up the process and to save labour costs. The effects of such stress are generally seen in the form of phase separation (whey and curd solids separate) which the consumer associates with a lack of freshness.

The fresh cheeses are very high in water content (80 g $100 \, g^{-1}$ for skimmed milk cheeses, and 85 g $100 \, g^{-1}$ of non-fat dry matter in full-fat cheeses). Water activity (A_w) is correspondingly high, at 0.99, explaining the need for a strictly limited display and 'use by' time limit, even when stored at refrigeration temperatures (at or below 6°C). The proportion of fat in dry matter is very variable amongst varieties of fresh cheese. Many are made with very low fat levels for the diet food market and use modern membrane concentration techniques or heat treatments to strongly increase the whey protein content of the cheese as a means of maintaining 'body' and texture in the absence of fat. As a result, many of these 'diet cheeses' have astringent off flavours. Here, we present mainly the products made with fat contents of about 70 g $100 \, g^{-1}$ of dry matter.

The pH of fresh cheese curd is about 4.6, though this is increased by adding nonacidified cream to certain 'double cream' smooth lactic cheese products. Lactose is usually about 3.0 g $100 \, g^{-1}$, and this should not change during storage of the product at 6°C. The strong acidification of fresh cheese by the lactic starter culture gives even lower calcium context in dry matter than is the case for soft ripened cheeses (about 0.4 g $100 \, g^{-1}$); the calcium ion to phosphate ion ratio is also very low.

Fresh cheese are usually unsalted, a factor adding to their limited shelf life and to the need for mechanised or automated production processes which are tightly controlled, hygienic and reliable.

9.3 The key phases in the process plant for soft cheese manufacture

For soft ripened and fresh cheeses, the vital syneresis (whey expulsion) stage of curd manufacture is heavily dependent on the acidification power of the lactic starter culture. This is particularly true in the case of fresh cheeses, in which little or no rennet is used to form the milk gel. In this sense, soft cheese manufacture is fundamentally less complex than hard cheese manufacture and, as such, is very amenable to mechanisation and automation. Figure 9.1 shows a comparison of the typical acidification kinetics of the main cheese types, illustrating the simple, strong sigmoidal acidification curves of fresh or soft cheeses compared will hard and semi-hard cheeses.

9.3.1 Soft ripened cheeses

All through the manufacturing process, two mechanisms of curd drainage and syneresis, driven respectively by acidification and rennet action on casein gel structure, will run concurrently, with opposing effects on the final properties of the curd. Rennet-mediated syneresis tends to produce a drier, less acid curd, whereas acid-mediated syneresis produces a softer, moister, 'de-lactosed', more

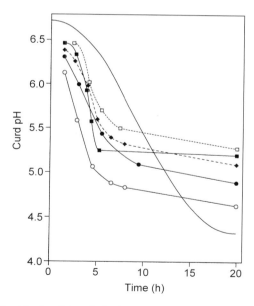

Figure 9.1 Acidification kinetics of the principal types of cheeses. Renneting occurred at time 0 h.—, fresh unripened cheese; ○, Camembert (large-scale industrial); ●, Carré de l'Est (stabilised curd); ■, Cheddar (hard); ◆, St-Paulin (semi-hard); □, Emmental (very hard). Source: Mietton (1991).

acidic and crumbly curd. Clearly, the moister curd will also mature faster (Vassal et al., 1984). An equilibrium must be found between the two kinetics of syneresis throughout curd manufacture, up to the 'dry' stage, when it is formed as cheese and salted.

Although relatively simple (compared with hard and semi-hard cheese in which rennet and temperature gradients play a critical role in curd formation, Lomholt and Qvist, 1999) the manufacture of soft ripened cheeses puts more constraints on the design of mechanical curd handling through its use of minimal cutting of the curd and the need for very gentle stirring in the whey, because the curd structure is more brittle. The extreme case is traditional Camembert, which is only cut with a single knife and is not stirred at all. It is ladled into moulds without disruption of the curd structure so that the curd is capable of reforming with a smooth surface, free from the 'pea' macrostructure which is so important in blue–green mould cheeses and which allows air passages for the germination of mould spores inside the cheese. Traditional Camembert must be smooth and free from such macrostructure.

In conclusion then, the syneresis which governs the final curd moisture and texture is dependent mainly on the acidification kinetics (which are, in turn, dependent on the starter inoculation rate and the type of culture) and on the curd handling process (the amount of cutting, the strength and duration of stirring and the state of the curd in the moulds). These factors are compared and summarised in the curves shown in Figures 9.2 and 9.3. In interpreting these curves, one must remember that in soft cheeses the final pH is not attained until 4 to 5 days after manufacture of the curd.

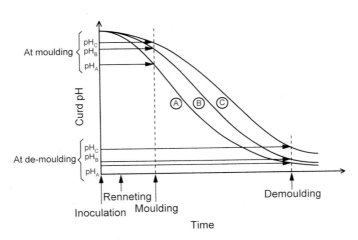

Figure 9.2 Acidification kinetics in relation to pH at moulding of soft ripened cheeses. A, Rapid acidification; B, normal acidification; C, slow acidification. Source: Mietton, 1991.

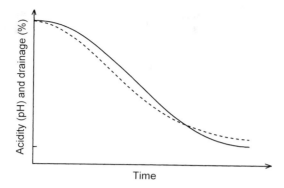

Figure 9.3 Relationship between acidification kinetics, curd pH and rate of curd drainage in soft cheese manufacture. —, slow draining; - - - -, rapid draining. Source: Mietton (1991).

Finally, it is important to note that, although mechanisation and automation must control acidification and syneresis if it is to be successful, the milk must also be standardised to avoid unpredictable variations in these processes arising from milk composition. The subject of standardisation is dealt with in detail in Chapter 3 of this volume and has been described specifically for soft ripened cheese by Mietton (1986). This standardisation achieves a true and verifiable 'mix' of cheesemilk components, including the starter culture as opposed to a raw material of unpredictable and variable composition.

The formulation of the 'mix' should have three different objectives.

a) The first objective is to fix a pre-determined, standard milk pH at the start of the process, when the starter culture and rennet are added—this can be done by 'pre-maturation' with a lactic culture, followed by pasteurisation, addition of D-glucono-δ-lactone (GDL), adding carbon dioxide.

b) The second objective is to standardise the dose rate of the lactic starter according to pre-acidification tests on the intended culture, so that acidification rate and lag phase are always the same. It is worth noting that modern automated plant often uses direct-to-vat starters (DVSs) supplied in guaranteed, checked, standardised form by the culture company as a cell concentrate which needs only to be thawed from frozen or rehydrated in the cheese vat (see Hoeier et al. (1999)). Standardisation of lactose content by membrane filtration can also be useful in this objective (see Chapter 3).

c) The last objective is to ensure constant syneresis potential (both acid-induced and rennet-induced) by standardisation of the casein:whey protein ratio by using ultrafiltration (UF) retentate or simply a low-concentration UF stage (see Chapter 11). This objective also ensures at the same time good management of the weight of individual cheeses.

After all this, it is also important to incorporate into automation the means of controlling the coagulation temperature, syneresis temperature and the rate and extent of curd cutting and gentle stirring.

Finally, it is worth making the following remarks about the soft ripened cheese process. First, because the manufacture of 'appellation de l'origine contrôlée (AOC) cheeses forbids the addition of ingredients to the milk, and clearly would not allow pre-maturation and pasteurisation to standardise pH, this sector is not suitable for mechanisation and automation. The low volume of production also mitigates against such developments. Second, because the transfer of curd to the moulds is followed by further acidification and syneresis, it is a long way from being the end of the cheesemaking process (unlike in Cheddar and Gouda manufacture, for example). Therefore, the automation and control of conditions in the mould and drainage stage is fundamental to ensuring the consistent high quality and weight of these soft cheeses.

9.3.2 Soft fresh cheeses

Here the concern is with a process in which the lactic fermentation and acidification entirely precedes syneresis. We will only deal with the production of very-low-fat curd prepared from skimmed milk, acidified by a lactic culture fermentation, recovered by centrifugation, then re-enriched with cream (sometimes cultured for aroma), sugar, flavours or fruit preparations. Figure 9.4 shows a flow diagram for the Westfalia smooth curd Thermo-Quark process. Production of cottage cheese is described in Section 9.3.3.

The maturation stage in this process is not, in this case, a means of standardisation, but is part of the method of increasing the flow rate and turnover of the process, incorporating milk protein at the least cost, with particular functional properties conferred by the maturation and pre-pasteurisation process. This maturation is an intermittent batch process, carried out in two parallel tanks of 100 000 l, each processing in turn $10 000 \, l \, h^{-1}$ and discharging well-stirred 'mature' milk to ensure homogeneity of the final product. From this point the process is essentially continuous, through the heat treatments and separators, until the curd storage tanks and cream enrichment process (plus other ingredients added to the curd). Unusually for a cheesemaking process, it is possible to take reliable samples during the different stages of this manufacturing stream to use in calculating the curd yield. This is precisely because the milk/curd transition is continuous and takes place in a homogeneous matrix. In fully automated plant (see below), continuous reliable sampling and precise automated compositional analysis is used both to calculate and to optimise the yield as a function of process parameters. Yield is simply based on a percentage calculation of skimmed milk curd minus the weight of whey as a proportion of milk weight minus the weight of whey. Percentage can be calculated at any time during the process flow, and when the milk flow, whey flow and curd flow are also constantly monitored

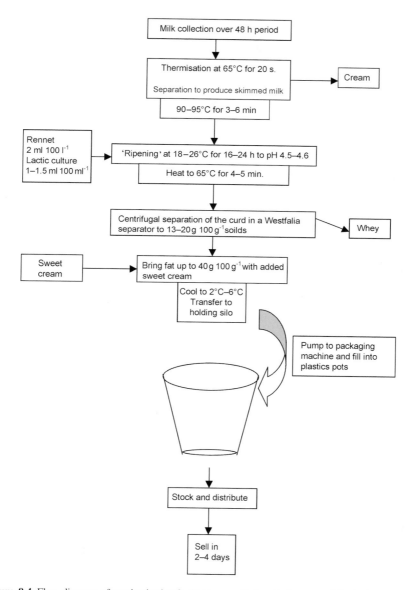

Figure 9.4 Flow diagram of mechanised soft cheese manufacture in the case of fresh cheese with a smooth texture (fromage frais). The process shown is the Thermo-Quark process of Westfalia, using specially designed 5000–6000 g centrifugal separators to achieve rapid syneresis and curd separation. Source: information supplied by Westfalia Separator AG, Germany.

any one of them can be adjusted to fit in with the other two to maximise yield, throughput or both yield and throughput.

The Westfalia curd separator is fully illustrated in the technical literature of the company, and here we describe it only to show examples of the key factors in mechanised and automated production of smooth soft cheese curd. The main working area of the plant is made up of outlet collection heads round the periphery of the centrifuge bowl. They expel the skimmed milk curd, which is heavier than the whey. The rate of outflow of curd per unit of surface area of the outlet surface is practically constant whatever its solids content (being about 520 kg mm^{-2} hour^{-1} for the model KDB 30). Therefore, for a particular outlet port diameter, regulation of the output of curd is simply effected by adjusting the inlet flow of coagulated milk, provided it has been homogeneously mixed beforehand. For example, with 12 outlet ports of 0.6 mm diameter (total surface area 3.396 mm^2), a skimmed milk input of 8.0 g 100 g^{-1} solids, and whey at 6 g 100 g^{-1} solids, the flow will be 1800 kg h^{-1} with a yield of 25%. Further calculation shows that if the flow rises by 1%, with a constant outflow of 1800 kg h^{-1}, the yield falls to 24.75%, and the solids content of the curd rises to 18.12 g 100 g^{-1}. This shows the importance of precise regulation of the throughput flow of the plant.

The first key to the automation of a fresh cheese line, therefore, is a combination of homogeneous milk coagulum input and a regulated flow precise to within 0.5%. Obviously, the subsequent addition of cream and/or other flavouring components must also be precisely metered and accompanied by efficient blending and, to avoid phase separation and shelf-life reduction of the final product, pumping to the pot-filling plant must be such that compression and expansion must be minimised.

Finally, to ensure the hygiene of the overall process, and the high microbiological quality of the product, an automatic cleaning-in-place (CIP) system is an essential component of the process. In particular, for the latest generation of completely watertight centrifugal plant, the cleaning and sterilising cycle must include an internal steam sterilisation applied rigorously. A guaranteed high microbiological quality in these fresh cheese products is an absolute necessity to ensure that no microbiological spoilage or growth of pathogenic bacteria happens between their manufacture and consumption.

9.3.3 Cottage cheese

Without automation, Cottage cheese is made from pasteurised milk (with or without standardisation) in open rectangular vats (Robinson and Wilbey, 1998) using a well-proven recipe which creates a relatively highly structured curd with a distinct granular structure, unlike the fresh soft cheese discussed in the previous section. The curds are formed by acid synersis by means of a mesophilic lactic starter culture, but a low level of rennet is also used (0.0002%) with calcium chloride (0.01–0.02 g 100 g^{-1}) to give strength to the granular texture. The milk is fermented down to pH 4.4–4.7 before cutting, then the curds are strengthened ('firmed up') and further syneresed in the whey by stirring and scalding to about 49°C. This process also prevents further action by the starter. The whey is drained and washed three times with water, which must be of very high quality to avoid contamination with psychrotrophic bacteria, which can easily spoil the cheese, even in cold storage. The curd at this stage is very bland, and is usually flavoured by the addition of fruit, herbs, vegetables or a dressing prepared by fermenting buttermilk or cream with an aroma-producing lactic culture.

There are plenty of opportunities for mechanisation and automation of this process; indeed, the major equipment manufacturers offer plant specifically for this purpose, and this will be described Section 9.4.2.2.

9.4 Mechanisation and automation solutions

Mechanisation replaces human work and physical effort with an ordered mechanical procedure, but an operator is still needed to start and control it. Automation takes over some or all of the mechanised process, including the start command to launch a series of continuous or discontinuous stages, often controlled by sensors (measuring flow rate, temperature, pH, dry matter or moisture, etc.), linked to the control system. The whole process is thus coordinated, checked and supervised automatically.

Automation is only possible if the cheesemaking process (particularly acidification and syneresis) has been modelled. In turn, accurate modelling is only possible if the milk is standardised at the outset (for soft ripened cheese) or if one has at one's disposal a large quantity of homogeneous coagulated milk (for smooth-curd fresh cheese).

9.4.1 Soft ripened cheeses

9.4.1.1 Mechanisation

Mechanisation is an indispensable means of ensuring uniformity and of cutting the cost of making the typical small-unit, individual soft ripened cheeses by a process that necessarily involves many sequential unit operations of curd formation, drainage and packaging. In large production units it is possible to reduce direct human work to a very small proportion of the total work of the production line. Even for AOC raw milk Camembert, whose characteristics and production process are carefully defined, concern for the moulding process in ladles (*moulage à la louche*) has led to some mechanisation in units converting more than 10 000 l per day, by using *robot mouleurs* (Figure 9.5) developed in the Co-operative d'Isigny sur Mer, France. These robot moulds are an adaptation of the small curd vats with perforated bottoms used in factory Camembert making

Figure 9.5 Mechanisation of curd production for traditional Camembert cheese with use of mechanical *moulage à la louche* equipment (see text). Reproduced by courtesy of Tecnal, France.

(see Figure 9.5) and, at the expense of some loss of material, this system reproduces the manual moulding operation.

All mechanisation of this type of cheesemaking demands at the outset that the concept of moulding, drainage and ripening of small-unit cheese is kept intact with the retention of equipment such as unit moulds, support trays, screens and supporting structures. The investment and maintenance costs are increased for the mechanised plant, but throughput and precise uniformity of the product is increased dramatically, cutting the unit cost of cheese manufacture.

The main stages where mechanisation is being most rapidly installed are as follows:

- the transfer of moulds, trays, cutting knives and drainage filter screens to the heart of the moulding and draining operation; gentle movement is vital in this area, without shock to the forming coagulum and soft curds
- the efficient loading and unloading of empty and full moulds boxes and draining screens
- moulding operations with weighing of coagulation vats, partial drainage and uniform levelling (to regulate unit weight of each cheese) of the curd in the moulding box
- a de-moulding operation to remove well-moulded curd from mould boxes to the draining tray prior to transfer to the ripening room.

Elements of this level of mechanisation are illustrated in Figure 9.6.

9.4.1.2 *Automation*

Automation does not yet use in-line sensors to measure, for example, curd firmness and pH. At present it is based simply on the processing time of a known mix of milk, coagulant and starter culture, on the amount of curd to be distributed before moulding and on the number of units for handling and

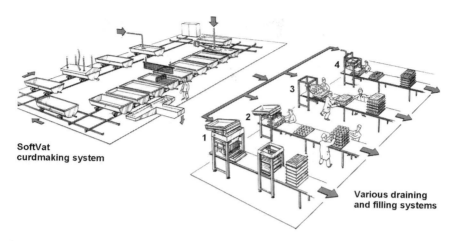

Figure 9.6 Typical mechanised production layout for ripened soft cheese manufacture involving different types of curd draining and mould filling equipment designed to take production as far as individual cheeses ready for salting and ripening. 1, Automatic ContiFiller with draining belt (with optional vibration strainer); 2, semi-automatic filling directly from the draining belt; 3, mould filling by shute straight from the curd-forming vats; 4, draining and filling from the vats via a vibration strainer. Reproduced by courtesy of APV Cheese AS, Denmark.

recycling (cheeses, moulding boxes, draining screens, etc.). Particular features of automated production lines are found in the 'levelling' of the curd in moulding boxes to make the weight of the finished cheeses almost uniform and in the automatic transfer of the coagulating milk to moulding boxes in a continuous production line, using very smooth mechanical movement systems and stirrers to avoid damaging the fragile gel. There are two main commercial systems in use, based on different mechanical layouts.

The first method involves the transfer of coagulating milk along the process line in small individual plastic vats, usually carried on a smoothly moving aerial line to avoid mechanical shock at this critical stage. The curd coagulum is cut automatically within these moving vats and stirred gently during syneresis before being weighed into box moulds, levelled (curd evening) to give constant cheese weight, drained while stacked on draining trays (screens), salted and transferred to the ripening room. This system is shown diagramatically in Figure 9.7 and is generally used in factories which need to be able to make several different varieties on the same line. It is very versatile in this respect and permits very tight control over the size and weight of the cheeses produced. The second alternative automated curd production line is a coagulator in the form of a long vat having a semicircular cross-section and equipped with moveable partitions and cutting knives, which can be deployed or retracted automatically according to the stage of the process (Figure 9.8).

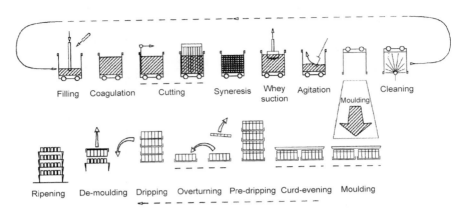

Figure 9.7 Automated production of soft ripened cheeses by using a system of individual coagulation vats (*bassines*) moving along a shock-free production line from vat filling, through coagulation by the rennet and acidifying culture, cutting, stirring and syneresis, moulding and drainage in multiple block moulds and de-moulding ready for salting and ripening. Reproduced courtesy of Tecnal, France.

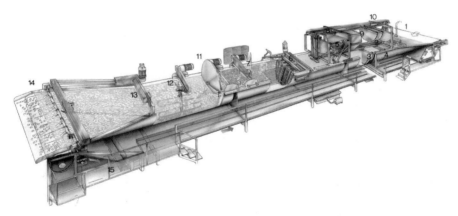

Figure 9.8 Automated production of curd for soft ripened cheese using a continuous, integrated coagulating machine. Note that the curd output can be drained or lightly pressed on leaving the machine so that its moisture content is adjusted to the type of soft cheese required. In the simplest form, the curd would be filled into moulding boxes without pressing and formed into individual cheeses for salting and ripening. 1, Milk inlet; 2, coagulation zones; 3, through-shaped conveyor belt with water film; 4, vat; 5, electrostatic curd detaching device; 6, curd cutter (longitudinal); 7, curd cutter (transverse); 8, spacing plate (return transport); 9, spacing plate (cleaning); 10, spacing plate (preparation); 11, dividing worm; 12, stirring device; 13, stirring device; 14, curd discharge device; 15, belt cleaning device. Note that this machine can also be used to make cottage cheese curd which would be fed into a cooling and washing vat prior to 'dressing' and filling into tubs or pots (see Section 9.4.2.2 and Figure 9.9). Reproduced courtesy of Alpma, Germany.

The milk coagulum is moved down the vat at a speed which can be regulated between 8–60 m h^{-1}. In large factories the vat can be almost 20 m long. The partitions, knives and stirrers are cleaned in a constant cycle of use, cleaning and return to use. This type of equipment is easier to clean than the lines which use individual coagulating vats and has a higher throughput of very uniform curd. It is favoured by the larger cheese plants with product flows of 20 000 l h^{-1}. The machine effectively takes in the 'mix' (milk plus coagulant plus starter culture) at one end and delivers curd to the moulds at the other. It is less versatile but more efficient than the individual box system.

In both systems the launch and flow through the line largely depends on the rate at which the curd in the moulding boxes can be drained, salted (either in brine or by surface dry-salting machines) and the moulds washed and returned automatically for refilling. Automated emptying, washing, disaffection, rinsing and return of the vats, moulds, partitions, knives and draining stacks is an essential and integral part of the overall automation of soft ripened cheese production, and the suppliers of the automated production lines also supply equipment for this purpose [e.g. Ultralav systems by Tecnal, France; Tetra Pak (Sweden) cleaning-in-place (CIP) units].

9.4.2 Soft fresh cheeses

9.4.2.1 Smooth-curd cheese

We have already seen how the production of soft fresh cheese with a smooth curd can be produced in specially constructed separators after formation of lactic curds (the Westfalia Thermo-Quark process; Figure 9.4); here we will consider the possibilities for automation of the plant, based on continuous sampling and on-line analysis of the composition of the product during production. The reader will recall that the process depends on the centrifugal (5000–6000 × g) separation, and almost instantaneous syneresis of lactic curd to make a smooth cheese product which is packaged in plastic pots and eaten fresh. Automation of the throughput parameters can be achieved by feedback adjustments to the 'mix' composition, its flow through the machine and the operating mode of the machine itself (speed, temperature, etc.) based on, for example, measurement of the viscosity of the emerging cheese as a measure of its solids content, or on-line near-infrared (NIR) sensors to measure moisture, fat, lactose and protein content. The viscosity, gelling properties and water retention properties of Quarg can be controlled and improved to some extent by processing the milk at high temperatures (Kelly and O'Donnell, 1998).

Obviously, an automatic CIP unit must be installed to take full advantage of this automated production line.

9.4.2.2 Cottage cheese

A typical mechanical, semi-automated cottage cheese line is shown in Figure 9.9. This particular line uses the Tebel O-Vat to make the cottage cheese curd. The

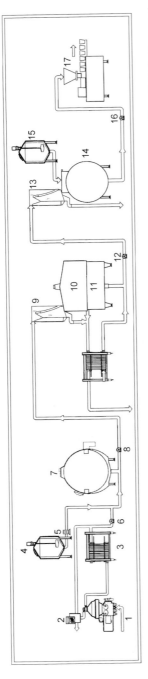

Figure 9.9 Flow diagram of a Cottage cheese production line based on a separate enclosed curdmaking vat (see Section 9.4.2.2 and Figure 9.10) feeding a curd cooling and washing vat and a 'dressing–creaming–flavouring' plant before the filling machine. 1, Separation; 2, standardisation; 3, pasteurisation; 4, bulk starter; 5, dosing pump; 6, feed pump; 7, cheesemaking vat (see Figure 9.10); 8, curd pump; 9, whey drainer; 10, cooling and washing tank; 11, water recirculation and cooling system; 12, curd pump; 13, water drainer; 14, creamer; 15, tank for dressing cream (flavours, etc. may be added at this point); 16, cottage cheese pump; 17, packaging machine. Reproduced by courtesy of Tetra Tebel-MKT, The Netherlands.

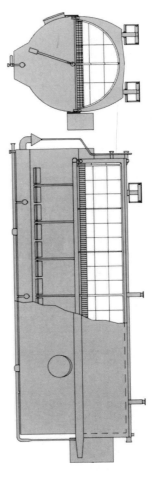

Figure 9.10 An enclosed-type curdmaking vat for soft cheese: the O-Vat. Reproduced courtesy of Tetra Tebel-MKT, The Netherlands.

vat is a fully automatic self-contained unit into which is pumped the inoculated, standardised pasteurised milk. Rennet addition is optional or minimal and can be done aseptically at any time after vat filling. All stirring and cutting equipment is contained within the vat and can be deployed automatically according to a time regime or acidification curve. A vat equipped with a viscometer or other curd strength-measuring device could be programmed to cut at a particular stage of coagulation. The initial curd cut is longitudinal, followed by automatic rotation of the shaft bearing the cutting blades through 180° to provide radial cutting into the curd cubes ready for stirring. Next, the stirrer blades are deployed with an oscillating action matched to the filling level on the vat. The mechanical cutting method which programmes the cutting time exactly (between 40 s and 2 min.) gives very uniform curd grain size (7-12 mm) with low wastage. A sectioned illustration of the O-Vat is shown in Figure 9.10.

As mentioned previously, Cottage cheese curd is washed and cooled before 'dressing' with cultured cream, flavour or vegetable preparations. In modern high-throughput cheese lines, the washing and cooling cycle, taking between 1 and 2 hours, is controlled automatically. If the curd has been washed and cooled by whey removal from the vat, followed by successive refillings with water, the curd is passed directly to the 'creamer' tank prior to packing in individual pots or tubs. The equipment manufacturers also supply whey-draining screens and pumps to transfer the curd to a separate washing and cooling tank (Figure 9.10) if this arrangement fits in better with fermentation vat emptying and refilling logistics. The whole process is controlled by a central processing unit (CPU), and connected to a proprietary CIP unit.

References

Bockelmann, W. (1999) Secondary cheese cultures, in *Technology of Cheesemaking* (Ed. B.A. Law), Sheffield Academic Press, Sheffield, UK, pp. 132-62.

Gripon, J.-C. (1997) Flavour and texture in soft cheese, in *Microbiology and Biochemistry of Cheese and Fermented Milk* (Ed. B.A. Law), Blackie Academic & Professional, London, pp. 193–206.

Kelly, A.L. and O'Donnell, H.J. (1998) Composition, gel properties and microstructure of Quarg as affected by processing parameters and milk quality. *Int. Dairy J.*, **8** 295–301.

Lomholt, S.B. and Qvist, K.B. (1999) Formation of cheese curd, in *Technology of Cheesemaking* (Ed. B.A. Law), Sheffield Academic Press, Sheffield, UK, pp. 66–98.

Mietton, B. (1986) La préparation des laits de fromagerie en technologie des pates molles. *Rev. I.A.A.*, (October) 951-63.

Mietton, B. (1991) *Cours de Fromageric*, ENIL, Poligny.

Robinson, R.K. and Wilbey, R.A. (1998). Selected cheese recipes, in *Cheesemaking Practice, R. Scott* (Eds. R.K. Robinson and R.A. Wilbey), Aspen, Rockville, MD, USA, pp. 327–437.

Vassal, L., Monnet, V., Le Bars, D., Roux, C. and Gripon, J.-C. (1984) Relation entre le pH, la composition chimique et la texture des fromages de type camembert. *Le Lait.* **66** 341-51.

10 Pasta Filata cheeses

O. Salvadori del Prato

10.1 General introduction and basic classification

Pasta Filata cheeses (*filata* literally meaning 'spinning or stretching the curd') constitute the main contribution of Italy to the world of cheesemaking in recent years. The term 'Pasta Filata' covers varieties of cheese, characterised by the special production process whereby the extensively soured curd is drained, heated and then stretched to modify the protein fibres, using mechanical tension to obtain the required texture and shape (Emaldi, 1968; Salvadori del Prato, 1998).

Mozzarella and Pasta Filata cheeses have a long Italian history dating back to mediaeval times. The area where they originate lies in the plain of Naples. Here, herds of buffalo produced very white milk which, because of the primitive production and transportation conditions, was already very sour by the time it arrived at the processing site. This acidity gave the curd its malleable qualities. The name of Mozzarella probably derived from the word *mozzare* which means 'to slice', a reference to the action performed by the cheesemakers when they used their hands to cut off a piece of cheese (the Mozzarella itself) from a long section of Pasta Filata they had produced. However, other researchers (Niccoli, 1902; Delforno, 1985; Ammirati, 1996) suggested that the name may have been derived from the Greek word *muzao*, meaning 'to milk', or from the French word *maison*, which dates back to the Middle Ages, when it referred to the barns used by buffalo shepherds near Naples.

Varieties of Pasta Filata include cheeses that are soft (such as Mozzarella), semi-soft (such as Caciocavallo) and hard (such as Provolone). These and other varieties are illustrated in Figure 10.1. Mozzarella can be produced from either cows' or buffalos' milk, and around 5 million tonnes of milk are used per year in Europe for the manufacture of this cheese; almost the same amount is utilised in the USA. Consequently, Pasta Filata cheeses represent one of the more important types of cheese made around the world, and their high consumption is a result of the popularity of pizza (Kindsted, 1993).

Typical Italian Mozzarella is in the shape of a ball, varying in weight from less than 100 g to over 300 g. It is off-white and milky in colour, very soft, with a delicate flavour of milk or cream, never sour, and with a slight buttery aroma. There are other types and shapes of Mozzarella in existence around the world, for example blocks of 'drier' cheese which are usually called Pizza cheese. A good Mozzarella should never be slimy, coarse or rough, and the rind

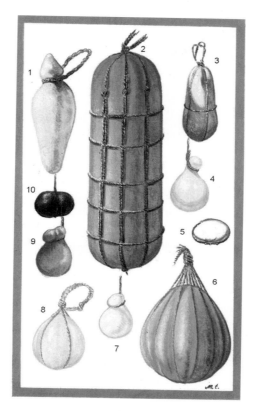

Figure 10.1 Traditional shapes and forms of Pasta Filata cheeses. 1, Caciocavallo of Naples; 2, Provolone; 3, Caciocavallo Silano; 4, Burrata; 5, Mozzarella; 6, Provolone 'Mandarone'; 7, Scamorza; 8, Provolone 'Boccia'; 9, smoked Scamorza; 10, smoked Provoletta. Source: Salvadori del Prato (1998).

should be very thin, soft and edible, and easy to peel off. The composition of different types of Italian Mozzarella and other Pasta Filata cheeses are shown in Tables 10.1 and 10.2.

In recent years, many technical innovations have been introduced throughout the world in Pasta Filata technology. In brief, they are:

- introduction of mechanical stretching machines (1950–60)
- production of cheese with milk acidified chemically rather than by means of starter cultures (1960–70)
- fully mechanised production lines (1970–80)
- attempts to make Pasta Filata with ultrafiltration (UF) of the milk (1980)
- production of 'imitation' Pasta Filata cheeses (1970–80)
- computerised control of cheesemaking parameters and mechanisation (1990)

Table 10.1 Proximate composition (all units 100 g^{-1}) and mean energy values of Italian Mozzarella cheese made from different mammalian milks

Constituents	Milk	
	buffalo	cow
Water (g)	55–60	54–62
Proteins (g)	19–20	18–22
Fat (g)	22–27	18–24
Carbohydrates (g)	0.6	0.7
Salt (g)	0–1	0–1.5
Energy (kcal)	280	265
Iron (mg)	0.2	0.4
Calcium (mg)	400	160
Phosphorus (mg)	240	350
Thiamine (mg)	0.05	0.03
Riboflavin (mg)	0.51	0.27
Vitamin A (mg)	190	300

Source: Salvadori del Prato (1998).

Table 10.2 Range of proximate composition of differents Pasta Filata cheeses

Cheese	Moisture[a]	Fat[a]	Proteins[a]	Ash[a]	pH
Kashkaval	40–45	27–33	19–25	2–5	4.8–5.3
Provolone	36–44	29–31	27–29	4–7	4.8–5.2
Pizza	45–55	15–25	18–24	1–3	4.9–5.3
Caciocavallo	34–42	30–36	23–28	4–6	4.8–5.2
Scamorza	40–48	25–30	22–28	3–6	4.8–5.3

[a]Expressed in g 100 g^{-1}
Source: Salvadori del Prato (1998).

The physico-chemical principles which govern the stretching properties of the cheese depend upon the presence of acid produced by the starter culture and/or added acid to the milk. The presence of acid results in a series of reactions which may be described as follows (Salvadori del Prato, 1998):

- coagulant

    calcium caseinate + rennet → calcium *para*-caseinate
    (milk pH 6.2)   (insoluble curd,
                                                      precipitate)

- fermentation

    calcium *para*-caseinate + lactic acid → mono-calcium-*para*-caseinate
    (milk pH 5.6)   + calcium lactate
                                                                     (soluble in hot water plus salt;
                                                                     fibrous; soft; malleable)

- Acidulant

    calcium *para*-caseinate + citric acid → mono-calcium-*para*-caseinate
    (milk pH 5.6)   + calcium citrate
                                                      (soluble in hot water plus salt;
                                                        fibrous; soft; malleable)

The technology used in Pasta Filata and Mozzarella cheeses is based upon the following characteristic stages:

- sufficient souring of the milk in a cheese vat (i.e. using microbial fermentation or added chemicals)
- draining and souring of the curd to pH 4.9–5.2
- 'stretching' (i.e. mechanical traction of the heated soured curd in order to obtain a fibrous and malleable consistency)
- cooling of the Pasta Filata, once it has been formed during the stretching process, into the required shape

However, the methods used to produce souring of the curd vary according to the technique used. Souring may be achieved by:

- the action of natural microflora in raw milk
- the addition of selected lactic starter cultures to the milk, including direct-to-vat inoculation (DVI; see Figure 10.2)
- chemical acidification through the addition of citric or other organic acids to the milk
- combined souring methods (i.e. chemical and biological acidification)

Once soured to stretching point the curd is either processed by hand or, nowadays by special machines called *filatrici*, or 'stretchers', which work on the principle of: (a) continuous screws that are juxtaposed and rotate in opposite directions and (b) 'diving arms' that work in a similar way to industrial kneading machines (see later). The two main types of cochlear-type mechanical stretchers are the inclined and horizontal screw models. Both work well, but the inclined type enables the curd to be stretched more thoroughly, although the curd may break up if it is too sour. The horizontal screw type stretches the curd less thoroughly but uses less water and a lower temperature and allows for the possibility of dry salting of the curd, even if it is very sour (Salvadori del Prato, 1993a).

10.2 Technology of Pasta Filata cheeses

10.2.1 *Mozzarella and soft Pasta Filata cheeses*

There are two main varieties of Mozzarella cheese in existence: the 'soft' Italian variety and the Pizza types. Production techniques are slightly different for each

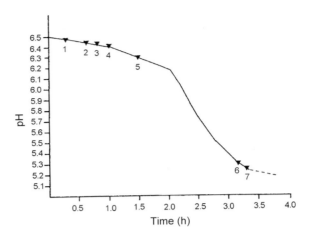

Figure 10.2 Acidification curve during the manufacture of Mozzarella cheese, with use of *Streptococcus thermophilus* as DVI starter culture. Note: whole milk (3.4 g 100 g^{-1} fat) is pasteurised and fermented at 37°C. Direct inoculation of starter into milk at 0 h; 1, addition of rennet; 2, first cutting of curd; 3, second cutting of curd; 4, extraction of whey; 5, beginning of maturation of curd on boards; 6, cutting and turning of blocks of curd; 7, stretching. Source: Stafanini (1991).

variety. Soft Mozzarella is produced by using whole pasteurised milk (with a protein:fat ratio of 0.88–0.92) which is soured by acidulants and coagulated at 35–38°C without further heating; draining is limited and processing of the curd is rapid with minimal or no salting. In contrast, Pizza cheese (66% of Mozzarella production in the USA; see Table 10.3) is usually obtained by using pasteurised milk which is partially skimmed, depending on the amount of fat required in the finished product. For example, a typical fat level in such cheese is 15–20 g 100 g^{-1} because at higher fat content the product 'oils off'. The milk fat is standardised to 1.7–2.3 g 100 g^{-1} to give a protein:fat ratio of 1.4–1.6. Coagulation occurs at lower temperatures than soft Mozzarella (30 to 35°C) and the curd is cooked up to 37–38°C. The curd is allowed to drain longer than

Table 10.3 Compositional standards of US Mozzarella cheeses

Type	Moisture, M (g 100 g^{-1})	FDM^a g 100 g^{-1}
Full fat	$52 < M < 60$	FDM > 45
Partly skimmed	$52 < M < 60$	$30 <$ FDM < 45
Low moisture	$45 < M < 52$	FDM > 45
Low moisture and partly skimmed	$45 < M < 52$	$30 <$ FDM < 45

^{a}Fat in dry matter.
Source: Anon. (1989).

is the 'soft' variety and, after spinning or during stretching, the cheese is salted (Larson and Olson, 1970; Kindsted, 1993).

10.2.1.1 Technology using natural milk flora for souring the milk
For both varieties of Mozzarella, the processing methods used in the cheese vat and later during the maturation period are very important. It is important to maintain the highest possible moisture level in the cheese when producing the 'soft' Italian, variety. Coagulation must, therefore, come mainly from the action of rennet, and the souring period of the milk in the vat must be shorter. This means that the milk will be coagulated at higher temperatures (35–38°C). The coagulum is cut into larger pieces (walnut size), compared with pizza cheese, after 35–40 min, and the curd pieces are left to rest under the whey until the acidity reaches 0.13–0.14 ml 100 ml^{-1} lactic acid (Salvadori del Prato, 1993a, 1998; Ghitti *et al.*, 1996). No heating is applied for soft Mozzarella, and the temperature rise must be restricted to a maximum of 2–4°C for Pizza cheese. The curd during the souring stage (i.e. on the draining tables) must be kept at a constant temperature to achieve the desirable level of acidity (higher pH for soft Mozzarella, lower pH for Pizza cheese). If the curd is kept soft it can be easily stretched even at a pH of 5.2. Curd which may have soured too quickly (less than 2 h on the draining tables) will have a fibrous structure which is too weak to stretch; curd which may have soured too slowly (more than 4 h on the draining tables) will contain less moisture and will produce Mozzarella which is hard and fibrous (Ghitti *et al.*, 1996; Salvadori del Prato, 1998).

The correct level of acidity of curd at time of stretching is essential in order to obtain a good Mozzarella; acidic curd will sequester the calcium, and the protein matrix will disintegrate, causing significant fat and protein losses. However, 'sweeter' curd proves difficult to stretch, and the cheese becomes hard, with significant fat losses in the whey (Lawrence *et al.*, 1987). Nevertheless, to keep the cheese soft the temperature of the stretching water must be as low as possible to ensure softening of the curd; this should be compatible with the type of machine used and the mechanical resistance of the curd fibres. It is important at this stage to control the composition and acidity of the white water: the fat level should be low, the protein content should be 1 g 100 g^{-1} or less and the water should not be too acidic (Table 10.4). The curd of 'soft' Mozzarella should not be salted in the machine during stretching because the cheese will become too dry and fibrous. The temperature of the Pasta cheese when it leaves the machine should not exceed 60–65°C. If the cheese is not sour enough, raising the temperature of the water by 2 or 3°C is recommended. If, on the other hand, the curd is too sour, the water temperature should be reduced.

10.2.1.2 Acidification of milk with use of lactic starter cultures
Choosing the right blend of starter cultures and the type of coagulant are very important in producing a good Mozzarella cheese. For the production of fresh

Table 10.4 Typical chemical composition of hot 'white' water circulating in the stretching machine during Mozzarella cheesemaking

Constitutent	Amount (g 100 g^{-1})
Water	97.50
Dry matter	2.30
Fat	0.90
Protein	0.80
Lactose	0.12
Ash	0.20

Source: Salvadori del Prato (1998).

Mozzarella balls, which are intended for consumption within a short period of time, a mixture of 50:50 of bovine rennet and microbial substitutes are suitable. They are used at different rates depending on the composition of the milk, but never exceed 30–35 ml 100 l^{-1} for standard rennet preparation (1:10 000). The use of starter cultures makes it possible to achieve controlled acid development during the manufacturing stages, and the cheese will have a distinctly better flavour. In addition, the production time is reduced, with a significant increase in the yield of the cheese, and the product has a longer shelf-life compared with a similar cheese made with use of the chemical acidification technique. A typical production process for Mozzarella cheese made with a starter culture is shown in Figure 10.3. This process is widely used at present in the industry and is based upon advanced technology together with the use of multipurpose vats and modern stretching and shaping machines.

The choice of starter culture blends is dependent on the type of Pasta Filata being made (Rossi and Leal, 1975; Stefanini, 1991; Addeo *et al.*, 1996). For example, for the production of fresh Mozzarella (i.e. shaped into balls) *Streptococcus thermophilus* is used as a single strain or, alternatively, a blend of *Lactococcus lactis* subsp. *lactis* and subsp. *cremoris* may be added. However, a yoghurt starter culture (*S. thermophilus* and *Lactobacillus delbrueckii* subsp. *bulgaricus*) is widely used for the production of Pizza cheese.

Starter cultures for Mozzarella or Pizza cheese may be used in form of DVI or as liquid. In both cases it is possible to obtain excellent products. When using the DVI culture, it is evident that there is no immediate rise in acidity in the milk, but souring of the curd will occur normally and more quickly than with a liquid culture. When processing the milk and the curd, soured by DVI starter in the cheese vat, it is recommended to extend the processing parameters (e.g. coagulation time, cutting, heating and/or stirring) to encourage drainage, which is only partially assisted by souring of the milk. The curd on the draining tables must be monitored closely to avoid fast acid development, especially when using mesophilic starter cultures.

Thermophilic starter cultures must be regulated in a different way depending on whether a pure culture of *S. thermophilus* (for fresh Mozzarella) is used

Pasteurise the milk and cool to 32–36°C (fat g 100 g⁻¹) for Pizza cheese, 1.8; for 'soft' cheese, 3.6
↓
Add starter culture
(1–2 ml 100 ml⁻¹ *Streptococcus thermophilus* and soured to 0.36 – 0.5% lactic acid)
↓
Coagulate at 32–36°C for 20–30 min with 30 ml liquid rennet
↓
Cut the coagulum
First cut: cut large cubes (4 cm³) and rest for 10–15 min
Second cut: cut walnut-sized pieces
↓
Remove 30–50% of the whey in 10 min
↓
Mature curd in the whey (traditional method), or

Transfer curd to draining table (pH 6.0–6.2)
↓
Cut and turn the blocks of curd at ambient temperature (20–26°C) for 2–4 h
or until pH reaches 4.9–5.2
↓
Stretch using hot water (72–82°C)
↓
Cool using cold water (8–10°C) for 30–40 min
↓
Brine (12–14 g 100 g⁻¹ salt) for 30 min to 1 h (no brining for 'soft' or mild varieties)
↓
Package, cold store and dispatch

Figure 10.3 Flow chart of the manufacturing stages of Mozzarella cheese with use of starter cultures. Source: Salvadori del Prato (1993a).

or whether a combined starter culture of *S. thermophilus* and *Lb. delbrueckii* subsp. *bulgaricus* (for Pizza cheese) is used. In the former case, the incubation temperature of the bulk culture is 42–44°C, and the final acidity is around 0.6–0.9 ml 100 ml^{-1} lactic acid. In the latter case, it is best to grow the culture at 44°C–45°C, and the acidity should be at 0.8–1.0 ml 100 ml^{-1} lactic acid. Inoculation of the milk, for the fresh variety, is usually 1 ml 100 ml^{-1} or enough to raise the acidity of the milk in the cheese vat by around $0.005\,\text{ml}\,100\,\text{ml}^{-1}$ lactic acid; however, for Pizza cheese a $1.5\,\text{ml}\,100\,\text{ml}^{-1}$ inoculum is used and the anticipated rise in acidity of the milk is 0.01 vol% lactic acid (Salvadori del Prato, 1993a, 1998; Ghitti et al., 1996). Such information on starter cultures is important during the traditional production method of Mozzarella and Pizza cheese. This method is still in use today, but the dairy sector has witnessed major technological innovations including the chemical acidification of milk.

10.2.1.3 Souring the milk with added acidulants

This technique of cheesemaking was first developed in the USA for the production of Mozzarella cheese using inorganic or organic acids to sour the milk (Breene et al., 1964; Larson et al., 1967; Olson, 1970). The process has spread rapidly around the world because of its practicality and the good results which can be achieved in terms of the keeping qualities of the product. However, use of a microbial rennet substitute combined with chemical acidification of the milk lowers the yields and gives rise to the development of a bitter taste (Carini and Todesco, 1974; Corradini, 1977). The principle of this method in cheesemaking is based upon replacing the slow and natural fermentation of the milk (i.e. on achieving sufficient demineralisation of the curd to increase its malleability so it can be stretched) by instant souring to a level similar to those attainable with use of the traditional process. The organic acids that have been used are citric, lactic, acetic and mineral acids; the preferred choice is citric acid diluted to $10\,\text{ml}\,100\,\text{ml}^{-1}$ concentration before addition to milk. The advantages of this procedure are: first, to eliminate the prolonged time caused by the slow souring action of starter cultures and, second, to obtain a product with improved keeping quality, acceptable organoleptic properties and standardised structural properties. A flow diagram of cheesemaking with use of citric acid is shown in Figure 10.4.

In many countries, including Italy, it is obligatory to package the Mozzarella cheese for direct consumption in waterproof plastic bags containing liquid with preservatives, for example, salt, sorbates, hypochlorite (Borin and Emaldi, 1987; Salvadori del Prato, 1998). When Mozzarella is produced with use of starter cultures the preserving fluid usually consists of the water coming out of the stretching machine, which may still contain high residual counts of starter culture microorganisms so that the product preserves well. When chemical acidification of the milk is used, the product may lack protection; thus it becomes necessary to use preservatives to extend the shelf-life of the Mozzarella without presenting a health risk.

10.2.1.4 Drainage of the curd

With Pizza cheese, the draining of the curd must be extended to obtain a more compact and durable curd which is also more acidic. Hence, processing in the vat is longer and the cooking temperature is increased by $2-4°C$, partly because the coagulation of the milk has occurred at lower temperatures (30–35°C). Cutting

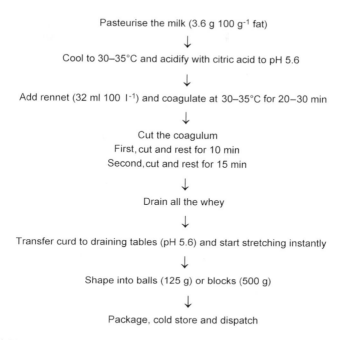

Figure 10.4 Flow chart of the manufacturing stages of Mozzarella cheese with use of chemical acidulants (citric acid). Source: Salvardori del Prato (1993a)

time is also longer, and the curd may be left in the whey for a longer time to develop more acid.

Souring of the curd on the draining tables to pH 5.0 may take up to 4 h and is markedly different when compared with the process employed for the 'soft' variety, and the curd will be firmer.

10.2.1.5 Salting of the cheese

Cooling the Mozzarella when it leaves the shaping machine (i.e. as balls) must be done as quickly as possible; slow cooling causes a grainy and chalky core in the middle of the cheese. It is therefore best to use refrigerated water (Mongiello, 1984). If the balls of Mozzarella have to be salted, this should be carried out by using a cool brine (maximum temperature of 12–14°C) and a salt concentration of of ca. 15–17 g $100\,g^{-1}$ NaCl. Immersion of the cheese in the brine should be rapid to allow just sufficient uptake of salt; for example, fresh Mozzarella normally contains ≤ 1 g $100\,g^{-1}$ of salt or less. The product is then ready for packaging. If the brine is highly concentrated with salt or the cheese is immersed for too long, apart from causing loss of weight, the system creates a rind on the surface of the cheese which is hard and excessively salty; however, the inside of the cheese will often remain insipid (Addeo et al., 1996; Ghitti et al., 1996).

The outside portion of the cheese also tends to form a scum and, if the salting process is prolonged further in a diluted brine, the Mozzarella becomes slimy on the outside and 'mushy' in texture.

Pizza cheese contains on average 1.5–1.7 g 100 g^{-1} salt in the final product. Above or below such salt levels the melting properties of the cheese on top of pizzas will be adversely affected (Olson, 1979; Nilson, 1982; Mongiello, 1984).

10.2.2 Provolone and hard Pasta Filata cheeses

Provolone is the most representative type of hard Pasta Filata cheese. It is made from whole cows' milk and is soured by the microflora naturally present in the milk. It has a smooth consistent skin with a white creamy colour, has few or no holes and is typically bound with ropes. It has a sharp and tangy flavour which is the result of the lipolytic activity of the lactic starter cultures and enzymes originating from the rennet (e.g. rennet paste derived from goat or lamb stomachs; Salvadori del Prato and Messina, 1991).

Hard Pasta Filata cheeses (Provolone, Caciocavallo or Provole) are produced from raw milk (if the cheese is matured more than 60 days) or pasteurised milk inoculated with a thermophilic lactic starter with high souring power (Bottazzi and Battistotti, 1986). Researchers have shown that Provolone's natural microflora is dominated by *Lactobacillus helveticus*, *Lb. delbrueckii* subsp. *bulgaricus*, *S. thermophilus* and *Lactobacillus delbrueckii* subsp. *lactis* (Gatti *et al.*, 1993). These species are able to grow at a temperature between 32 and 53°C, with optimum growth at 40–50°C, and produce great quantities of lactic acid. When a single species of starter culture is used, or even when a blend of several strains is used, it becomes difficult to obtain a close-texture cheese because the fibrous mass of the stretched curd is unable to form a proper skin or rind. This is especially true when *Lb. delbrueckii* subsp. *lactis* is used. Moreover, there are significant differences between starter cultures with respect to fat retention in the curd during stretching. For example, *Lb. delbrueckii* subsp. *bulgaricus* creates more favourable conditions than *Lb. delbrueckii* subsp. *lactis* or *S. thermophilus*. Comparative tests with the same milk and using equal quantities of the same rennet have demonstrated that the time required to liberate butyric acid during ripening of the cheese is regular and quantitatively significant when mixed starter cultures are used whereas a shorter time is required when *S. thermophilus* is used alone. Similar observations were reported when the quantity of free glutamic acid was used as an index for protein degradation. It is therefore recommended that starter cultures for Provolone cheese consist of a mixture of cocci and rods using several species of lactic acid bacteria at a ratio of 1:1 if the culture is composed of just two species, or 1:2 in favour of the rods if the culture is composed of three or more species.

For this type of cheese, the lipid content of the milk and curd is not only essential to guarantee good body and close texture characteristics, but also for the formation of a sharp flavour due to the liberation of free fatty acids from the milk triglycerides during ripening. In addition, the lipolytic activity of the coagulant is important to develop the desirable flavour characteristic of Provolone cheese. However, in order to achieve such taste in the cheese, rennet paste is used at a rate of 40–50 g $100\,l^{-1}$ of milk, and lipase powder is used at a rate between 5 and 15 g $100\,l^{-1}$ of milk (Christensen, 1965; Harper, 1965; Shahani et al., 1971; Corradini, 1973; Moskowitz, 1980; Salvadori del Prato, 1993b). Liquid vell rennet is used at a rate of 20–30 ml $100\,l^{-1}$ of milk for the production of sweet Provolone or Caciocavallo cheeses.

10.2.2.1 Traditional method of production

The raw milk is ripened with a whey starter culture (1.5–2 ml $100\,ml^{-1}$) and set in the vat at 35–38°C; the total coagulation time is around 20 min (Salvadori del Prato and Messina, 1991). Once sufficient consistency of the coagulum has been reached, the curd is cut into hazelnut size pieces in ca. 10–15 min. After cutting, the curd is then stirred in the vat for several minutes to help whey expulsion, followed by removal of two thirds of the whey, which is heated to 72°C and added back to the cheese vat to cook the curd (the amount of whey used is 30 l $100\,l^{-1}$ of milk used). This is done in two stages to increase the temperature of the curd/whey mixture gradually to 44°C and then up to 53°C, respectively; inbetween each stage, some whey is removed, reheated and added back to the cheese vat. All these stages are carried out under constant, high-speed stirring of the curd/whey mixture.

After the cooking stage, the curd and whey is poured onto a draining table where further acidification takes place and light pressure is applied to the curd mass. Acidification of the curd takes place with the temperature maintained at between 40 and 50°C (maintaining the room temperature at 22–26°C) to achieve the sharp-flavoured cheese; slightly lower temperatures are used for the milder varieties. The acidification period may require up to 15 h or more, but on average is only 3–4 h (pH range 4.9–5.2) to avoid excessive drying of the curd (Lawrence et al., 1983; Salvadori del Prato and Messina, 1991). Stretching takes place in machines and under similar working conditions to those of Mozzarella, but with a higher water temperature and lower mechanical speeds. The stretched mass must be malleable and of silky consistency, and perfectly stretched fibres must be capable of being separated, with use of the fingers, into very fine stretchable and coherent filaments. This is known in Italy as 'spiderwebs', and Figure 10.5 illustrates the physical characteristics of the stretched curd.

The stretched curd is then moulded into different shapes, allowed to rest at room temperature for a few minutes and then cooled by immersion in cold water to harden; this is followed by brining (20–24 g $100\,g^{-1}$ salt; 10–12°C) for 8–10 h per kilogram of cheese. The salt content in the final product will vary from 1.5 to 3.0 g $100\,g^{-1}$ (Salvadori del Prato and Messina, 1991; Salvadori del Prato, 1998). After brining, the cheeses are usually tied with ropes or are caged

Figure 10.5 Illustration of the 'spiderweb' formation of a typical stretched curd during the manufacture of Provolone cheese. Source: Salvadori del Prato and Messina (1991).

in nylon net, and hung to dry. During the drying and ageing process the cheeses are hung vertically, suspended by the ropes or nets, and are hung side-by-side in a manner which is a distintive characteristic of these cheeses.

Once salted, tied and dried the Provolone cheese is first kept in hot chamber at 22–28°C for around 24–48 h; this process is known as 'stewing'. Here the cheese 'sweats', and fat coats its rind to give the cheese its characteristic creamy yellow colour. Afterwards, the cheese is moved to ageing chambers at 12–14°C to mature (i.e. for 2–3 months if consumed fresh or more than 6 months for sharp cheese). In the maturation rooms, pairs of cheeses are hung together on special racks. Forced air ventilation is avoided to prevent the rind from wrinkling, despite the fact that good air circulation helps to prevent mould growth and optimises the maturation process. The relative humidity of the stewing room is less than 80%, and it should vary according to ageing stage; the relative humidity is increased to more than 85% in the ageing stage to minimise moisture loss from the cheese as a result of evaporation (Salvadori del Prato and Messina, 1991; Salvadori del Prato 1998).

10.2.2.2 Mechanised Provolone cheesemaking

Among the most important process parameters, where modifications have been introduced compared with the traditional methods, are:

- substitution of rennet paste with lipase powders (2–15 g 100 l^{-1})
- monophase cooking at low temperatures rather than in two-stages
- an increasing in the moisture content of the curd by minor draining of the whey during the cooking process in the cheese vat
- different maturation methods, including types of packaging

At present, the tendency is to make high-moisture Provolone in order to accelerate its maturation. Therefore, the cooking of the curd is achieved by adding water or whey only once, and the final temperature of 48°C is reached in

20–25 min. This method of cooking, combined with faster working of the curd, is done by using slow rather than high-speed stirring in order to produce high moisture and softer cheese. Furthermore, accelerated maturation of the cheese is achieved by (Salvadori del Prato and Messina, 1991; Salvadori del Prato, 1998):

- the addition of lipolytic enzymes
- ageing the cheese at higher temperatures (15–18°C) and with higher relative humidity for 4–8 weeks, followed by eight or more weeks at lower temperatures (10°C) and lower relative humidity

Another important technological development is fast cooling of the cheese in baths using continuous cold-water circulation (10–14°C). Cooling after stretching must be done as quickly as possible, especially for big cheeses, to avoid any defects in the product. An efficient cooling process should cool the centre of a 30 kg cheese to less than 20°C in less than 30 min. Also, Provolone cheese may be wrapped in a permeable or semipermeable, plastic coating material (based on polyvinyl acetate) or in multilayer plastic bags to allow good maturation of the product (Giroletti and Bartels, 1993).

10.3 Mechanisation and control of Pasta Filata cheese production

Mechanisation of the production of Pasta Filata cheese has made a positive contribution to the technology of manufacture while maintaining the typical artisanal characteristics of the products and improving their hygienic quality; it has also helped indirectly to increase dramatically world production figures (Kindsted, 1993). Fully automated lines, which are capable of handling large volumes of milk, are now available for the manufacture of soft and hard Pasta Filata cheese. The production lines come in different sizes and with different engineering designs, but all are based on the traditional stages of manufacture of Pasta Filata cheese. Mechanisation has developed primarily with respect to the:

- use of specially designed cheese vats
- stretching of the curd
- forming and moulding of the stretched curd
- cooling of the cheese
- salting
- packaging

10.3.1 Coagulators or cheese vats

Successful production of Pasta Filata cheese has been achieved by using a specially designed cheese vat with a revolving cradle in which the curd is cut, purged of some of the whey and left until it has reached the pH necessary for stretching. At this point, the curd can be collected by capsizing the vat over a draining table or a curd maturator (Figure 10.6) (Anon., 1995). Such vats have distinct advantages such as:

- operation under highly hygienic conditions
- suitability for cleaning-in-place (CIP)
- low energy requirement
- continuous control of pH of the milk and curd with use of sensors
- all the operations in the vat, including the connections of pH sensors, can be programmed in a programmable logic controller (PLC) (see Chapter 1)

The coagulator (i.e. the cheese vat) consists of a curd-setting unit which also contains a special stirring device and slanting knives. The cut is obtained by rotating the knives to produce a three-dimensional cut. The curd or whey is drained by reversing the rotation of the knives and is then pushed out of the coagulator through a butterfly valve located on the bottom of the vat. The vat is emptied completely when the rear part of the vat is automatically lifted toward the outlet. The complete processing cycle can be controlled either automatically or manually. The discharged curd is acidified on a table where it is manually cut to feed the cutter (Figure 10.7) which later feeds the curd to the stretcher machine. There are many different models of curd cutters, but the most common type has a funnel of stainless steel which supports a motor that drives a horizontal rotating blade; the curd is fed into the upper part of the funnel and drops to the bottom where the blades cut it into small pieces. Incidentally, the Alpma continuous coagulator can also be used for the production of Pasta Filata cheese;

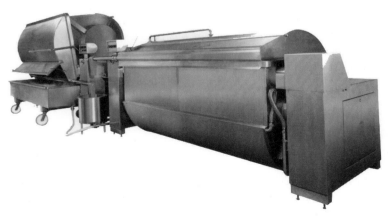

Figure 10.6 A Pasta Filata cheese coagulator filled with a revolving cradle and a maturator. Reproduced by courtesy of CMT SpA, Peveragno, Italy.

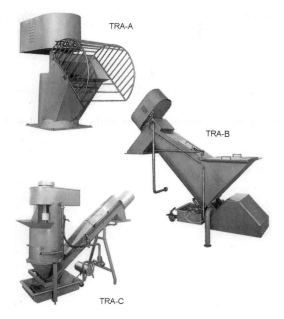

Figure 10.7 Three different types of matured-curd cutters which are used in Pasta Filata cheesemaking. Reproduced by courtesy of CMT SpA, Peveragno, Italy.

this system will not be reviewed in this chapter, but for further details refer to Chapter 9.

In some modern equipment, the curd, which is acidified by lactic starter cultures, is discharged into specials tanks called 'maturators'. These large metal tanks are made of sections of 'cradles' capable of containing all the curd from the cheese vat plus part of the whey. The cradle slowly tilts on its axis like a pendulum, and the curd is kept warm and wet in the whey as it moves sideways. This operation lasts until the curd has matured and then, at the end of the cradle, two augers convey the matured curd to the cutting device and moulding plant (see Figure 10.6).

10.3.2 Filatrici and moulding machines

The curd is stretched in hot water in a special machine called a *filatrice*, which has two screws that are juxtapositioned and rotate in opposite directions. Other types of stretching machines, known as 'diving arms', resemble industrial kneading or stirring machines; the mechanical arms are controlled by epicyclic gearing and move in opposite directions, cutting across each other and picking up the softened curd from the bottom of the container. Sometimes the screw-type and 'diving-arms'-type of the stretcher are combined in one machine to obtain a softer curd (Figure 10.8).

Figure 10.8 A modern stretching and moulding machine for the production of Provolone cheese. Reproduced by courtesy of CMT SpA, Peveragno, Italy.

As mentioned earlier, the main two types of cochleate mechanical stretchers are the inclined and the horizontal screw models. In a modern combined stretching and moulding machine, the curd is loaded into the machine after cutting by the pasta cutter, which is placed upstream. The product is then pre-stretched into the feeding shoot tanks, which are located towards the outer rotary movement of the two augers. The augers convey the curd in a round mixing vat with a rotary paddle, which stretches the cheese and gives it a 'whisked' structure. In both sections, i.e. in the feeding tunnel and in the mixing vat, there are inlets for hot water, with adjustable capacity, an exhaust outlet to remove the fat-rich water through waste pipes, and movable pipes that allow the operator to regulate the level of liquid in the machine. The two feeding augers are extended for a short distance out of the circular mixing vat in order to push the stretched pasta through a Teflon-coated pipe that feeds the moulding unit. Since the curd enters the machine at about 30°C and must be raised to 60–70°C (with a temperature difference of about 35°C), the proportion of curd to water (if the hot water temperature is at 80–90°C) must be around 1:3–1:4 (Addeo *et al.*, 1996). The quantity of water added can be regulated automatically by using a temperature sensor.

During the stretching operation in the *filatrice*, the curd can be salted in two ways: first, by distributing dry salt on the stretched curd (the quantity of salt is automatically regulated by the speed of the advancing curd), or second, by adding salt to the hot water directly (the quantity of salt is regulated continuously by salt sensors inserted into the water tank). The former system of salting is more precise because the dosing of the salt is normally made by a special machine (not illustrated), made by double truncated cone which contains the salt and is kept in constant movement by the continuous rotation of the tank. Inside the tank, there are several spoon-shaped units which take the salt and place it inside a conveyor tube fitted with a distributor at the end. A special chute for product expansion is assembled at the outlet manifold from the stretching machine. At this stage, the salt is distributed according to the speed of the curd, which in turn is controlled mechanically by the rotation of the cylinder positioned on the chute; the rotation cylinder also regulates the quantity of the salt distributed. The system is very accurate and allows precise dosing of the salt in the curd. The curd is then conveyed to the kneading unit consisting of an auger within a sloping and jacketed tunnel. In this part of the machine, the Pasta Filata cheese is mixed with salt and conveyed forwards to the moulding unit.

Options of automation controls during the stretching operation are:

- the temperatures of the water and curd
- the pH of the water
- the salt level in the curd and/or in the water

The stretched curd can be mechanically shaped into different forms depending on the type of cheese. The machine is known as a moulding or forming unit, and some examples of the final shapes are: ball-shaped for Mozzarella, brick-like for Pizza cheese, sausage-like or oblong for Provolone, and aubergine-like for Caciocavallo. A moulding machine, which is used during the manufacture of Caciocavallo cheese, is shown in Figure 10.9, and a complete automated line for Pizza cheese is illustrated in Figure 10.10. Basically, the moulding machine for Pizza cheese consists of three main sections assembled together to make up one unit; a curd feeding and advancement system, a moulding section and a system for removing the product from the moulds. The curd is loaded into the hopper to feed the moulding section nonstop with use of an auger conveyor. At the end of the augers, the curd passes through the desired 'shape' distributor made of polyethylene and fills the moulds which are mounted on a rotary carousel with vertical axle containing a series of cylindrical moulds that are welded on the spokes of a wheel (Figure 10.10). After moulding, the product undergoes cooling and pre-hardening by passing the blocks of cheese through a shower of cold water. The moulding carousel turns intermittently so that the curd is delivered into each mould on the lower part of the machine and is then turned over by a pneumatic system after making a complete circle through the pre-hardening section. A self-priming pump circulates the water between the carousel and the plate heat exchanger (PHE) that cools the water. A level regulatory system is installed in the moulding unit to produce brick-shaped Mozzarella of different heights or sizes, but using the same moulds on the carousel.

10.3.3 Hardening and brining

The moulded form of Pasta Filata cheeses must be hardened as soon as possible in order to retain their shape. This is achieved by dropping the moulded cheese

Figure 10.9 An automatic moulding machine for Caciocavallo cheese. Reproduced by courtesy of CMT SpA, Peveragno, Italy.

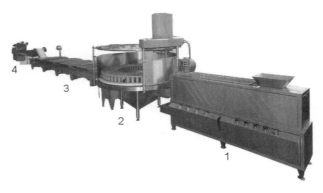

Figure 10.10 CMT production line for the manufacture of brick-shaped Pizza cheese. 1, Stretching machine; 2, mould former; 3, hardening–salting tunnel; 4, packaging machine. Reproduced by courtesy of CMT SpA, Peveragno, Italy.

into cold water in special vessels called hardening vats (see item 3, Figure 10.10). The length of the vat is calculated on the basis of daily output and the type and size of cheese being made (Mozzarella balls, aubergine-shaped Caciocavalli, Pizza cheese bricks). The product from the moulding machine is delivered to the hardening vat through a loading hopper inside the vat. A series of floating 'carpets', made of stainless steel, are activated by a connecting rod and a crank shaft mechanism that moves the product forward. Also, a series of water jets coming out from nozzles in pipes assembled along the side of the vat assists in moving the cheeses. At the end of the hardening vat, the product is automatically conveyed onto a stainless steel or plastic belt which is driven by a geared motor

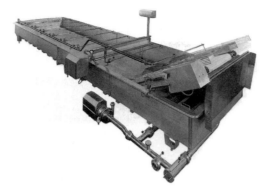

Figure 10.11 A brining vat fitted with 'jetting' nozzles to move the Pasta Filata cheese forward. Reproduced by courtesy of CMT SpA, Peveragno, Italy.

unit. The centrifugal pump, which recirculates the cooling water to the vat, is also used during the CIP cycle.

After the hardening, the Pasta Filata cheeses are usually salted by submersion in brine (if they have not been dry salted during stretching). A mechanised salting vat, suitable for Mozzarella balls and Pizza cheese bricks (Figure 10.11) is made of stainless steel and its length and capacity can be adjusted (i.e. by connecting more sections together to extend its length) to suit production requirements; water jets from nozzles assembled in a series of pipes placed along the vat advance the cheese. At the end of the salting process, the product is automatically conveyed out of the machine on a stainless steel or plastic belt; the excess brine drips off the cheese which is then ready for packaging.

10.3.4 Packaging

Fresh Pasta Filata cheeses, such as Mozzarella balls or Pizza cheese bricks, are sold packed in plastic pouches, whereas Caciocavalli and Provoloni are generally sold as individual whole cheeses ready to be cut and packed in the store. Form–fill–seal packaging machines are used to make pouches by heating the flat plastic sheet; the sheet is sealed longitudinally to form a tube and transversely to provide the bottom closure. The pouch is filled with a Mozzarella ball and a metered quantity of 'preserving' liquid; it is then transported downwards a length. At the next stroke, transverse sealing jaws close the top of the filled pouch and, at the same time, the bottom seam of the next pouch is made; the two pouches are separated by a cutting knife (Figure 10.12a). For Pizza cheese, packaging in pouches is usually done under vacuum using a continuous form–fill–seal machine; a salami-shaped cheese can be formed by extruding the pasta into a plastic tube before sealing (Figure 10.12b).

(a)

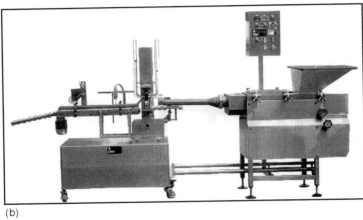

(b)

Figure 10.12 Packaging machines for (a) Mozzarella balls and (b) salami-shaped Pizza cheese. Reproduced by courtesy of PFM SpA, Vicenza, Italy and of DIMA SpA, Modena, Italy.

10.3.5 Miscellaneous systems

Ultrafiltration (UF) technology has been used in the USA since the late 1980s for the production of Mozzarella cheese, and one such system has been used commercially by Ridgeview Food (Elliott, 1984). The system consists of a series of plastic vertical cylinders with a capacity of around 70 l each; each cylinder is filled with UF skimmed milk which is mixed with cream, starter culture and rennet. The cylinders move forward within an insulated chamber at constant temperature to a discharge station. The coagulated milk is removed by revolving the cylinders and pouring the coagulum into the cutting machine. The curd is cut into cubes, transferred to draining tables for souring and is subsequently stretched. However, the results obtained with the UF technique for Mozzarella cheesemaking are poor because of the poor melting property of the cheese

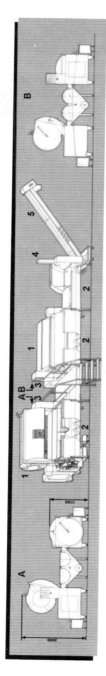

Figure 10.13 CMT mechanised production line of Mozzarella cheese. 1, Cheese vat; 2, trough; 3, control panels; 4, guillotine; 5, automatic auger or feed conveyor. Reproduced by courtesy of CMT SpA, Peverango, Italy.

(Ernstrom and Anis, 1985; Lawrence, 1989; Hooker, 1990), and this method of manufacture is not widely used for the production of Pasta Filata cheese.

Today, fully automatic lines for Pasta Filata cheeses are made by many equipment suppliers throughout the world including for the use of chemical acidification or microbial fermentation of the milk and curd.

A typical line for the continuous production of Pasta Filata is illustrated in Figure 10.13; the curd is matured at a controlled temperature in four, continuously filling vats (each 6 500 l). It is then poured into a whey draining vessel through a rotating grill before being moved forward by augers towards a guillotine blade and into an automatic loader for stretching. Depending on the processing method used, the system can yield from 1000 to 2000 kg of curd per hour, with a minimum requirement of 2 or 3 people to operate the system.

10.4 Quality control of Pasta Filata cheese processing

10.4.1 Rheological properties

Most Mozzarella and Pizza cheeses are used for cooking purposes and it is important, therefore, to assess the melting properties of the cheese subjectively as pizza topping baked under commercial conditions, or in an oven in the laboratory. Several methods are used to evaluate the meltability of the product, including heating a standardised cylindrical plug of cheese in a oven or in a boiling water bath under defined conditions (Arnott *et al.*, 1957; Breene *et al.*, 1964; Olson and Neu, 1964; Kosikowski and Mistry, 1997). Meltability can be expressed as a function either of decreased height or of increased area of the cheese sample. Another meltability test which could be applied to Mozzarella cheese involves melting the sample in a test tube and measuring the distance over which it flows (Olson and Price, 1958). Furthermore, Olson and Nelson (1980) developed a melting test to measure the behaviour based on the tendency of a viscoelastic material to climb up a rotating rod; a modified version of this method involved replacing the rod by a helical screw (Kindsted, 1993).

Other versions of meltability tests have used other parameters, and one such example is free oil formation in melted Mozzarella (Breene *et al.*, 1964; Nilson and LaClair, 1976). Many researchers have characterised the reheological behaviour of Mozzarella by using the Instron Universal Testing Machine, where the consistency of good Mozzarella shows a high degree of elasticity and a low level of hardness when subjected to compression response under various test conditions. Chen *et al.* (1979) and Lee *et al.* (1978) reported strong correlations between rheological measurements by the Instrom test and the sensory assessment of the product.

Moisture and fat in dry matter (FDM) contents in low-moisture Mozzarella cheese affect hardness and chewiness. In chemically acidified Mozzarella, the moisture content is inversely related to the firmness of the cheese (Keller *et al.*,

1974). Cervantes *et al.* (1983) investigated the effect of salt content on the texture of Mozzarella. Firmness was not significantly affected by the salt level in fresh cheese, however, after a month in cold store, high salt content (2–4 g $100\,g^{-1}$) was firmer than low salt content (0.3 g $100\,g^{-1}$) Mozzarella. Casiraghi *et al.* (1985) and Masi and Addeo (1986) have demonstrated how cheese firmness decreases with increasing ratio of FDM: non-FDM with increasing moisture content or with decreasing pH level (which demineralises the pasta).

Determining the right moment for stretching the curd during production is crucial for obtaining Mozzarella of acceptable organoleptic characteristics and good melting behaviour when cooled on pizza. Traditionally, the decision to stretch the curd was determined through a manual test of gradually stretching a piece of heated curd in hot water. At present, special laboratory instruments (Figure 10.14) are used to evaluate two typical properties of the curd (i.e. stretching and spreading of the melted curd; Cavella *et al.*, 1992; Addeo *et al.*, 1996). Samples of Pasta Filata cheese are tested for extrusion and stretching; the temperature and stretching ratio has been calculated by changing the fusion temperature of the curd at point P (see Figure 10.14) to give different lengths of curd and different resistance in the curd. Figure 10.15 illustrates a typical performance curve for curd at different fusion temperatures. An increase in fusion temperature tends to increase the force required to break the fibres; maximum force is recorded at 70°C. At higher temperatures, a lower force is needed to break the fibres (Cavella *et al.*, 1993; Addeo *et al.*, 1996).

Variations in the stretching capability of the curd at different temperatures are dependent on the melting point of the curd (Figure 10.16; Cavella *et al.*, 1993; Addeo *et al.*, 1996). The extensibility of the melted curd fibres increases from a minimum of 120% at 63°C to a maximum of 220% at 72°C, (i.e. it approximately doubles over a temperature range of 10°C).

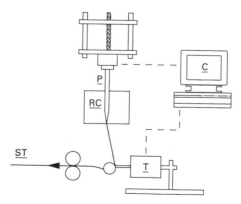

Figure 10.14 Schematic drawing of a laboratory instrument for measuring the stretchability of curd. P, piston; RC, rheometer; C, computer; T, transducer; ST, traction system. Source: Addeo *et al.* (1996).

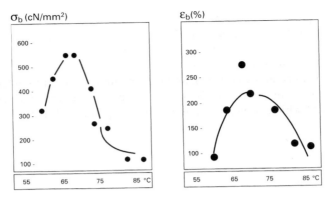

Figure 10.15 (a) Force required (σ_b) to break the curd and (b) percentage of elongation (ε_b) as a function of fusion temperature. Source: Addeo *et al.* (1996).

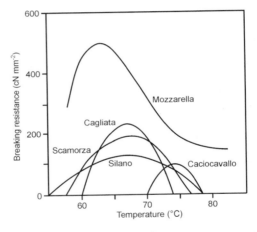

Figure 10.16 Variation in breaking resistance (cN/mm^2) of the stretched curd for various cheeses, as a function of temperature. Source: Addeo *et al.* (1996).

An additional problem in the evaluation of processing technologies capable of giving Pasta Filata of optimal quality is the correlation between the functional properties of the curd (i.e. fusion and stretching) and the moisture and fat content of the curd. Mozzarella made from whole milk or partially skimmed milk, as long as it contains a high percentage of moisture (more than 52 g 100 g^{-1}) is considered a soft product. It has a shorter shelf-life and is not always suitable for pizza topping as it is difficult to shred and the shredded particles tend to stick together. Mozzarella made from partially skimmed milk with reduced moisture content (ca. 47 g 100 g^{-1}) is more suitable for pizzas. All Pasta Filata products require a minimum temperature of fusion ca. 55°C; however, the resistance of the fibres to stretch decreases with increasing temperature of fusion, which varies according to the type of cheese (see Table 10.5). Such data suggest that the highest temperatures of fusion are found in cheeses with low moisture contents, and vice versa, and that the fibres in high moisture cheeses have the highest tensile strength. Therefore, each product has a temperature range beyond which stretching cannot properly proceed (Cavella et al., 1993; Addeo et al., 1996). Based on these results, it is safe to conclude that, by increasing the moisture content in the cheese, continuous fibres tend to form at lower temperatures. Such physical effects of the curd reflect the ability of nonbiological polymers in water to plastify and act as a lubricant between the protein molecules. The stretchability of the curd is also dependent on the pH level, but not on the moisture level, which helps to form long and resistant fibres. Similar tests can give good results regarding the minimum temperature recommended for the stretchability of the curd of each variety (Table 10.6).

10.4.2 Microstructure

The microstructure of Pasta Filata in relation to its malleablity has been studied by many researchers (Kalab, 1977; Altiero et al., 1984; Masi and Addeo, 1986; Paquet and Kalab, 1988; Addeo et al., 1996). The overall structure of the curd is granular. During the stretching stage, the structure consists of parallel fibres

Table 10.5 Temperature ranges (ΔT) for stretching different Pasta Filata cheeses

Cheese variety	Temperature(°C)		ΔT (°C)
	minimum	maximum	
Curd	63	73	10
Caciocavallo	70	77	7
Mozzarella	58	83	25
Scamorza	59	76	17

Source: Addeo et al. (1996).

Table 10.6 Optimum temperature for stretching the curd fibres, ε_{max}, in different Pasta Filata cheeses

Cheese variety	Temperature (°C)	ε_{max} (%)
Curd	68	173
Caciocavallo	74	165
Mozzarella	70	217
Scamorza	68	114

Source: Addeo et al. (1996).

containing aggregates of fat globules. The fibrous structure of the curd does not become evident until the pH is 5.8 or less (Addeo *et al.*, 1996). The gradual increase in the micellar strength and number of bonds that causes contraction of the casein aggregates to expel the interstitial whey is dependent on pH level and the decrease in colloidal calcium phosphate. Some of the residual moisture is distributed within the structure of the protein matrix, which is a typical property of the structure of the cheese curd.

According to Kimura *et al.* (1992), when milk coagulates the caseine micelles aggregate to form a continuous matrix of protein. Such a protein matrix holds within it the fat globules, free moisture and starter culture bacteria (Addeo *et al.*, 1995). When the coagulum is cut, the release of whey causes the micellar structure to become more compact and the micelles tend to fuse or interact together to form a denser core. However, on heating and stretching the curd, the casein matrix changes to a fibrous structure, which also holds within it the whey, fat globules and microorganisms. The microstructure of the curd, visualised by a scanning electron microscope, reveals a 'beehive' structure or large void spaces, when the milk has been chemically acidified; this is in contrast to the fibrous structure with very few void spaces seen in the protein matrix when microbial fermentation is used (Addeo *et al.*, 1996). Furthermore, scanning electron microscopy makes it possible to distinguish the differences in the microstructure of the curd which result from the use of different stretching machines. For example, in Pizza cheese, vertical stretchers give well-formed fibres whereas horizontal stretchers form less well orientated fibres (Addeo *et al.*, 1996).

10.4.3 Hazard analysis critical control points

Pathogenic microorganisms have been isolated from Mozzarella cheese (Irvine, 1978; Donnelly, 1990; Eckner *et al.*, 1990; Persano *et al.*, 1995), and current European Union (EU) regulations have prescribed the application of hazard analysis critical control points (HACCPs) in Pasta Filata production. These regulations and HACCP control of Mozzarella production have been recently reviewed by Marino *et al.* (1995). The critical controls points (CCPs) for a mechanised production line can be summarised as follows:

- CCP 1, the milk after storage and pasteurisation
- CCP 2, the milk in the cheese vat after the addition of citric acid or starter culture and rennet
- CCP 3, the curd before the stretching stage
- CCP 4, the curd after stretching and moulding
- CCP 5, the cheese after packaging

It is safe to conclude that proper control at the above-mentioned manufacturing stages will ensure the microbiological safety of Pasta Filata products.

References

Addeo, F., Emaldi, G.C. and Masi, P. (1995) Tradizione e innovazione nella produzione della Mozzarella di bufala campana. *Bubalus bubalis*, **3** 46-62 (in Italian).
Addeo, F., Chianese, L., Masi, P., Mucchetti, G., Neviani, E., Schiavi, C., Scudiero, A. and Strozzi, G.P. (1996) *La Mozzarella*. Morfin, Novara (in Italian).
Altiero, V., Masi, P. and Addeo, F. (1984) Influenza dell'acidificazione della cagliata al momento della filatura sulla qualità e sulla struttura della Mozzarella di bufala. *Latte*, **9** 764-70 (in Italian).
Ammirati, L. (1996) *Il Formaggio Mozzarella e le sue Origini*, Thesis, Dairy and Cheese Technology School of Lodi, Italy (in Italian).
Anon. (1989) *Code of Federal Regulations*, parts 100–69, US Department of Health and Human Services, Washington, DC.
Anon. (1995) Nuovi orizzonti per la produzione della cagliata in continuo. *Ind. Alimentari*, **7**(34) 18-20 (in Italian).
Arnott, D.R., Morris, H.A. and Combs, W.B. (1957) Meltability test for Pizza cheese. *J. Dairy Science*, **40** 957-62.
Borin, F. and Emaldi, G.C. (1987) Problemi technologici e di conservazione del formaggio 'Fior di Latte' prodoctto con acidi organici. *Latte*, **12** 126-33 (in Italian).
Bottazzi, V. and Battistotti, B. (1986) Fermentazione lattica e demineralizzazione della pasta per Provolone. *Sci. Tecn. Latt. Cas.*, **37** 117-25 (in Italian).
Breene, W.M., Price, W.M. and Ernstrom, C.A. (1964) Manufacture of Pizza cheese without starter culture. *J. Dairy Sci.*, **47** 1173-80.
Carini, S. and Todesco, R. (1974) The proteolytic and curd-forming properties of microbial rennets and their use in the manufacture of cheese. *XIX Int. Dairy Congress*, New Delhi, India, **IE** 702.
Casiraghi, F.M., Bargley, E.B. and Christianson, D.D. (1985) Behaviour of Mozzarella, Cheddar and processed cheese spread in lubricated and bonded uniaxial compression. *J. Texture Studies*, **16** 281-301.
Cavella, S., Chemin, S. and Masi, P. (1992) Objective evaluation of stretchability of Mozzarella cheese. *J. Texture Studies*, **23** 185-91.
Cavella, S., Chemin, S. and Masi, P. (1993) Objective evaluation of stretchability of Pasta Filata cheeses. *Ind. Agri. Alim.*, **110** 11-15.
Cervantes, M.A., Lund, D.B. and Olson, N.F. (1983) Effects of salt concentration and freezing on Mozzarella cheese texture. *J. Dairy Sci.*, **66** 204-11.
Chen, A.H., Larkin, J.W., Clark, C.J. and Irwin, W.E. (1979) Texture analysis of cheese. *J. Dairy Sci.*, **62** 901-07.
Christensen, W.V. (1965) *Flavor development and control in Provolone and Romano cheese*, in Proceedings 2nd Annual Marschall International Italian Cheese Seminar, Marschall-Miles, Madison, Wi, p. 3.
Corradini, C. (1973) I vari tipi di caglio nella produzione del Provolone. *Sci. Tecn. Latt. Cas.*, **24** 24-34 (in Italian).
Corradini, C. (1977) Coagulanti diversi dai tradizionali nella produzione dei formaggi Italiani. *Sci. Tecn. Latt. Cas.*, **28** 345-51 (in Italian).
Delforno, G. (1985) Il formaggio Caciocavallo. *Mondo Latte*, **6** 361-64 (in Italian).
Donnelly, C.W. (1990) Concerns of microbial pathogens in association with dairy foods. *J. Dairy Sci.*, **73** 1656-61.
Eckner, K.F., Roberts, R.F., Strantz, A.A. and Zottola, E.A. (1990) Characterization and behaviour of *Salmonella javiana* during manufacture of Mozzarella-type cheese. *J. Food Protection*, **53** 461-4.
Elliot, R. (1984) Hi-tech system ensures quality. *Dairy Field*, **167**(11) 26-8, 31.
Emaldi, G.C. (1968) Il Provolone. *Latte*, **6** 24-30 (in Italian).

Ernstrom, C.A. and Anis, S.K. (1985) Properties of products from ultrafiltered whole milk, in *New Dairy Products via New Technology*, Proceedings of International Dairy Federation Seminar in Atlanta, GA, USA. IDF Publishing, pp. 21-30.
Gatti, M., Fornasari, E. Giraffa, G. and Neviani (1993) *Moderne Strategie Lattiero: Casearie.* ILC, Lodi (in Italian).
Ghitti, C., Salvadori-Bianchi, B. and Rottigni, C. (1996) *Il Formaggio Mozzarella.* CSL Publications, Milan (in Italian).
Giroletti, A. and Bartels, H.J. (1993) La maturazione sottovuoto dei formaggi. *Latte*, **10** 1015-18 (in Italian).
Harper, W.J. (1965) *The effect of starter micro-organisms on the ripening of Provolone and Romano cheese*, in Proceedings 2nd Annual Marschall International Italian Cheese Seminar, Marschall-Miles, Madison, Wi, p. 23.
Hooker, C. (1990) Can Mozzarella be made with UF? *Dairy Field Today*, **173**(4) 30-2.
Irvine, D.M. (1978) *Survey on the bacteriological quality of Canadian variety cheeses*, in Proceedings 15th Annual Marschall International Italian Cheese Seminar, Marschall-Miles, Madison, Wi, p. 30.
Kalab, M. (1977) Milk gel structure: VI. Cheese texture and microstructure. *Milchwissenschaft*, **32** 449-58.
Keller, B., Olson, N.F. and Richardson, T. (1974) Mineral retention and rheological properties of Mozzarella cheese made by direct acidification. *J. Dairy Sci.*, **57** 174-80.
Kimura, T., Sagara, Y., Fukushima, M. and Taneya, S. (1992) Effects of pH on submicroscopic structure of string cheese. *Milchwissenschaft*, **47** 547-52.
Kindsted, P.S. (1993) Mozzarella and Pizza cheese, in *Cheese: Chemistry, Physics and Microbiology, Volume 2*, (Ed. P.F. Fox), 2nd edn, Chapman and Hall, London, pp. 337-62.
Kosikowski, F.V. and Mistry, V.V. (1997) *Cheese and Fermented Milk Foods, Volume 1*, 3rd edn, F.V. Kosikowski LLC, Westport, CT, pp. 174-203.
Larson, W.A. and Olson, N.F. (1970) Continuous direct acidification system for producing Mozzarella cheese. *J. Dairy Sci.*, **53** 664-71.
Larson, W.A., Olson, N.F., Ernstrom, C.A. and Breene, W.M. (1967) Curd-forming techniques for making Pizza cheese by direct acidification procedure. *J. Dairy Sci.*, **50** 1711-12.
Lawrence, R.C. (1989) The use of ultrafiltration technology in cheesemaking. Document 240, *International Dairy Federation*, FIL-IDF, Bruxelles, pp. 2-15.
Lawrence, R.C., Creamer, L.K. and Gilles, J. (1987) Texture develoment during cheese ripening. *J. Dairy Sci.*, **67** 1748-53.
Lawrence, R.C., Gilles, J. and Creamer, L.K. (1983) Review: the relationship between cheese texture and flavour. *N Z J. Dairy Sci. Technol.*, **18** 175-90.
Lee, C., Imoto, E.M. and Rha, C. (1978) Evaluation of cheese texture. *J. Food Sci.*, **43** 1600-05.
Marino, M., Butta, P., Cocolin, L., Comi, G. and Duratti, G. (1995) Controllo microbiologico di punti critici nella filiera di produzione di Mozzarella. *Ind. Alimentari*, **10**(**34**) 1027-31 (in Italian).
Masi, P. and Addeo, F. (1986) An examination of some mechanical properties of a group of Italian cheeses and their relation to structure and conditions of manufacture. *J. Food Eng.*, **5** 217-19.
Mongiello, A. (1984) *State of the arts of salting and cooling pasta filata cheese*, in Proceedings 21st Annual Marschall International Italian Cheese Seminar, Marschall-Miles, Madison, Wi, p. 80.
Moskowitz, G.J. (Ed.) (1980) *The Analysis and Control of Less Desirable Flavors in Food and Beverages*, Academic Press, New York.
Niccoli, V. (1902) *Saggio Storico Bibliografico Sull'Agricoltura Italiano*, UTET Publishers, Turin (in Italian).
Nilson, K.M. and LaClair, F.A. (1976) *Free oil formation and Mozzarella composition*, in Proceedings 13th Annual Marschall Italian Cheese Seminar, Marschall-Miles, Madison, Wi, p. 64.
Nilson, K.M. (1982) *Quality and yield of Mozzarella manufactured from pasteurized milk in prolonged storage*, in Proceedings, 19th Annual Marschall International Italian Cheese Seminar, Marschall-Miles, Madison, Wi, p. 188.

Olson, N.F. (1970) *Direct acidification system for Mozzarella production*, in Proceedings 7th Annual Marschall International Italian Cheese Seminar, Marschall-Miles, Madison, Wi, p. 28.

Olson, N.F. (1979) *Direct injection of salt in Mozzarella cheese*, in Proceedings 1st Biennal Marschall International Italian Cheese Conference, Marschall-Miles, Madison, Wi, p. 3.

Olson, N.F. and Nelson, D.L. (1980) *Testing for melting behaviour in Mozzarella cheese*, in Proceedings 7th Annual Marschall International Italian Cheese Seminar, Marschall-Miles, Madison, Wi, p. 109.

Olson, N.F. and Neu, J. (1964) *Mozzarella cheese composition and meltability characteristics*, in Proceedings 1st Annual Marschall International Italian Cheese Seminar, Marschall-Miles, Madison, Wi, p. 16.

Olson, N.F. and Price, W.V. (1958) A melting test for pasteurized process cheese spread. *J. Dairy Sci.*, **41** 999-1000.

Paquet, A. and Kalab, M. (1988) A view of cheese microstructure using a scanning electron microscope. *Food Microstructure*, **7** 99-107.

Persano, A.E., Tantillo, G. and Augelli, R. (1995) Qualità igienica dei formaggi freschi a pasta filata confezionati. *Ind. Alimentari*, **6**(27) 377-80 (in Italian).

Rossi, J. and Leal, B.G. (1975) Gili agenti microbiologici nella maturazione della Mozzarella. *Latte*, **31** 378-87 (in Italian).

Salvadori del Prato, O. (1993a) Italian Mozzarella. *Dairy Ind. Int.*, **58**(4) 26-9.

Salvadori del Prato, O. (1993b) Le lipasi e il loro impiego Caseario, in *Aggiornamenti Technologici*, Milc, Parma (in Italian).

Salvadori del Prato, O. (1998) *Trattato di Technologia Casearia*, Edagricole Publishers, Bologna (in Italian).

Salvadori del Prato, O. and Messina, G. (1991) *Il Provolone e Altre Paste Filate*, ILC Publishers, Lodi (in Italian).

Shahani, K.M. (1971) *Lipase and flavor development*, in Proceedings 4th Annual Marschall International Italian Cheese Seminarp, Marschall-Miles, Madison, Wi, p. 18.

Stefanini, G. (1991) La vera Mozzarella Italiani. *Process*, **9** 78-82 (in Italian).

11 Membrane processing

H.C. van der Horst

11.1 Principles of membrane processes

A major activity in the processing of dairy liquids is the separation of components from the milk into dairy components and products. In the beginning of the 1970s, membrane filtration processes were introduced and were recognised by the dairy industry to be a major breakthrough in the processing of milk and whey into valuable components and dairy products. Nowadays membranes are widely applied in the dairy industry (Figure 11.1).

Membranes are molecular sieves. They are made of polymeric [poly(ether) sulfone, polyacrylonitrile, polyamide] or ceramic (zirconium oxide, aluminium oxide) materials. The liquid is driven through the membrane by applying a pressure difference (Figure 11.2). The membrane rejects components much larger than the pore size and permeates components much smaller than the pore size. Components that are rejected by the membrane cause plugging and thus, high resistance for permeation. Therefore, the feed is pumped along the membrane at a certain speed in order to prevent what is called concentration polarisation and fouling.

The principle of the fractionation is based on size difference. Membranes are folded into membrane modules. Four different module configurations are

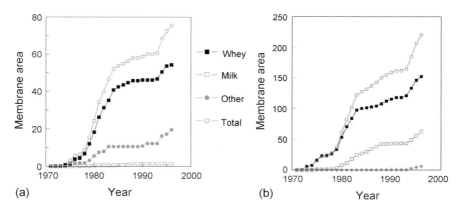

Figure 11.1 Worldwide use of membrane surfaces for (a) reverse osmosis, and (b) ultrafiltration of dairy liquids. Membrane areas in use are expressed in thousand m^2.

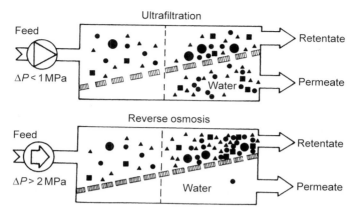

Figure 11.2 The principle of membrane processes: schematic view of ultrafiltration and reverse osmosis during concentration of cheese whey to a volume reduction of 67%. ΔP, pressure difference. ●, 1 whey-protein molecule; ■, 10 non protein nitrogen (NPN)-particles; ▲, 100 lactose molecules; •, 100 saltions.

common: tubular, hollow fibre, plate and frame, and spiral wound (Figure 11.3). In Table 11.1 properties of the different module configurations are ranked. The spiral wound configuration is nowadays the most widely used configuration because of its good cost/performance ratio.

Polymeric membranes are available in many different module configurations and with a variety of pore sizes. Ceramic membranes are commercially available in the tubular configuration. They are not available for reverse osmosis applications.

The volumetric throughput of the membrane surface per hour per meter-squared is called the flux ($1\,h^{-1}\,m^{-2}$) and is dependent on the applied pressure and the total membrane resistance. During the process two product streams are produced: a retentate or concentrate, and a permeate.

$$J = \frac{dP - \Pi}{\eta(R_m + R_p + R_f)} \quad (11.1)$$

where

- J = flux ($m^3 m^{-2} s^{-1}$)
- Π = osmotic pressure at the concentrate side ($10^5\,N\,m^{-2}$)
- dP = pressure difference between feed and permeate side ($10^5\,N\,m^{-2}$)
- R_m = resistance to permeation of the clean membrane (m^{-1})
- R_p = resistance to permeation as a result of concentration polarisation and adsorption (m^{-1})
- R_f = resistance to permeation as a result of long-term fouling (m^{-1})
- η = dynamic viscosity of the permeate (Pas)

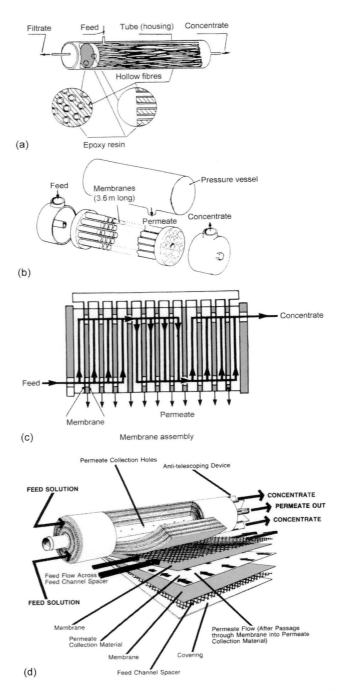

Figure 11.3 Different ultrafiltration module configurations: (a) hollow fibre, (b) tubular membrane, (c) membrane assembly and (d) spiral wound.

Table 11.1 Features of various types of membrane modules

Module type	Packing density (m² m⁻³)[a]	Process costs[b]	Sensitivity[c]
Tubular	Low	High	Low
Plate and frame[d]	↓	↑	↓
Spiral wound			
Hollow			
fat fiber			
fine fiber	High	Low	High

[a] Surface area per module volume.
[b] Energy, cleaning agents, start–stop losses.
[c] Sensitivity to plugging of flow channels.
[d] Also called 'flat sheet'.

Rejection, R, is a measure of the permeability of the membrane for a certain component and is dependent on membrane characteristics such as pore size, membrane charge, hydrophobicity, viscosity, etc. In particular, the pore size distribution of a membrane is an important factor in rejection. All commercially available membranes have a pore size distribution. The effect of this is that components much larger than the pore size can still permeate the membrane, decreasing the selectivity. The adsorption of proteins and precipitation of minerals increases fouling, resulting in a higher rejection.

Rejection, R, is defined as follows:

$$R = 1 - \frac{C_p}{C_r} \qquad (11.2)$$

where

- C_p = concentration of component in the permeate
- C_r = concentration of component in the retentate

At $R = 1$ the membrane rejects all components, and at $R = 0$ all components are permeated and concentration in the permeate and retentate are the same: no separation is achieved.

The advantage of the process is that simultaneous separation and concentration is achieved: the liquid as well as some specific components are removed from the feed. The process can be applied at low temperatures (less than 55°C) which can have a positive effect on the product characteristics; the removal of water is an energy-efficient process because there is no phase change as in evaporation and drying. Furthermore, no solvents are needed for obtaining fractionation on a molecular level. In addition to all other advantages it is a technology that enables the dairy industry to develop new products that otherwise could not have been made cost effectively.

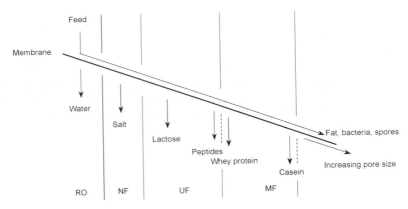

Figure 11.4 Permeation of dairy components. RO, reverse osmosis; NF, nanofiltration; UF, ultrafiltration; MF, microfiltration.

Depending on the pore size of the membrane, four different membrane processes are distinguished that are widely applied in the dairy industry

- reverse osmosis (RO)
- nanofiltration (NF)
- ultrafiltration (UF)
- microfiltration (MF)

In Figure 11.4 the milk components that permeate the membrane, dependent on the pore size of the membrane, are shown. The higher the pore size the lower the pressure during operation.

Typical process pressure are as follows:

- RO of whey, 3–4 MPa
- NF of whey, 2–3 MPa
- UF of whey and milk, 0.3–0.8 MPa
- MF of whey and milk, 0.0–0.2 MPa

11.2 Process control and automation of membrane processes

Membranes are folded into membrane modules, and these modules are built into a membrane system. In principle, two ways of operation are possible: batch and continuous. In a batch system the feed is circulated over the feed tank until the desired concentration factor is obtained; permeate is continuously produced. The advantage of the batch mode is the generally higher flux that can be obtained because the system operates, most of the time, at a low concentration factor; flux usually declines with increasing concentration factor. The disadvantage is that the feed is circulated in the system for as long as the process takes. This can effect the (microbiological) quality of the concentrate. Therefore, this mode of operation is used when permeate is the product. In continuous operation the feed is continuously fed to the system, and concentrate and permeate are continuously removed. At start-up, the retentate outlet is closed, the feed circulates in the system and permeate is continuously removed until the desired concentration factor in the system is obtained. The retentate outlet is then opened. The retentate and permeate outlet are then controlled. Usually, the feed flow and permeate flow are measured and used as control system. The concentration of the concentrate can be measured (e.g. by measuring the refraction index).

Most systems in the dairy industry operate in the continuous mode. At a low concentration factor one stage is used. The flux during filtration depends on the actual concentration in the membrane stage. At high concentration factors it is economical to apply more stages, each stage operating at its own concentration factor. With an infinite number of stages the batch system is approached (Figure 11.5).

A system can be controlled in constant flux operation or variable flux operation. The driving force is the difference in pressure applied between the retentate and the permeate side. Many membrane installations have to operate at a constant capacity in order to match the operation characteristics of other installations, (e.g. evaporators and dryers). For realising this objective, membrane installations are started up at a pressure level below the maximum permitted pressure. A constant flux is realised by raising the process pressure in order to compensate for

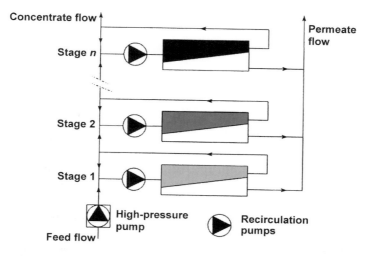

Figure 11.5 Example of the configuration of a multistage membrane system.

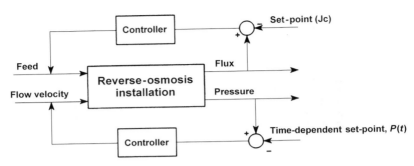

Figure 11.6 Example of a reverse osmosis system with constant flux operation control. Source: van Boxtel (1995).

the increased permeation resistance due to fouling (equation 11.1). This procedure is continued until the maximum permitted pressure is attained (Figure 11.6; van Boxtel, 1995).

Membrane fouling can be distinguished on two different time scales. The first one concerns fouling, which takes place in the first seconds after the start-up of a membrane system. This (short-term) fouling is a combination of concentration polarisation and adsorption of product components at the membrane surface resulting in a basic fouling layer. The second, long-term fouling, is a gradual fouling caused by precipitation of product components such as proteins and salts (especially calcium phosphates and calcium citrates) which takes place during the operation and results in a gradual decrease of the production capacity (van Boxtel, 1991; 1995).

Membrane fouling is a major problem, reducing the performance of membrane systems. In research on membrane fouling much attention has been paid to the description of the mechanisms and sources of membrane fouling. Today, this knowledge has resulted in methods to reduce fouling by membrane surface modification, improved module design and better module arrangement and optimal product pre-treatments as described above. All these methods precede the actual operation of membrane systems; that is, the information is used for the design of membrane equipment and in the pre-treatment. In addition to these common methods, membrane fouling can be controlled during the operation of membrane filtration units, resulting in lower operational costs. The controlling variables are: process pressure, flow velocity at each stage and feed temperature. There is a new development introducing static and dynamic process optimisation in membrane operation.

Another example with respect to the effect of control of the cross-flow velocity in a multistage tubular RO system is shown in Figure 11.7. The cross-flow velocity in each stage is adapted to the flux at each moment during processing and in each stage, in this case achieving minimal process costs.

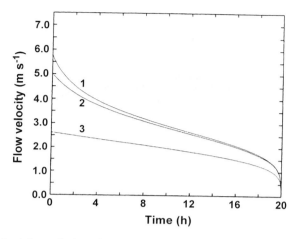

Figure 11.7 Optimal flow velocity trajectories for stages 1, 2 and 3 (see Figure 11.5): three-stage installation for reverse osmosis of cheese whey, with concentration increasing with each stage. At the start of the process the flux is high in all stages, but is highest in stage 1 (lowest concentration and osmotic pressure) and lowest in stage 3 (highest concentration). Cross-flow velocity is therefore highest in stage 1 in order to prevent high fouling rates. Moderate flow velocities suffice in the later phases of the operation. Source: van Boxtel, 1995.

The best example of successful control of process pressure is found in the uniform transmembrane pressure (UTMP) system of Tetra Pak and APV Nordic.

11.3 Membrane applications for milk

With respect to the use of membrane filtration processes in the processing of milk, different applications can be distinguished.

11.3.1 Milk concentration by reverse osmosis

Since the introduction of RO for dairy applications, the concentration of milk by RO at the farm has been mentioned as an interesting application. It would have the advantage of lower transportation costs from farm to factory, and the farmer could use his or her own water production. This is yet further an advantage in countries where the distance from farm to factory is large. However, there are many disadvantages such as losses of components, possible damage to fat globules, microbiological contamination and more handling costs at the farm. Furthermore, legislation does not permit milk components to be taken out of the milk, and last, but not least, RO would require a complete change in the logistics of milk processing, both at the farm and in the factory. In Europe, therefore, there is no incentive as yet to introduce such a system.

11.3.2 Demineralisation by nanofiltration

Milk could be de-salinated and concentrated simultaneously by NF. In particular, sodium, chloride and potassium ions would be removed during the process. A milk concentrate produced by NF has interesting gelling properties, for instance for deserts and cheeses. On an industrial scale this application has not yet been applied to my knowledge.

11.3.3 Milk protein standardisation by ultrafiltration

During the year the protein content in raw milk varies. Also, there is a difference in protein content depending on the feed, the climate and the variety. Standardisation of the protein content of the milk by UF results in a consistent milk composition all year round, independent of seasonal variations.

Advantages of protein standardisation of consumer and cheese milk are:

- uniform milk quality over the year
- a more constant process
- constant product quality over the year
- better utilisation of existing equipment (pasteurisers, cheese vats)
- increased yield
- possibly better cheese quality (structure) for low-fat cheeses

It is a growing application for cheesemilk (used for Camembert in France and for Feta in Denmark), milk powder and market milk. Milk is concentrated by UF to a constant protein concentration, usually higher than the natural protein content [The concentration factor (CF, the ratio between feed volume and concentrate volume) for Brie can be as high as 2.0–2.5]. This concentrated milk can be used for cheese production, increasing the production capacity per cheese vat, or can be used for protein-enriched milk products. For the production of Brie, an increased yield of 5% is reported (IDF, 1992). For the production of hard cheese, the CF of the protein is limited to approximately 1.3, owing to the simultaneous inclusion of whey proteins. For Gouda cheeses, higher protein concentrations result in inferior cheese quality.

The permeate that results from UF of cheesemilk can be used to standardise market milk to a constant (lower) protein level. Standardised milk can, furthermore, be processed into powder with a constant protein concentration, the level of which is dependent on the application.

An important bottleneck for the implementation of milk standardisation is the legislation in each country. It is not permitted to 'dilute' milk. Implementation of protein standardisation in a processing line for cheese or for market milk places some high demands on process control and automation: high-volume-throughput systems with constant capacity are needed despite membrane fouling. Furthermore, the protein content needs to be controlled precisely, independent of the incoming protein content of the feed.

11.3.4 Milk protein concentration by ultrafiltration and microfiltration

As an alternative for production of casein, caseinate and co-precipitates by precipitation with acid, heat or other precipitants, the total milk protein fraction consisting of casein and whey proteins can be fully concentrated by UF (cut-off molecular weight, less than 25 000 Da). The ratio of casein to whey proteins can be altered by application of MF (IDF, 1992). It is even possible to produce whey protein free milk or milk concentrate by MF by using a ceramic membrane with a pore size of 0.2 or 0.5 μm. The MF permeate of such a process is a high-quality whey, free of bacteria, nitrate, casein macropeptide, colorants and fat. The flux for this process is about $20–40 \, l \, m^{-2} \, h^{-1}$, which means the process is not yet feasible for large-scale application.

During production of milk protein concentrate (MPC) preferably a high protein concentrate is achieved (ca. $\sim 16 \, g \, 100 \, g^{-1}$ protein) suitable for direct spray drying, hence avoiding protein damage during evaporation. Tailor-made MPC powders can be produced in this way with different functional properties especially with respect to viscosity and heat stability. Important factors affecting the quality and functional properties of MPC are:

- microbiological quality and heat treatment of the raw milk
- protein concentation and fractionation
- fat content of the skimmed milk
- process conditions such as temperature, pH, and method used during diafiltration
- quality of the water (microbiological and chemical) during the diafiltration stage (the dilution of the feed or concentrate by the addition of [usually] water to remove more permeable components like minerals and lactose)
- treatment of the concentrate (e.g. heating, evaporation and/or drying)
- storage conditions

It is clear that it is possible to produce MPC according to specific demands. However, competition with widely applied caseinates is fierce, especially with respect to the costs.

11.3.5 Removal of bacteria, spores and somatic cells from raw milk by microfiltration

One of the most important processes in dairy industry is the conservation of milk by heat treatment. In order to ensure safe products with a long shelf-life, pasteurisation or sterilisation is used to inactivate pathogenic bacteria and spores.

In the 1980s, Tetra Pak developed a microfiltration bactocatch system able to achieve a 3 log unit bacteria reduction in skimmed milk (van der Horst, 1990), while performing almost complete permeation of all other milk components. Fat globules were concentrated by MF, decreasing the flux and limiting the concentration factor. Therefore, skimmed milk was used. The bacteria reduction decreased with decreasing initial bacterial load and depended on the organism concerned. The membranes used in the system were tubular ceramic (α-alumina), with a pore size of 1.4 μm (Membralox, Société Ceremiqûes Techniques, Bazet, France). The system was able to operate continuously with a flux of more than 500 $l\, m^{-2}\, h^{-1}$. This was achieved by application of a uniform transmembrane pressure (UTMP) along the membrane flow channel. In MF of milk for the removal of spores, large pores were used (larger than 0.8 μm). This resulted in a very high flux, especially at start-up of the process when the membrane was yet not fouled. The feed was pumped at a high-speed (6 m s^{-1}) along the membrane channels in order to avoid concentration polarisation. The flow of components to the pores in a MF membrane is not in balance with the cross-flow applied. A complicating factor here is that with increasing cross-flow the pressure drop along the membrane channel increases, hence increasing the driving force and, thus, the flow of components to the pores. In other words: the flow to the membrane needs to be in balance with the flow along the channel. Tetra Pak solved this by decreasing the driving force along the length of the channel by application of a cross-flow along the permeate channel filled with plastic balls. In an uncontrolled system the pressure drop at a cross-flow velocity of typically 6 m s^{-1}, a channel diameter of 4 mm and a length of 0.85 m would be about 0.2 MPa. Only one third of the channel would be effectively used for permeation. By controlling the membrane pressure along the membrane channel at a UTMP of approximately 0.05 MPa, the whole length was used for permeation, resulting in a better performance (see Figure 11.8).

The process was originally developed for producing milk with an extended shelf-life (ESL), to give so-called ESL milk. The shelf-life of pasteurised milk is restricted by the growth of spores of *Bacillus cereus* at refrigerator temperatures. Sterilisation or ultra heat-treatment affects the taste and nutritional value of the milk. Removal of the spores by microfiltration at low temperatures (< 55°C) gives milk with an extended shelf-life and good taste (for further details, see Chapter 3; IDF, 1997).

The process has also been used for the production of cheesemilk free from spores of *Clostridium tyrobutyricum*, which are responsible for the phenomenon of late blowing in semi-hard cheeses (Figure 11.9; IDF, 1992). A limiting factor here is the reduction in spores that can be achieved. Especially with the development of cheese varieties with low salt, low-fat and high-moisture content the sensitivity for late blow increases, thus demanding very low concentrations in butyric acid-forming bacteria. A competing technology is bactofugation. At this moment MF performs better than one cycle of bactofugation but performs worse than two cycles of bactofugation.

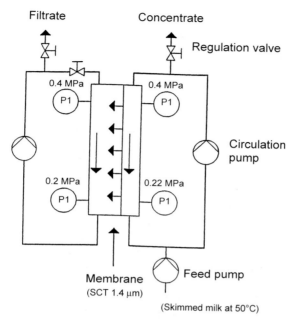

Figure 11.8 The principle of uniform transmembrane pressure (UTMP). Source: IDF (1997). PI, pressure indicator; SCT, Societe Ceramiques Techniques (Bazet, France).

A disadvantage of using MF is that a relatively large part of the milk, including the cream (inactivating the xanthine oxidase, active against *C. Tyrobutyricum*), is sterilised, thus resulting in reduced cheese quality. APV Nordic developed a system for treating cheesemilk in which the fraction of milk to be heat-treated was reduced. The retentate of the MF process is discarded or returned to the raw milk, which is then skimmed by centrifugation. Spores are also removed during this step, increasing the overall spore reduction. The module construction of APV Nordic to ensure UTMP is different: the membranes are placed into stainless steel tubes facilitating easy membrane exchange and cleaning.

MF for production of cheesemilk is more or less comparable in cost to the use of bactofugation. One disadvantage of MF is that all cheesemilk needs to be skimmed. In The Netherlands only the part needed for standardisation of the milk is skimmed.

Developments in this area are membranes with higher spore retention owing to a controlled narrow pore size distribution and membranes with smaller channel diameters (3 mm) hence increasing the membrane surface per module and compacted systems, reducing energy consumption. Another trend in favour of MF is the increase in production capacity. The maximum capacity of bactofuges is limited, whereas the capacity of MF systems can be increased easily by adding extra modules (Figure 11.10).

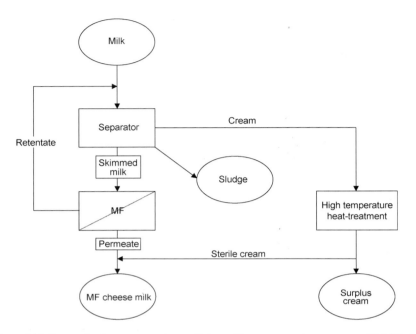

Figure 11.9 Process for the treatment of raw milk is a uniform transmembrane pressure (UTMP) plant. MF, microfiltration.

Figure 11.10 Example of a microfiltration plant for the treatment of milk and whey. The membrane area installed is ca. $9\,m^2$, having a capacity ranging from $200\,l\,h^{-1}$ to $4000\,l\,h^{-1}$.

11.4 Applications to cheese

11.4.1 Soft and hard cheese varieties

In principle UF is a unique process for the continuous production of fresh, soft and hard cheese types such as Feta, Camembert, Quarg, Ymer, Cream cheese,

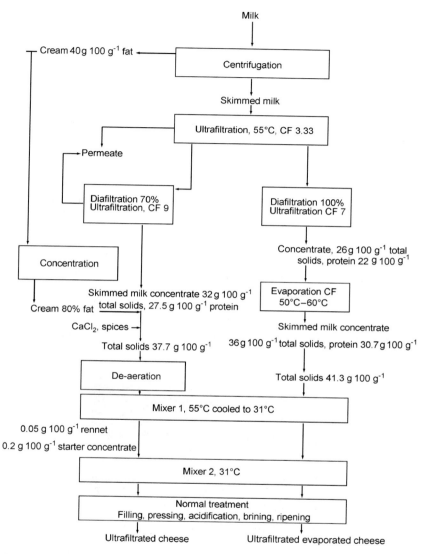

Figure 11.11 Manufacturing process of cheese (including low-fat cheese 20 g 100 g^{-1} fat in dry matter) from ultrafiltrated milk. CF, concentration factor. Source: de Boer (1980).

Cheese base and Gouda cheese. Advantages of using UF for the production of cheese are:

- continuous fully automated production instead of batchwise production
- higher flexibility with respect to cheese variety (cheese composition, addition of starter, colorants, herbs, brine)

- shorter processing cycle (removal of coagulation and the brining step in the production of hard cheese types)
- higher yield owing to the inclusion of whey proteins
- reduction of waste (removal of the brining step)
- reduction in water consumption (removal of the washing step)
- the creation of a product with constant product quality

Why is it, then, that UF is so far only a success in the production of high-moisture fresh cheese such as Ricotta, Quarg and Cream cheeses? In the first place, any attempt to produce a traditional cheese with a new technology always results in a cheese that has properties different from the traditionally produced cheese. Although these differences may be very small in texture, in taste they are significant for consumers who are used to the traditional taste. Major differences result from the inclusion of whey proteins in the cheese, changing the water-holding capacity, from outgrowth from the lactic acid bacteria and from ripening. Furthermore, the high total solid concentration needed for semi-hard cheese is at the limit of what is possible with UF at an industrial scale.

This explains why, for example, Feta produced by UF is a success in countries where the product is new (the Middle East), whereas in countries where people are used to the traditional product the product has failed (e.g. Greece; the type of milk used in Greece is ovine or caprine as compared with bovine).

For fresh acid curd cheese (Quarg and Cream cheese) UF enables a high yield and improved product texture as compared with the use of the traditional process or the Thermo-Quark process in which whey proteins are already included into the product (see Chapter 9).

In Figure 11.11, an example is given of a possible general UF process for the production of low-fat cheese. All kinds of variations on this process are applied, depending on the cheese type.

11.5 Applications for whey

11.5.1 Concentration of whey by reverse osmosis (RO)

Concentration of whey by RO is an attractive alternative technology for evaporation. RO has low energy consumption compared with evaporation (Figure 11.12). The maximum concentration that can be obtained by RO is $28\,g\,100\,g^{-1}$; the osmotic pressure of the concentrate then exceeds the maximum pressure of commercially available RO systems. Economically, RO is competitive with evaporation up to a concentration of approximately $16\,g\,100\,g^{-1}$ total solids. For RO (4 inch) spiral-wound composite membranes are most widely used. The process temperature should be lower than 15°C in order to prevent severe irreversible fouling by precipitation of calcium phosphate and calcium citrate during concentration. The selection criterion for the membrane is the retention

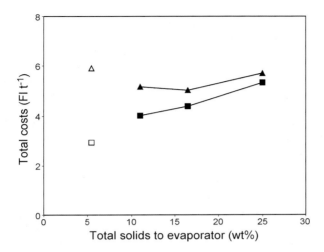

Figure 11.12 Costs for concentration of Gouda whey [in Dutch florins (Fl) per g 100 g^{-1} tonne of water removed] at 50 g 100 g^{-1} total solids, by evaporation with thermal vapour recompression (TVR, △), mechanical vapour recompression (MVR, □), and a combination of reverse osmosis (RO) with TVR (▲) and of RO with MVR (■). Source: unpublished NIZO data.

for lactose and sodium chloride. Permeating lactose pollutes the permeate and restricts its use. More and more the RO permeate from whey concentration is further treated and re-used in the factory.

11.5.2 Demineralisation of whey by nanofiltration

For whey products intended for human or animal consumption, demineralisation enhances the nutritional value of the product. In industrial processes, whey is concentrated by evaporation and subsequently demineralised by electrodialysis and/or ion exchange. NF is an alternative for the partial (approximately 30 g 100 g^{-1}) demineralisation of whey (Figure 11.13). It has the advantage of simultaneous concentration (approximately 15 g 100 g^{-1} total solids) and demineralisation of whey. This eventually reduces processing cost (Figure 11.14). During NF monovalent ions such as sodium, potassium, chloride and nitrate are preferentially removed. This results in a favourable mineral composition of a product that is 30% demineralised *vs* a product that is demineralised by electrodialysis. Also, improved taste is claimed in certain applications. NF is also applied for the demineralisation of salt cheese whey with the purpose to 'sweeten' the whey. Even more important than for RO, the selection criterion for NF of whey is the lactose rejection of the membrane: it should be greater than 0.99. Lower rejection results in permeate with high wastewater load. As with RO, spiral-wound polymeric composite membranes are generally used.

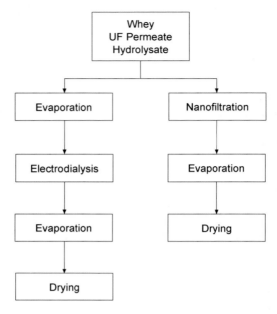

Figure 11.13 Flow chart for concentration and demineralisation of Gouda whey UF, ultrafiltration. Source: Timmer and van der Horst (1997).

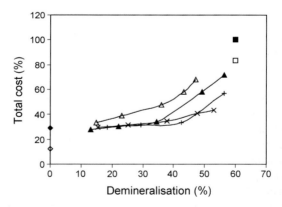

Figure 11.14 Costs for demineralisation of sweet Gouda whey by combinations of nanofiltration (NF) and evaporation [thermal vapour recompression (TVR) and mechanical vapour recompression (MVR)] and by electrodialysis (ED). Note: M1 and M3 are different NF membranes. ◆, TVR; +, NF (M1), pH 4.6; ×, NF (M1), pH 5.8; ▲, NF (M1), pH 6.6; △, NF (M3), pH 5.8; ■, ED+TVR; ◇, MVR; □, ED+MVR. Source: Timmer and van der Horst (1998).

The process is operated at temperatures lower than 15°C. The flux during NF of whey is much higher than during RO (IDF, 1997).

11.5.3 Whey protein concentrate production by ultrafiltration

By far the most successful application of membrane processes in the dairy industry is the production of a whole new product from whey: whey protein concentrates. Whey is concentrated by UF to produce whey protein concentrate: total solids ratio of 35 (skimmed milk equivalent), 50, 60, 70, 80 or even higher (termed WPC 35, WPC 50, WPC 60, etc.) The ratio is dependent on the concentration factor during UF. WPCs can replace (part of) the milk base in yoghurts, desserts, bakery products, soup and sauces, etc. After a treatment they can also replace part of the fat in the production of cheese (Zoon, 1993). They have special gelling, foaming and stabilising properties. The functional properties of WPCs depend very much on the composition. As with MPC, the composition of the final product depends on many parameters such as (see also IDF, 1992):

- whey quality and history (method of whey generation, microbiological factors, cheese type, heat treatment, age, storage conditions, concentration)
- pre-treatment prior to UF (lipid removal, pH, heating)
- process conditions during UF (temperature and diafiltration, diafiltration water, pH, concentration factor)
- final product processing and storage (evaporation and drying conditions)

An example of the difference in composition of various commercial WPC powders is shown in Table 11.2. Small differences in, for example, mineral composition or fat content can lead to quite large differences in functional properties. Furthermore, high protein concentrates are used as nutrients in infant formulae or as a basis for clinical products. Particularly for this type of product, the spore content and total count can be high because the concentration factor may be as high as 80 or more. Use of MF for the removal of spores is an option. For specific products, the low concentration of fat and denatured proteins in centrifuged whey can give negative functional properties and removal is required. Several processes have been developed for the clarification and fat separation of whey (Fauquant *et al.*, 1985; Rinn *et al.*, 1990; Pearce *et al.*, 1992).

UF of whey is performed mainly in multi stage spiral-wound systems with polysulfone membranes. Measuring the refractive index and coupling this to the concentration factor allows control of the protein content of the concentrate. Processes are operated between 50°C and 60°C, requiring a pre-treatment by heating or acidification in order to avoid severe membrane fouling during operation (Figure 11.15) (Hanemaaijer, 1998). Flux at 50°C is about twice as high as flux at 10°C. This was a major incentive for operation at such higher

Table 11.2 Comparison of gross composition of different commercial whey protein concentrate (WPC) powders

	Powder			
	WPC 35 A	WPC 35 B	WPC 50 A	WPC 45 B
Composition g $100\,g^{-1}$ pH	6.77	6.52	6.54	6.94
Dry matter	96.0	96.3	96.3	96.4
Total protein[a]	37.2	34.9	48.9	44.8
Nonprotein nitrogen	4.1	3.6	4.2	8.3
NSI, pH 4.6 (rel%)	85	85	82	72
β-Lactoglobulin	19.2	20.0	25.6	17.9
α-Lactalbumin	4.35	4.1	6.02	3.8
BSA	0.583	0.69	0.617	0.72
Immunoglobulin	0.856	1.3	1.02	0.13
Casein nitrogen[b]	5.6	5.2	8.8	12.5
Casein macropeptide	4.61	1.75	6.35	2.69
Noncasein nitrogen	31.6	29.7	40.1	32.3
Lactose	45.9	52.5	34.5	49.5
Fat	4.2	2.36	5.7	3.44
Ash	7.23	5.95	5.94	2.23
Na^+	0.568	0.38	0.392	0.17
K^+	1.71	1.92	1.36	0.72
Ca^{2+}	0.932	0.46	0.906	0.12
Mg^{3+}	0.114	0.10	0.104	0.04
P (organic)	0.176	0.30	0.206	0.93
P (inorganic)	0.629	1.10	0.504	0.09

[a]Calculated as total nitrogen content multiplied by 6.38.
[b]Calculated as the total nitrogen content minus the total noncasein nitrogen content.
NSI, nitrogen solubility index; BSA, bovine serum albumin.
Source: NIZO data. (non-published)
Note: WPC 35 is a powder with a whey protein concentrate: total solids content ratio of 35% WPC A and WPC B are WPC from different sources (manufacturers).

temperatures. Lately, prices for membrane systems have decreased rapidly, enabling new systems to operate at temperatures below 15°C. With respect to the microbiological quality of the end product this is to be favoured. Furthermore, the costs for pre-treatment can be avoided.

11.5.4 Whey protein fractionation

A next step in the valorisation of whey proteins is the fractionation of whey proteins using membrane filtration. In particular, WPC 70 and higher are used as gelling agents and for use in infant formulae. WPC enriched in, for example, α-lactalbumin mimic the composition of mother milk and, therefore, different processes have been developed to produce WPC enriched in α-lactalbumin. However it, seems that WPCs enriched in (native) β-lactoglobulin have improved gelling properties. α-Lactalbumin and β-lactoglobulin do not

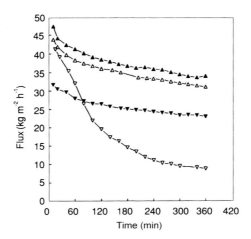

Figure 11.15 Effect of pre-treatment of whey on flux rate during ultrafiltration for the manufacture of whey protein concentrate WPC 35. △, 30 min, 55°C at pH 6.6; ▲, 30 min, 55°C, pH 7.5; ▼, 30 min, 55°C pH 5.8; ▽, untreated. Source: van der Horst et al. (1992).

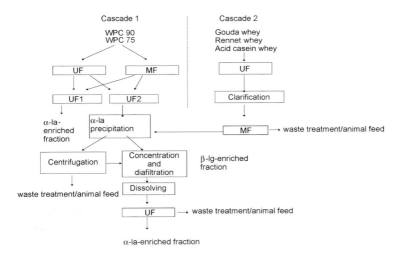

Figure 11.16 Tentative process for fractionation of whey protein. MF, microfiltration; UF, ultrafiltration, WPC 75, WPC 90, whey protein concentrates with a concentration factor of 75 and 90, respectively; α-la, α-lactalbumin; β-lg, β-lactoglobulin.

differ very much in molecular weight (ca. 14 000 and 36 000 Da, respectively) and are, therefore, difficult to separate from each other by membrane filtration. With pre-treatment a difference in solubility can be achieved enabling a separation by membrane filtration. An example of a process for the separation of whey proteins is given in Figure 11.16 (IDF, 1992, 1995).

11.6 Miscellaneous processes

11.6.1 Clarification of brine

During brining of Gouda cheese, water and components from the cheese migrate into the brine causing dilution. Part of it therefore needs to be discharged daily. Furthermore, particles from damaged cheese pollute the brine and, in certain cases, it is contaminated with yeast or salt-resistant lactobacilli which can give rise to defects in the cheese. Such microbiological contamination can be solved by pasteurisation. This will not, however, solve the problem of particle pollution and the need to discharge overloads of brine.

Microfiltration can clarify the brine without the need for pasteurisation. Indeed, microfiltration followed by evaporation, in which surplus water is evaporated, solves there quality problems (Zoon *et al.*, 1991).

11.6.2 Recycling of cleaning solutions

During evaporation of milk and whey, the fat, proteins and minerals are precipitated. Evaporators are cleaned with caustic and acid cleaning agents. A major part of the sediment is collected in the first cleaning fraction, which is discharged to a municipal waste collection service, in some countries giving rise to high wastewater costs. This fraction can be cleaned by using tubular NF membranes resistant to caustic. The caustic can be re-used for cleaning, and the retentate of the process containing protein and fat can be used for fodder after mixing with other waste streams. The mixing with other streams is necessary in order to reduce the pH and the mineral (NaCl) content of the concentrate. This is applicable to all detergent solutions used for cleaning dairy equipment.

References

de Boer, R. and Nooy, P.F.C. (1980) Low-fat semi-hard cheese from ultrafiltrated milk. Nord eruopesk mejer tidssckrift **3** 52-61.
van Boxtel, A.J.B. (1991) Strategies for Optimal Control of Membrane Fouling, PhD thesis, Department of Chemical Engineering, University of Twente Enschede, The Netherlands.
van Boxtel, A.J.B. (1995) Fouling control during the operation of membrane filtration units (RO and UF), in *Fouling and Cleaning in Pressure Driven Membrane Processes*. IDF, Brussels, Belgium, ch 6, pp. 93-117.
van Boxtel, A.J.B., Otten, Z.E.H. and van der Linden, H.J.L.J. (1991) Evaluation of process models for fouling control of reverse-osmosis of cheese whey. *J. Membr. Sci.*, **58** 89-111.
van Boxtel, A.J.B., Otten, Z.E.H. and van der Linden, H.J.L.J. (1992) Dynamic optimisation of a one-stage reverse-osmosis installation with respect to membrane fouling. *J. Membr. Sci.*, **65** 277-93.
Fauquant, J., Vieco, E., Brule, G. and Maubois, J.L. (1985) Clarification des lactosérums doux par agregation thermocalcique de la matière grasse residuelle. *Le lait*, **65**(1) 1-20.
Hanemaaijer, J.H., Robbertsen, T., van den Boomgaerd, T.H., Olieman, C., Both, P. and Schmidt, D.G. (1988) Characterisation of clean and fouled ultrafiltration membranes. *Desalination*, **68** 93-108.

van der Horst, H.C. and Hanemaaijer, J.H. (1990) Cross-flow microfiltration in the food industry: state of the art. *Desalination*, **77** 235-58.

van der Horst, H.C., Hols, G. and Teerink, S. (1992) Production of whey protein concentrates. *Voedingsmiddelentechnologie* **25**(19) 13-15.

Pearce, R. J., Marshall, S.C. and Dunkerley, J.A. (1992) Reduction of lipids in whey protein concentrates by microfiltration: effect on functional properties, *Special Issue Int. Dairy Fed.* **9201** 118-29.

Rinn, J.C., Morr, C.V., Seo, A. and Surah, J.G. (1990) Evaluation of nine semi-pilot scale whey pretreatment modifications for producing whey protein concentrate. *J. of Food Sci.*, **55**(2) 510-15.

IDF (International Dairy Federation) (1992) New applications of membrane processes. *Special Issue Int. Dairy Fed.* **9201**, Brussels, Belgium.

IDF (International Dairy Federation) (1995) Fouling and cleaning in pressure driven membrane processes. p*Special Issue Int. Dairy Fed.* **9504**.

IDF (International Dairy Federation) (1997) Implications of microfiltration on hygiene and identity of dairy products. *Bull. Int. Dairy Fed.* **320** 8-40.

Timmer, J.M.K. and van der Horst, H.C. (1997) *Whey processing and separation technology: state of the art and new developments*, in Proceedings of the Second International Whey Conference, Chicago, USA, October 27–29, 1997. IDF, Brussels, Belgium.

Zoon, P., Straatsma, J. and Allersma, D. (1991) Concentration of cheese brine by evaporation and its effect on cheese quality. *Voedingsmiddelentechnologie*, **24**(11) 13-16.

Zoon, P. and van den Berg, G. (1994) Increasing cheese yield by incorporation of whey proteins. *Voedings middelen technologie*, **27**(1/2) 12-14.

12 Nonproduct operations, services and waste handling

L. Robertson

Dairy technology has changed dramatically over the past two decades, from an industry in which large numbers of relatively small plants were operated in a labour-intensive mode to the huge and highly automated plants that are common today. These changes are not only changes of scale but also changes of basic operating philosophy. We have moved from an essentially manual industry to a highly mechanised and automated industry. Although mechanisation and automation of the processing functions are essential for this transition, no product will be produced unless the nonproduct functions, the services and the waste-handling facilities are capable of supporting the processing plant.

12.1 Nonproduct operation and maintenance

Nonproduct operation and maintenance covers all operational aspects of dairy plant when the plant is not actually making product. This section of the chapter discusses the period between completion of construction and first production, the maintenance functions and the periods between completion of maintenance and recommenced production. The normal (nonproduct) start-up, product change-over and shut-down periods are also covered, as are the regular cleaning operations that are a vital part of automated dairy plant design. Emphasis is placed on the automation and mechanisation aspects of these nonproduct operations and on their design, planning and execution.

12.1.1 Plant commissioning

In the context of dairy technology, the term 'commissioning' has an engineering aspect and a contractual aspect. In engineering terms, commissioning refers to the activities that take place after plant (re)construction is completed and before the plant is ready to process product. Contractually, the commissioning period is commonly the period when the plant owners are given access to the plant and when the contractors' and the owners' representatives work together to bring the fully constructed plant to a state where it can actually carry out its intended functions.

The largest commissioning tasks generally occur after a plant is first constructed. Although construction may be complete, it is likely that pipelines will contain debris, valve actuators cannot be relied upon to be connected correctly and three-phase motors are as likely to rotate in the wrong direction as in the right direction. All of these and many more issues must be identified and rectified before the plant can be put into production. The planning of the commissioning stage is commonly very detailed and may include separate stages in which mechanical and electrical problems are rectified, leaks (air and water) are identified and fixed, and lines are flushed out to temporary drains. Commissioning will also include the supervisory control and data acquisition (SCADA) systems, sensors and other electronic equipment.

After a major maintenance task, many of the same issues of (re)commissioning occur as after initial construction, though on a reduced scale.

12.1.2 Start-up and shut-down

Once commissioning is completed a plant is capable of carrying out its function—in the case of processing plant, it is capable of making product. For a food processing plant (unlike a chemical processing plant), however, there are normally significant periods when the plant is not actually processing product. Apart from the shut-downs caused by lack of raw material, this may be for several reasons.

For plants such as pasteurisers, evaporators and spray driers it is common to start-up using water as the feed. This allows the plant to be brought up to running temperatures, allows stable operation to be achieved and also ensures that all product-contact surfaces are sterilised. This warm-up and sterilisation step can take 20 min or more, depending upon the size of the plant.

Having achieved stable operation on water, and also having sterilised the plant, the operators then need to change the evaporator and spray drier over to product. This is not a simple operation because the thermal properties of the product are different from those of water and the operating parameters will need to be adjusted. Particularly with large plant, it is very important to achieve stable operation on product as quickly as possible, to minimise the quantity of quality-downgraded product.

The transition off-product at the end of a run is somewhat less critical to product quality. The operator will want product to flow through the plant at equilibrium conditions, and some small temperature excursions as product is replaced by water are not too critical.

As plant size grows, the issue of change-over from one product specification to another grows in importance. There may be times when the milk supply is limited or when market demand suggests relatively small quantities of a number of product specifications. In such cases, a high-capacity processing plant may be forced to make relatively short runs, resulting in the time between initial change-over and stable on-specification operation becoming a significant proportion of the total run time. There is a large incentive to install control systems that are capable of getting the equipment operating on-specification in the absolute minimum time.

Off-product running, for whatever reason, implies that the full energy requirement of the plant is being utilised, even though no product is being produced. This can have a very large effect on the total energy used. In particular, lack of coordination between the evaporator operators and the spray-drier operators, which results in the spray drier operating on water for extended periods, will result in large and expensive energy bills.

12.1.3 Maintenance, including predictive or planned maintenance

The dairy industry places some particular and peculiar constraints on maintenance operations; being a natural product, supply commences whether the plant is prepared to receive it or not, and the supply continues regardless of the availability of the plant. Many countries place severe limits on the quantity of organic material that can be released to the environment and, therefore, the dairy plant operator faces a severe problem if the plant either is not ready on time or suffers unplanned outages while the milk supply is available. It is, of course, possible to carry out some maintenance during the regular cleaning cycles of the processing plant. It is also possible to take advantage of below-peak raw milk supply rates to fit in some maintenance tasks.

Recognition of the importance of maintenance has led to the development of a number of 'maintenance philosophies': these include breakdown maintenance, run to destruction, pre-planned maintenance, reliability-centred maintenance, total productive maintenance (TPM), condition-based maintenance and 'campaign maintenance' (Stoneham, 1998). Only the last five in this list have been found to be applicable in the dairy industry. It is common practice to benchmark maintenance costs by calculating them as a percentage of equipment replacement value (ERV).

Because of the peculiar constraints of the industry, staff of a large dairy processing plant make every effort to plan all maintenance tasks and to plan them in great detail. An emphasis on plant reliability leads almost inevitably to an emphasis on scheduled maintenance, implying the replacement or servicing of all items on a fixed schedule (regardless of whether the item appears to need replacement or not). Under a scheduled maintenance regime, the two vital planning activities are identify every item that needs maintenance as well as to the maintenance period. It is the identification of the maintenance period that is the more difficult and which inevitably leads to some balancing of replacement costs, assessment of mean time between failure (MTBF) and cost consequences of unplanned failure. Maintenance planning requires a knowledge of the statistical risks of each failure mode and an assessment of the costs of maintenance and the costs of failure; together these allow staff to develop a schedule of planned maintenance that balances all known risks. However, the schedule is statistically

NONPRODUCT OPERATIONS, SERVICES AND WASTE HANDLING 321

based; this inevitably means that some very expensive maintenance tasks will be undertaken, only to find that the part is in perfect condition (and the task unnecessary) and, conversely, that there may be unplanned failures despite the maintenance schedule having been followed.

As plants grow in size and complexity, as the available downtime decreases and as the need to remove any unnecessary costs increases there is a need to consider alternative approaches to maintenance planning. A common alternative to scheduled maintenance planning is predictive maintenance planning, in which characteristics of plant deterioration are monitored, allowing staff to predict when these characteristics will reach unacceptable levels and to arrange the maintenance activity.

Examples of measurements include scheduled oil sampling (analysing and measurement of the buildup of contaminants or of the results of wear within oil reservoirs) and vibration monitoring of critical bearings. There are many indicators of performance that can be monitored for signs of deterioration. A recognition that failure is often not a simple event has led to the use of such software approaches as 'expert systems' and 'artificial neural networks' to help the predictive maintenance task.

A key characteristic of dairy plant maintenance is the uneven distribution of effort during the year. This uneven distribution presents a problem of personnel utilisation. To cope with this uneven personnel requirement, it has become common to let maintenance contracts for annual shut-down work.

12.1.4 Cleaning-in-place operation, control and automation

In dairy plant, cleaning has two purposes: the control of microorganisms and the removal of deposits. Dairy processing plants deal with a natural raw material that is not sterile. Despite strenuous effort to prevent the ingress of microorganisms to the processing plant, it is almost inevitable that some do gain access and, having done so, find conditions that are ideal for rapid multiplication. The typical result is that, after beginning processing, there will be a period in which bacteria counts remain low, followed by a period in which there is an exponential growth in live bacteria numbers. For any processing run in a particular plant (i.e. for a particular bacterial growth environment) there is a stage beyond which processing cannot continue; the plant must be stopped and cleaned. This period can vary considerably: for a spray drier operating at high temperatures the period can be several days; for an evaporator operating at 40–85°C, or a pasteuriser, the period can be as short as a few hours. Different microorganisms react differently to a particular environment, and one of the problems faced by operators of automated dairy processing plant is 'thermophilic' (heat-loving) bacteria, which survive and thrive at surprisingly high temperatures.

Cleaning-in-place (CIP) is a feature that most sharply distinguishes the modern automated and mechanised dairy plant from older, smaller and more manually operated plants. Manually operated plants are disassembled and manually washed. For large, complex and high-throughput plants, manual washing would be impractical and prohibitively expensive. The procedures, design and materials necessary to achieve adequate CIP have been the subject of a great deal of research; the following is a very brief introduction to the topic, and readers are referred to texts by Romney (1990), Maddox (1994) and Gutzeit (1997) for further information.

When considering CIP from an engineering viewpoint, it is possible to distinguish three main physical aspects of a system being cleaned: the (metallic) surface to be cleaned, the deposit and the cleaning solution. The cleaning process is a series of processes: diffusion of the cleaning solution to the surface, wetting of the surface, penetration of the cleaning solution into the deposit, reaction between the cleaning solution and the deposit, and removal of the deposit. Note that some of these steps depend on mass transfer between the solution and the deposit (and hence on the velocity and coverage of the cleaning solution), others depend on diffusion through deposits and other steps depend on the nature of the deposit and of the cleaning solution chemistry.

Adequate CIP depends critically on plant being designed to facilitate CIP. The design features that allow adequate CIP are quite different and distinct from the design features that allow adequate or efficient manual cleaning. In particular, it is essential that plants (including pipework, vessels, valve gear, etc.) to be cleaned have absolutely no crevices or other design features that will allow product to lodge and remain throughout a CIP cycle. The most important, internationally accepted, standards for construction detail of a dairy processing plant are the '3-A' standards of the International Association of Food Protection (IAFP, 1999). As with all codified aspects of design, adherence to the 3-A standards does not necessarily mean that all possible sources of contamination are removed. A problem that requires particular attention is that of 'dead legs' of pipework between valve shut-off elements and the pipe manifold into which the valved pipe drains.

Proteinaceous material adheres strongly to metal surfaces (particularly in areas of high heat flux), and removal is not a simple task. The chemicals that are effective are also quite aggressive, and it is essential that materials of construction are chosen to ensure that corrosion is not caused by CIP chemicals. Apart from damage to the mechanical properties of materials, corrosion has another major effect; even very small pits created in product-contact surfaces are very likely to harbour bacteria and are unlikely to be successfully cleaned by CIP processes. The selection of material combinations and CIP chemicals that will avoid the formation of corrosion pits is an essential part of the specification of automated dairy processing plant. In general, stainless steel is not corroded by high-pH cleaners but is very susceptible to corrosion by chloride ions. The reader should note that corrosion affects not just the product pipelines and vessels; the high-pH solutions resulting from alkali flushes can cause severe damage to the materials used in drainage systems, flooring, valve seals and other components.

A typical CIP process involves a pre-rinse with water to remove loosely bound material, one or more cleaning cycles (involving the circulation of a cleaning liquid), a clean water flush and a final rinse to remove traces of cleaning chemicals.

Two systems are in common use for the cleaning cycle:

- two-stage acid–alkali cleaners in which an initial alkali clean with 0.1–2 g 100 g^{-1} NaOH is carried out to remove the proteinaceous material; after a water flush, an acid cleaning cycle with 0.1–2 g 100 g^{-1} nitric acid is used (followed by a rinse cycle) to remove the mineral layer adjacent to the metal surface
- single-stage commercial cleaning formulations, commonly based on NaOH, plus detergents, plus (optionally) chelating agents such as ethylenediaminetetra-acetic acid (EDTA) to improve the suspension of solids in the flush liquid and (optionally) surfactants or wetting agents

CIP solutions are commonly prepared in vessels in a CIP room, so that the various flushing solutions are ready for circulation as soon as the product run finishes. All stages of the CIP cycle use moderately high fluid velocities to ensure good scrubbing action of the CIP solutions. Romney (1990) and Bylund (1995) provide good flow diagrams of several common types of CIP system configuration.

CIP chemicals represent a very significant cost for the dairy plant, and their disposal represents a further major cost. There is a large incentive to re-use CIP chemicals, and there is also an incentive to minimise water use. These two considerations have encouraged plant designers to change from single-use CIP systems to re-use CIP systems. In re-use CIP systems chemical use is reduced by storing and reusing the alkali and acid rinses. Assuming that the initial rinse is effective in removing all loose dairy product it should be possible to re-use an alkali or acid wash solution for up to two weeks provided sludge is removed from the stored solution. It is also possible to use membrane processes to reconcentrate CIP solutions and recover pure water.

In some cases it is possible to use an abbreviated CIP process part way through a product run, followed by a full CIP at a later time. This approach allows longer effective run times, with adequate cleanliness. 'Cleanliness' is rather an imprecise term; it could be taken to mean that a surface is physically clean (appears to be clean), that a surface is chemically clean (free of cleaning chemicals) and/or that a surface is microbiologically clean. In the dairy industry, 'microbiologically clean' generally means that there is less than a certain number of viable microorganisms per square centimetre.

A significant portion of the control system and control logic in a mechanised or automated plant will be devoted to the control of the CIP process, ensuring that every part of the plant is thoroughly cleaned. It is still common to find CIP processes controlled by simple timers that set the length of the initial, alkali, acid and final rinse runs. Although simple and robust, timers tend to be set conservatively and often produce much longer total CIP times than are strictly necessary for adequate cleaning. In modern plants there is a trend towards replacing timers with sensor-operated controls that end each CIP component when the fluid condition shows that its task is completed.

12.2 Supply and control of services

The operation of the mechanised and automated dairy plant is very strongly dependent on the automation and control of services to the plant. These services including electricity, steam, hot water, chilled water, compressed air and other services. This section emphasises the automation and mechanisation of the supply and control of these services. As service supply is a moderately specialised topic, reference is made to some standard texts.

12.2.1 Water quality

One of the most important services required by the dairy plant is water. Several grades of water are commonly reticulated within a food processing plant. These grades are quite distinct and not interchangeable. For example, potable water for product-contact applications, is specified by the World Health Organization (WHO; web page: http://www.who.int/dsa/cat98/water8.htm). Water treatment is quite a specialised topic and is the subject of many texts and several journals, some of which are referenced at the end of this chapter. Generally, treatment includes a filtration step to remove suspended solids, treatment to remove excessive amounts of hardness-forming salts and finally disinfection using chlorine or ozone injection, or ultraviolet (UV) light. Each disinfection method has specific advantages and disadvantages. Chlorine injection leaves residual chlorine in the water which resists bacterial growth during reticulation; however, chlorine also leaves a distinctive taste. Ozone leaves no residue in the water. UV light also leaves no residue, but its effectiveness is dependent on careful lamp maintenance and no possibility of water escaping the required UV dosage.

12.2.2 Electricity

Of the energy services used in the automated dairy plant, electricity is the most ubiquitous. Electricity supply is almost invariably three-phase. For large loads, it becomes economic to use higher voltage power supplies to reduce conductor sizes. In the largest food processing plants it is not uncommon to find 11 kV circuits in use. Electricity costs for a large processing plant can be several million dollars per year and, together with steam costs, can be among the higher-cost components of the processing operation. Electricity supply authorities generally have several charging regimes for their power; each charging regime will commonly include a base cost per 'unit' (kW h) and a surcharge for usage during times of peak demand. It is also common to find 'peak demand penalties', which are calculated quarterly according to the peak load drawn during that quarter, and also 'power factor' penalties calculated according to the level of reactive power drawn.

Although food processors do not have complete control over their times of operation, they do have some degree of control, and they have control over many of the 'background' electricity users such as bore water pumps, compressors, etc. 'Demand-side management' (DSM) refers to the practice of shifting electricity loads so as to match best the purchase tariff and to minimise the total price paid by the processor. Effort invested in DSM can be extremely cost effective; DSM often involves no capital equipment and can show very significant cost savings. For smaller processors, consultants are able to recommend DSM measures and quantify savings available. For larger processors, there is the even more attractive possibility of negotiating a tariff specifically to fit their demand profile. Processors should realise that good DSM benefits both the supplier and the consumer.

As power factor departures from unity are almost certainly a result of inductive loadings by electric motors, capacitors are commonly used for power factor correction. Power factor correction equipment is readily available. If the penalties are high, installation of correction equipment becomes justified.

Variable speed drives (VSDs) commonly operate by rectifying the alternating current (AC) supply and then 'reconstructing' a variable frequency AC by high-speed switching of a small number of voltage levels. This approach can generate very significant harmonic power levels, and poorly designed VSD equipment can allow harmonics of the supply frequency to appear on the supply lines. Supply companies commonly have limits to the level of total harmonic power on their lines. Problems of excessive harmonic levels are somewhat easier to avoid (by careful specification of VSD equipment) than to cure.

An issue with large food processing plants is control of electricity supply vulnerabilities. An unexpected electricity outage can lead to a very expensive cleanup and restart exercise as well as lost production and the possibility of environmental problems resulting from the dumping of raw material. With these severe consequences from an outage, modern plants commonly arrange for two or more separate electricity feeders and automated protection systems to allow the electricity supply to the plant to continue uninterrupted if one feeder unexpectedly fails. The keys to supply security are multiple redundancy of supplies, good protection and changeover systems and care to ensure that the multiple supplies are truly independent (otherwise, a failure on one supply feeder can lead to a 'domino effect', resulting in the failure of other feeders).

12.2.3 Steam

Site steam pressure is dictated by plant temperature requirements, and specific requirements (relating to the chemicals used in treating the boiler feedwater and to the filtering of the steam) exist for steam that will come into contact with product.

Steam-making boilers have changed significantly as factories have become more automated: manually-stoked solid-fuel boilers are still to be found but are increasingly being replaced by unattended or limited-attendance boilers. The level of attendance possible is determined primarily by local regulations, because there is little doubt that it is now possible to design boilers for unattended and automatic operation while retaining levels of safety and reliability at least as good as for attended boilers. The keys to achieving these levels of reliability and safety are redundancy of systems, design for fail-safe operation and backup power supplies.

Fuel sources for boilers can generally be categorised as

- natural gas (treated or untreated)
- liquid fossil fuels (of grades ranging from heavy residual oil to light motor spirit)
- solid fossil fuels (ranging in grade from anthracite to brown lignite)
- biomass fuels (wood chip, 'hog fuel', planer and sander dust and sawdust)
- local fuel sources, including bagasse (sugar cane residue) and alcohol fermented from local biomass

The need for reliable, automatic and efficient boiler operation affects fuel selection. As a plant becomes more and more automated, the need for consistent fuel quality and high levels of supply reliability becomes critical. Supply reliability is particularly important when considering a 'local' fuel that is available from only one source. Although it is certainly possible to design automated boilers for bulky, fibrous or otherwise 'difficult' fuels, such fuels make it more difficult to install a boiler within the confines of the processing plant.

The large quantities of carbon dioxide (CO_2) released from the fossil fuel reserve into the atmosphere have led to the 'greenhouse gas' effect. The reality of this effect and the resulting 'global warming' are gaining recognition among governments, who are seeking regulatory approaches to reducing net rates of 'greenhouse gas' emission. This topic is likely to be of major concern to large energy users such as dairy processing plants. If financial penalties for greenhouse gas emissions are introduced, this will significantly affect the economics of energy-saving measures, the selection criteria for energy-efficient plant and the selection of fuel for combustion plant. There will be pressure to select fuels with lower CO_2 production, and to move from fossil fuel to renewable energy sources such as biomass fuel. Biomass fuel generally does not have a zero greenhouse gas emission (because it is difficult to avoid the use of fossil fuel in collecting and processing it) but it does have a net greenhouse gas emission rate very much smaller than that of fossil fuels.

Feedwater for steam boilers must be treated in line with the specifications laid down by the boiler manufacturer; this is a specialised topic and is outside the scope of this text. Most industrial steam-users minimise feedwater treatment requirements by collecting and reusing condensate; in the case of dairy manufacturing plants the first effect of an evaporator is a particularly large source of such condensate.

12.2.4 Hot water

For some dairy processes, the highest temperature required is close to 100°C; for these processes it is possible to avoid the issues related to using pressurised steam systems and to use instead essentially unpressurised hot water systems.

Fired hot water boilers are commonly configured with a circulating supply and return system, supplying heat either directly to plant or via heat exchanger and secondary hot water circuits. Secondary circuits are invariably used when product-contact water is to be heated. Concerns over stress corrosion cracking of pipework have led to the adoption of duplex stainless steel for this service.

Hot water is required for washdown of nonproduct-contact surfaces (e.g. floors, road tanker exteriors). Water from another source can often be reused for washdown, and waste heat can also often be used for this service.

12.2.5 Chilled water

Chilled water is commonly produced centrally and reticulated around the plant by means of supply and return circuits. Although insulating chilled water lines may save some energy, the problems of condensation inside insulation are great and many chilled water lines are left uninsulated. Many types of refrigeration plant are used and a detailed treatment is beyond the scope of this text. Further information is available from texts such as Stoecker (1998). Refrigeration plant is simply a heat pump, and it is possible to consider other sinks (than atmospheric air) to take the heat removed from the chilling circuit.

Many 'traditional' refrigerants are being phased out to prevent further damage to the earth's ozone layer. This topic requires careful attention when specifying new plant.

12.2.6 Compressed air

Compressed air is commonly used for packaging equipment and process control systems. Although reciprocating compressors are still found, the large compressed air demand of highly automated plant has led to the widespread adoption of helical screw compressors. As with many service plants, compressors operate most efficiently when running steadily at close to rated capacity. Efficient operation under varying load can be achieved by combinations of air receivers (storage vessels) and by staged operation of several compressors operating in parallel. Compressor manufacturers offer detailed advice on this topic.

12.2.7 Dryer air

Hot air for drying dairy products is seldom identified as a 'service'; however, the air heater is commonly the largest single user of energy in the whole dairy manufacturing plant and its design and operation deserve close attention.

Although some direct-fired driers (in which products of combustion, plus dilution air, are used to dry products) are still used, the modern plant uses indirectly fired or steam-heated air heaters exclusively. As drier air contacts product intimately, it is always drawn through high-efficiency filters. Because the intake air is at ambient temperature, it is one of the relatively few low-temperature streams that can be used as a 'sink' for waste heat, to improve the overall plant energy efficiency.

12.2.8 Cogeneration

Dairy product manufacturing facilities are among a relatively small number of processing plants that require both large quantities of comparatively low-grade heat and large quantities of electricity. The second law of thermodynamics makes it inevitable that generation of electricity by thermal processes will also result in the discharge of large quantities of medium- to low-grade thermal energy. For a standalone generation facility, this discharge is inevitably wasted. However, for a generation plant located within or adjacent to a dairy processing facility there is the option of using the heat discharged from the power generation process to operate the dairy processing plant. This concept is known as 'cogeneration' or combined heat and power (CHP). Cogeneration offers a very efficient energy supply option for mechanised dairy processing plant.

Several general types of cogeneration systems can be broadly categorised according to the prime mover used. Among the oldest cogeneration systems installed were reciprocating steam engine systems. These beautiful pieces of machinery took medium-pressure steam (commonly at about 1 MPa) and expanded the steam through multi-cylinder reciprocating engines (often using several expansion steps) down to about atmospheric pressure. The steam was then ducted to a multi-effect evaporator. A very similar duty is performed by the steam turbine. For the same output, the steam turbine is substantially smaller than the reciprocating engine and, although lubrication and maintenance are both critical, both are significantly easier to arrange. Thermally, the steam engine or turbine has the lowest ratio of electricity produced to heat. A major advantage of steam-operated systems is that, unlike internal combustion engines

(ICEs) and gas turbines (GTs), it is possible to design the system for any combustible fuel.

Although less common than the steam turbines, ICEs deserve careful consideration as an energy source. Provided a good quality fuel is available at reasonable cost, they offer a good ratio of power to heat, good open-cycle efficiency (particularly in larger sizes) and the option of generating hot water directly. Unlike the GT, or the steam cogeneration systems, ICEs generate 'waste' heat in two distinct ways: from the cooling jacket and from the exhaust gases. In motors designed for cogeneration system use, it is possible to take water from the cooling jacket at up to about 120°C (i.e. to use water at pressures significantly above atmospheric pressure). The exhaust gases can exceed 500°C. In smaller sizes, the input energy is almost equally distributed between shaft power, exhaust and jacket heat. In very large sizes, the thermal efficiency of the engines increases and this distribution is changed.

The large range of unit capacities and the significant range of operating speeds also allow plant staff to consider replacing a large electric motor directly with an ICE; although such an arrangement is more complex, it does offer substantial overall energy and cost advantages. With care, it is possible to design an ICE to operate with poor quality liquid or gas fuel, keeping costs low.

For a larger plant, it is possible to use a GT as the prime mover in a cogeneration system. GT units are competitively priced, have low maintenance requirements (many tens of thousands of hours between major overhauls), have low vibration levels (leading to moderate mounting requirements) and have moderate to good open-cycle efficiency. They are also physically small and lend themselves to installation in confined spaces. Aero-derivative and 'industrial' GTs are available: the former are compact; the latter are larger, heavier and may allow more flexibility for installation of burner systems for an unusual type of fuel. GTs do require careful attention to noise emission levels. Although the heat output of a GT system could in theory be used directly as a heat source, this is not commonly done; the hot gases are usually routed to a heat recovery steam generator (HRSG), also known as a waste heat boiler. HRSGs can be comparatively simple, with a bank of (often finned) tubes stacked inside a rectangular duct and with a single-steam drum for separating steam from circulated water. HRSGs can also be extremely complex, with multiple pressure levels (seldom more than three) and separate circuits for economiser, evaporator and superheater passes within each pressure level and with supplementary firing arrangements. It is possible to design an HRSG for 'dry running', allowing the GT to be run without operating the boiler; it is also possible to design a gas bypass stack (blast stack) that not only will allow the GT to be operated without the boiler but also will enable maintenance to be carried out on the boiler at the same time. Many HRSGs are designed for supplementary firing; however, it is difficult to design an HRSG to operate efficiently without the GT operating. Many plants have chosen to retain at least some steam-raising capacity as well as the HRSG for this reason. Figure 12.1 shows the principal components of a combined-cycle cogeneration system; a typical dairy factory GT cogeneration installation is illustrated in Figure 12.2.

When cogeneration systems were first installed, the heat was invariably used within the plant and the electrical output was normally used within the plant (although many plants had the capability to export power this was very seldom done). As many countries have deregulated sections of their electricity supply industry, it has recently become possible for operators of large plants to design combined cogeneration/combined-cycle plants for a capacity much larger than the requirement of their plant. Such an installation is commonly managed as a separate enterprise, selling steam and power to the dairy company. Although valuable for all parties, this arrangement clearly needs very careful contractual arrangements to work satisfactorily. As well as contractual arrangements, the

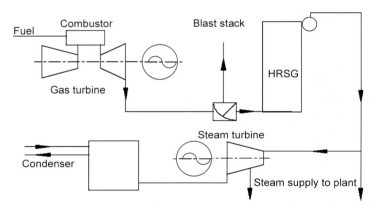

Figure 12.1 Principal components of a typical combined-cycle cogeneration system. HRSG, heat recovery steam generator.

Figure 12.2 Gas turbine cogeneration plant installed at New Zealand Dairy Group's TeRapa site. Reproduced by courtesy of Contact Energy, New Zealand.

plant management must carefully consider the engineering issues of integrating a cogeneration system into the plant; as reliability is essential for automated dairy plant, the effect of a failure of either the heat supply or the power supply will need to be considered. A much more detailed treatment of the subject is presented in the *Cogeneration Handbook for the Food Processing Industry* (PNL, 1984).

12.2.9 Waste heat recovery and re-use

If waste heat can be recovered and re-used, it effectively becomes another source of heat energy, reducing the net purchased energy requirement of the site. The essential key to identifying useful waste heat recovery projects is the process integration ('pinch') analysis technique, described by Linnhoff *et al.* (1982).

12.3 Waste handling

Minimisation and disposal of solid, liquid and gaseous waste products are important topics for large and automated plants in today's more environmentally conscious society. Practices that may have been acceptable in small production facilities with low levels of automation are no longer acceptable in larger plants, and practices that might have been tolerated by society many years ago will not be tolerated now. The dairy industry is a 'citizen of the world' and must act as a responsible citizen in managing its own waste.

12.3.1 Legal issues

All major dairy-producing countries have legal restrictions on the volume and nature of contaminants (liquid, solid and gaseous) that can be discharged to the environment, and these are backed up by penalties for failure. There is an increasing awareness that the producer of waste is responsible for controlling environmental damage caused by the waste. This section concentrates on the waste that is produced directly by the dairy processing plant. However, the reader should be aware that raw materials supplied to the dairy processing plant (ingredients, packaging and, indeed, equipment) all generate waste in the course of their manufacture and that bulk packaging and distribution activities also generate significant waste remotely from the processing site.

Central to all waste management systems is the identification and quantification of waste. There are several common elements and a great number of specific technologies involved here. One of the most pervasive topics is that of flow measurement. For liquid wastes, flow measurement approaches include measurement flumes (with bubblers) or magnetic flow meters. For gaseous wastes, pitot tube or annubar and manometer systems are commonly used.

12.3.2 Waste minimisation

Whenever waste handling is considered it is essential to first minimise waste generation, as all waste represents loss of saleable product as well as costs (capital and running) for waste treatment plant. Strictly speaking, the most important issue is to minimise the loss of product value; the distinction between loss of product value and loss of product can be illustrated by reference to a milk powder plant. Milk powder that escapes from the dryer cyclones and is recovered from the baghouse is commonly regraded as stockfood. Hence, although the product is not actually lost as a result of poor cyclone performance, product value is lost.

Recovery of specific by-products that would otherwise be wasted is largely a question of markets and treatment costs. Decisions on processing options must be made on the basis of product value *vs* capital and running costs of recovery equipment, and disposal costs.

Minimising product losses during processing depends critically on the ability of the plant operator to monitor product losses and to use the outputs of the monitoring to progressively reduce total product losses. Best results come from continuous monitoring; however, inadequate instrument reliability still limits the variables for which continuous monitoring is possible.

As well as minimising the loss of product in liquid and air streams it is also important to minimise the discharge of gaseous wastes. Gaseous wastes that must be minimised and retained within local regulatory limits include sulfur oxides (SO_x), nitrogen oxides, (NO_x), volatile organic compounds (VOCs, fortunately not common in dairy processing, the most likely source being fuel or lubricant tank vents) and refrigerants. In the future, there may be increasing pressure to control CO_2 and other greenhouse gases. The industry does discharge significant quantities of water vapour to the atmosphere, but this is a benign discharge and causes no environmental damage.

The generation of gas-borne particulate emissions is minimised by correct design and operation of the drying equipment and, more importantly, by the correct design, selection and operation of the primary product recovery devices (usually cyclones and baghouse filters).

An important factor in liquid waste treatment is wastewater volume; careful selection and operation of equipment will reduce net water use. Operators should aim to keep water use within the minimum specifications of the plant. The selection of casein washing and de-wheying equipment is an example of a field in which plant design can allow major reductions in water use and treatment volume.

Net reductions in water use are possible by reusing water. In cases where the quality of used water precludes direct re-use it is often possible to treat the water (taking care that each stream receives only the level of treatment that it requires) to bring it to a quality suitable for re-use.

Water streams that are likely to be available for re-use include cooling water, CIP final rinse water, ultrafiltration (UF) permeate and evaporator (particularly first-effect) condensate. Limited opportunities exist for using a technique similar to process integration (pinch) analysis to design plants for minimum water use. This technique is described by Rossiter (1995), among others.

12.3.3 Waste characterisation

Liquid-borne waste from dairy processing has several common characteristics:

- it consists largely of diluted milk or milk products
- commonly, significant quantities of cleaning compounds and sanitisers are present
- the organic contaminants result in the wastewaters having very high strengths compared with other typical waste streams such as domestic and municipal wastes
- there are marked variations in hourly, daily and seasonal composition and flow rate as well as major variations related to the product manufacturing mix in multiproduct plants
- it has high five-day biological oxygen demand (BOD_5) and chemical oxygen demand (COD) levels and a high initial rate of oxygen uptake

Dairy processing wastes commonly also feature high sodium ion contents, resulting from the use of sodium hydroxide for cleaning.

These characteristics of liquid-borne waste from dairy processes are quite distinct from the characteristics of municipal sewage, and the differences make it hazardous simply to extrapolate design principles used for municipal waste when designing waste treatment facilities for a large dairy plant.

Gaseous-borne waste solids from dairy processing plant fall into two distinct categories: product from dryers, and fly ash or incompletely burned solid fuel particles from boilers. Notwithstanding the danger of explosions, most countries regulate the allowable emissions of particulates, and most communities will be intolerant of significant deposits of milk powder on their cars or property. Isokinetic sampling of particulate flows within ducts is an essential tool for particulate emission control. Sampling equipment must normally comply with national standards.

Gaseous emissions can be characterised as follows:

- benign (e.g. water vapour)
- greenhouse gases (e.g. CO_x and NO_x from combustion, and methane)
- potentially corrosive compounds, e.g. SO_x, but (only if sulfur-containing coal is being burned)
- ozone-depleting substances [e.g. hydrochlorofluorocarbon (HCFC) and chlorofluorocarbon (CFC) refrigerants]; these are being rapidly phased

out internationally, in accordance with the 1987 Montreal Protocol [the Environ.com website (*http://www.environ.com/*) provides the following information regarding the Montreal Protocol: 'The "Montreal Protocol on Substances that Deplete the Ozone Layer" was agreed to in 1987 by Governments of the world, including the United States. The Protocol aims to reduce and eventually eliminate the emissions of man-made ozone depleting substances']

Mechanised dairy processing plants also generate significant solid waste such as packaging materials, and this must be disposed of responsibly.

12.3.4 Waste product and by-product treatment

Primary treatment of water-borne dairy processing waste begins with a screening operation to remove large debris that would otherwise cause damage to the treatment plant and impair the efficiency of the remaining waste treatment steps.

Dairy processing waste streams are generated not continuously but rather occur as discontinuous flows with widely varying COD, pH, etc. As treatment plants operate most efficiently under steady-state conditions, it is common to smooth the flow and characteristics of the waste stream by a combination of buffer and storage facilities. It should be noted that two considerations affect the selection of acceptable output pH: end-of-pipe environmental requirements and the operating requirements of the secondary treatment processes. The pH can be neutralised by mixing high-pH and low-pH streams, and then by adding either acid or alkali materials.

Sedimentation processes (which occur when gravity brings suspended waste to the bottom of a vessel) can involve flocs of individual waste material particles, lattices of flocs ('zone sedimentation') or unflocculated particles settling directly. The amount of suspended material in wastewater streams can be influenced by adjusting the pH of the stream. In all cases, sedimentation is a hydrodynamic process involving the actions of gravity, density difference, buoyancy and density difference, and viscosity acting together to affect settlement time and to determine the size and shape of the sedimentation facility required to achieve a given degree of separation. Whereas flocculation and sedimentation facilities encourage dense waste materials to accumulate on the bottom of a tank or vessel, fat traps are designed to cause low-density fats to rise to the surface of a 'trap' and coalesce for removal. Sedimentation and fat trapping processes together remove waste material that is suspended rather than dissolved in the waste stream. In dissolved air flotation (DAF) systems, large numbers of very small air bubbles are forced into the liquid waste, the air bubbles attaching themselves to suspended solids and causing these solids to float to the surface where they can be separated and recovered from the liquid waste. DAF systems achieve high levels of solids separation in relatively small plants and are becoming increasingly common in dairy waste treatment systems. Sedimentation and DAF systems are improved by coagulation, in which chemicals such as alum, ferric chloride, lime or various polyelectrolytes are added to a colloidal suspension of waste to allow flocculation to occur.

Final disposal of the solid waste recovered from suspension is commonly done by spreading the material onto land (land disposal), by use as an animal feed (many of the primary sludges are rich in fat and protein and, when suitably selected and treated, are acceptable feedstocks) and occasionally by composting (aerobic degradation). Alternatively, it is possible to dispose of solid waste in landfill sites. It is also possible to spray irrigate the liquid waste stream directly onto land after primary treatment, up to loading limits set by the soil hydraulic properties and allowable nutrient levels; this is an attractive option, because the cost is low, the land may well benefit from the application and there is no discharge directly to waterways or environmentally sensitive areas.

12.3.5 Nutrient and biological oxygen demand reduction

The primary treatment processes (screening, neutralisation, coagulation, sedimentation, fat trapping and clarification) all relate to the removal of suspended waste material. These processes do not affect the remaining dissolved waste material.

Dissolved material includes biological material (and, in particular, lactose— a simple sugar in milk) and nutrients (particularly nitrogen and phosphorus). Because lactose is a simple (low molecular weight) sugar, lactose-bearing waste has a very high initial rate of oxygen demand.

It is possible to remove nitrogen and phosphorus by chemical or physical treatments, however, it is very much more common to use biological treatment systems. Biological treatment systems utilise microorganisms (principally bacteria) to use the organic material and inorganic ions in wastewater to promote their growth and reproduction. In the process of growing and reproducing, the bacteria generate energy by oxidising the nutrients to gases and also convert biological material into bacterial mass.

Biological treatment processes are designed to create optimum conditions for the growth of BOD- and nutrient-converting bacteria in a continuous or batch flow waste process. Most modern continuous flow treatment systems are based on activated sludge processes in which treated waste from an aerated reactor vessel is passed through a clarifier and a portion of the biologically active sludge is recirculated to the reactor vessel to ensure vigorous bacterial growth.

Biological treatment systems can be broadly classified into suspended growth systems and biofilm systems. Suspended growth systems include plug-flow approaches, 'completely mixed' reactors and batch systems such as sequencing batch reactors. Suspended growth systems include trickling filters, rotating biological contactors and several other methods. It is important to note that treatment systems (e.g. activated sludge systems) may include both aerobic and anaerobic culturing phases or zones.

Dairy factory wastewaters contain substantial quantities of nitrogen, primarily in protein breakdown products and CIP nitrate. Nitrates are plant nutrients and, if released into waterways, will enable aquatic plants to grow, often to nuisance proportions. Aerobic treatment systems create an ideal environment in which aerobic microorganisms can use up the waste material to grow and reproduce vigorously, allowing the nitrogen (and also much of the phosphorus) to be removed from the waste stream to be separated and removed. The bacteria leaving an activated sludge system (waste activated sludge) contain up to about 12% nitrogen and 2–3% phosphorus, which will allow removal rates to be estimated.

Sequential nitrification–denitrification can be used to release organic nitrogen back to the atmosphere. It is also possible to remove organic nitrogen from wastewater by using ion exchange approaches.

When fed with organic waste, lagoons (man-made) develop high concentrations of bacteria and algae. Conditions can vary from aerobic near the surface to anaerobic close to the bottom. Lagoons are commonly operated on a flow-through basis and discharge organic material as sludge and as gases. Although lagoons are mechanically simple they require significant land area and are not recommended for the treatment of dairy waste.

Wetlands (swamps, fens, mangroves, marshes, etc.) are generally populated with hydrophyte plants and have anaerobic soil conditions. Bacteria attached to the submerged stems and roots of aquatic plants are largely responsible for wastewater BOD removal. In ideal climatic conditions, a wetland is able to achieve a high degree of denitrification of wastewater discharged from a biological treatment system. Wetlands are used for municipal wastewater; however, the volume and the strength of dairy waste make wetlands impractical for treatment of dairy waste.

It has recently become possible to design high-rate anaerobic digesters that convert organic wastes to methane and carbon dioxide. The use of anaerobic conditions reduces the energy demand of these systems. Adequate performance requires particular types of bacteria and particular pH conditions. Configurations include anaerobic lagoons, Imhoff tanks, anaerobic contactor processes, upflow anaerobic sludge blanket (UASB) digesters and anaerobic filters. Anaerobic treatment presents several problems, however, research and development are continuing to expand its applicability.

De-watered sludge is usually applied to land, although it is possible to incinerate it. In many industries, land treatment is becoming a preferred method of dealing with biological waste; as previously noted, the capital cost is moderate if (as is common in cooperatively owned dairy farm and processing organisations) land is available for waste application. The running costs of land application systems are also moderate. In a properly designed system the land is not harmed by the application and plant growth can be assisted by the nutrients. Before considering land treatment, however, it is essential that the staff of the dairy processing facility have a very clear understanding of the waste stream to be disposed of and a very good understanding of the soil and groundwater properties of the region proposed for the treatment. As soil response to waste application is not precisely predictable a land treatment regime will require monitoring to ensure satisfactory operation in the long term.

Gas-borne particulate emission rates are usually expressed in terms of grams per cubic metre of gas, corrected to a given temperature and (commonly) corrected to a 12 ml 100 ml^{-1} CO_2 concentration. Particulate emission limits vary between countries and possibly between regions. Particulate emissions must be measured by means of isokinetic sampling probes such as that illustrated in Figure 12.3, and many countries have detailed specifications for test procedures and equipment.

Technologies such as cyclones, baghouse filters (commonly pulse cleaned) and scrubbers are well suited to the recovery of product dust from air discharges. Plant operators should not forget that dairy product particulates are combustible and that even very dilute waste streams can accumulate explosive amounts of product in a poorly designed system. An adequate treatment of dust explosions is beyond the scope of this text (the reader is referred to texts such as Masters, 1991). However, the possibility of dust explosions is a particular

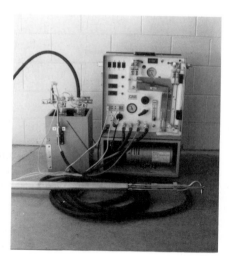

Figure 12.3 Equipment used to perform isokinetic particulate sampling according to USEPA (United States Environmental Protection Agency) Method 17: determination of particulate emissions from stationary sources (in-stack filtration method). Equipment manufactured by CAE Express, Palatine, IL, USA. Reproduced courtesy of NZDRI, New Zealand.

issue for automated plants, in which a design flaw may allow combustible dust to accumulate in some unsuspected location. Such an accumulation of product can be dispersed into a smoke (a technical description of solid particles suspended in a gas) of explosive concentration by many events, including an explosion in an adjacent area.

Boiler flue gas exhaust temperatures are commonly too high for baghouse use, and cyclones, scrubbers and even electrostatic precipitators can be considered for particulate recovery.

Many countries now set emission limits for flue gas SO_x, which cannot be met by coal-burning plants without flue gas desulfurisation (FGD) equipment. There are three main types of FGD equipment: dry injection systems, spray dryer scrubbers and wet (lime or limestone) scrubbers. Equipment may also be categorised as regenerable or nonregenerable. The design and the selection of FGD plants are beyond the scope of this text, and texts such as that by Herminé and Takeshita (1994) should be consulted.

The main approach to reducing NO_x is careful control of the combustion process and the design of the burner. Combined SO_2 and NO_x removal processes (regenerable and nonregenerable) are also possible.

It is technically possible to absorb CO_2 from flue gas and re-sequester it. However, all currently available processes are extremely expensive. A good overview of currently known sequestration technologies is provided by Herzog et al. (2000).

Emissions of VOCs, refrigerants, etc. are easier to prevent than to treat.

Acknowledgement

Assistance received from the staff of the New Zealand Dairy Research Institute's Environment Technology Program is gratefully acknowledged.

Further reading

IDF (International Dairy Federation) (1984) Dairy effluents. *Bull. Int. Dairy Fed.* **184**; web page: http://www.fil-idf.org/Welcome.html

IDF (International Dairy Federation) (1985) Nutrient sources in dairy effluent. *Bull. Int. Dairy Fed.* **195**; web page: http://www.fil-idf.org/Welcome.html

IDF (International Dairy Federation) (1990) Anaerobic treatment of dairy effluents. *Bull. Int. Dairy Fed.* **252**; web page: http://www.fil-idf.org/Welcome.html

McCabe, J. and Eckenfelder Jr, W.W. (Eds) (1956) *Biological Treatment of Sewage and Industrial Waste: Part IV. Aerobic Oxidation*. Van Nostrand Reinhold, New York.

Overcash, M.R. and Pal, D. (1981) *Design of Land Treatment Systems For Industrial Wastes: Theory and Practice*. Ann Arbor Science, Ann Arbor, MI, USA.

Pergamon Press (1992) *Modern Power Station Practice*. Pergamon Press, Oxford.

WHO (World Health Organization) (1989) WHO technical report 778: ALTH Guidelines for the use of Wastewater in Agriculture and Aquaculture; web page: http://www.who.int/dsa/cat97/ztrs.htm

The following journals, all published by Elsevier, also contain papers of relevance: *International Journal of Integrated Waste Management, Science and Technology.*
Journal of Loss Prevention in the Process Industries.
Water Quality International: A News Magazine of the International: Association on Water Quality (Formerly Iawprc).

References

Bylund, G. (1995) *Dairy Processing Handbook*, Tetra Pak Processing Systems A/B, Lund, Sweden.
Gutzeit, J. (1997) *Cleaning of Process Equipment and Piping: A Complete How-to Manual*, Elsevier.
Herzog, H., Eliasson, B. and Kaarstad, O. (2000) Capturing greenhouse gases. *Scientific American*, **282**(2) 54-61.
IAFP (International Association of Food Protection, Inc). (1999) *3-A Sanitary Standards*, IAFP.
Linnhoff, B., Townsend, D.W., Boland, D., Hewitt, G.F., Thomas, B.E.A., Guy, A.R. and Marsland, R.H. (1982) *A User Guide on Process Integration for the Efficient Use of Energy*. The Institution of Chemical Engineers, Rugby, UK.
Maddox, I.S. (1994) *Practical Sanitation in the Food Industry*. Gordon & Breach Science Publishers,
Masters, K. (1991) *Spray Drying Handbook*, 5th edn. Longman Scientific and Technical, Essex, UK.
PNL (Pacific Northwest Laboratory) (1984) *Cogeneration Handbook for the Food Processing Industry*, Pacific Northwest Laboratory; web page: http://www.eren.doe.gov/consumerinfo/refbriefs/ea6.html
Romney, A.J. (1990) *Cleaning in Place: In-place Cleaning of Dairy Equipment*, 2nd edn. Society of Dairy Technology, Cambridgeshire, UK.
Rossiter, A.P. (Ed.) (1995) *Waste Management Through Process Design*, McGraw-Hill, Elsevier Advanced Technology.
Soud, H.N. and Takeshita, M. (1994) *FGD Handbook*, 2nd edn. IEACR-65, International Energy Agency Coal Research, London.
Stoecker, W. (1998) *Industrial Refrigeration Handbook*. McGraw-Hill.
Stoneham, D. (1998) *Maintenance Management and Technology Handbook*, 1st edn. Elsevier Advanced Technology.

Index

accelerated maturation 279
acetaldehyde 161
acid syneresis 258
acidification 205, 218, 219, 255
acidification kinetics 253, 254
acidulant 269
aerobic treatment systems 336
air conditioning 248
AISY (automatic inoculation system) 204, 205
AlfOmatic 216, 218
Alpma continuous coagulator 280
AMF (anhydrous milk fat) 86, 88, 135, 145, 147, 171
anaerobic sludge blanket (UASB) digesters 336
anhydrous milk fat (AMF) 86, 88, 135, 145, 147, 171
AOC (appellation de l'origine contrôlée) 256, 259
appellation de l'origine contrôlée (AOC) 256, 259
APV-Sirocurd process 206, 289
aroma-producing lactic culture 259
artificial neural networks 321
aseptic
  buffer tank 131
  container 130
  filler system 100
  packaging 186
  product valve 90
ASR (automatic storage and retrieval) 177, 178
atomisation 98
augers 281, 282
  feeding 282
  speeds 141
automated
  bag loading 222
  curd drainage 206
  cluster removal 132
  detection of coagulation 198
  direct standardisation system 167
  inoculation system (AISY) 204-205

mould filling 220, 241
packaging machines 177
production lines 261
splicer 89
standardisation system 168
storage and retrieval (ASR) 177, 178

BAB (butyric acid bacteria) 228, 229
Bacillus cereus 306
Bacillus licheniformis 102
Bacillus stearothermophilus 115, 131
Bacillus stearothermophilus subsp. *calidolactis* 102
bacteriophage 228
bactofugation 229-31, 306, 307
BAF (butyric acid fermentation) 229
bagging and labelling machine 223
bagging machine 222
baghouse filters 337
Bifidobacterium adolescentis 154
Bifidobacterium bifidum 154
Bifidobacterium breve 154
Bifidobacterium infantis 154
Bifidobacterium lactis 154
Bifidobacterium longum 154
Bifidobacterium spp. 160-62
biological oxygen demand (BOD) 333, 335, 336
biomass fuel 326
black-box models 111, 113
blockformer systems 223
blue cheese 251
BOD (biological oxygen demand) 333, 336
bovine rennet 272
brine composition 243, 285
brining 229, 278, 316
buffer tanks 184, 204, 205, 235-37
bulk starter 204
butter 135, 137, 141, 147, 148
  churn 140
  oil 147
  serum 145
  spreadability 136
buttermaking, continuous 140

INDEX

buttermilk 158
buttermilk powder 99, 101, 143, 192
Butterschmalz 147
butyric acid 276
    bacteria (BAB) 228, 229
    fermentation (BAF) 229
    -forming bacteria 306

Caciocavallo 283, 285
calcium caseinate 268
calcium chloride 259
carbon monoxide 114
carton accumulator unit 90
α_s1-casein 156
β-casein 156
κ-casein 115, 159
casein 305
    macropeptide 305
    micelles 211, 236, 292
caseinate powder 99, 101
Casomatic® system 237, 240, 241
CCPs (critical controls points) 292
central data processing unit (CPU) 165
central processing unit (CPU) 265
centrifugal pumps 134, 235
CF (concentration factor) 170, 188, 191, 301, 304
CFD (computational fluid dynamics) 108
Cheddar cheese 208, 220
    plants 208
    production 204, 222
cheddaring 206, 208, 211
    effect 218
    machines 215, 218, 223
    tower 216, 218, 219
Cheddarmaster 216, 218
cheese
    handling 242
    handling systems 245
    loading station 245
    presses 241
    ripening room 245, 247
    tanks 231
    vat design 206
    whey 145
    yield 204, 212
cheesemaking 204, 205
    continuous 206
    process 209
    technology 212
chemical acidification 274, 288
chemical oxygen demand (COD) 333, 334

chemically acidified Mozzarella 288
chilled water lines 327
chipping mill 218, 219
CHP (combined heat and power) 328
churning 137, 141
chymosin 210, 225
CIP (cleaning-in-place) 110, 127, 130, 164, 165, 175, 184, 185, 204, 211, 220, 233, 240, 258, 263, 265, 280, 285, 321, 323, 324, 333, 336
CIP chemicals 322, 323
CIP room 323
citrate-fermenting bacteria 226
citric acid 274
cleaning-in-place (CIP) 110, 127, 130, 164, 165, 175, 184, 185, 204, 211, 220, 233, 240, 258, 263, 265, 280, 285, 321, 323, 324, 333, 336
Clostridium botulinum 131
Clostridium tyrobutyricum 228, 306
coagulation 205, 209, 210, 236, 256, 271
    probe 199
    time 197, 198, 272
    time detectors 196, 199
    vats 260, 263
Coagulite 211, 236
Coagulometer 210
coagulometer probe 193
coagulum 194
cochleate mechanical stretchers 282
COD (chemical oxygen demand) 333, 334
cogeneration systems 328, 330, 331
colloidial calcium phosphate 292
combined heat and power (CHP) 328
computational fluid dynamics (CFD) 108
concentrate flow rate 113
concentration 95
concentration factor (CF) 170, 188, 191, 301, 304
concentration of whey by RO 310
condensed milk 99, 100, 115
constant flux operation 301
continuous buttermaking 140
continuous cheesemaking 206
control charts 93
control of cheese weight 241
convective boiling 106
cooking 279
co-precipitates 305
cottage cheese 250, 258, 263
CPS (curd particle size) 212, 213
CPU (central processing unit) 165, 265

INDEX

cream 126-29, 131, 132, 134, 137, 140, 145, 168, 171, 256, 257
   enrichment process 256
   pump 125
   separator 143
   standardisation 124
critical controls points (CCPs) 292
cross-flow velocity 302
crystallisation
   of fat 136
   of the lactose 101
culture
   inoculation 204
   production 204
   technology 204
cultured
   butter 145
   buttermilk 166
   cream 265
curd 215, 217-23, 226, 233, 235, 237-40, 251-54, 275, 281
   cutters 256, 280, 282
   drainage 253
   fibres 289
   fines 235
   firmness 234
   gel strength 206
   handling 254
   maturator 280
   moisture 212
   particle size (CPS) 212, 213, 214
   particles (fines) 212
   shattering 213
   storage tanks 256
curd-draining belt 215
curd-making tank 235
curt texturisatior 205
cutting 205, 206, 209, 210, 213, 223, 232, 234, 240, 241, 254, 259, 272, 277
   knife density 213
   knives 211, 260
   time 211, 235

DAF (dissolved air flotation) 334, 335
dairy spreads 141-43
dairy waste treatment systems 335
Damrow (double-O) cheese vat 213, 215, 233
date print unit 89
de-aeration unit 126
deep-brining systems 244
demand-side management (DSM) 325
demineralisation 311, 312
de-moulding operation 260
density transmitter 126, 127
deposit formation 132
de-sludging 127
D-glucono-δ-lactone (GDL) 255
diffuse wave spectroscopy (DWS) 211, 236
direct-to-vat
   inoculation (DVI) 174
   starter system 204
   starters (DVSs) 255
disinfectant 324
dissolved air flotation (DAF) 334, 335
dosing system for salting 220
double-O vats 213-15, 233
drainage
   filter screens 260
   machine 239
draining 269
   belt 216, 217
   screens 261
   table 277
drinking yoghurt 193
dry
   salting machines 251
   stirring 218
drying 95, 312
   technology 192
DSM (demand-side management) 325
dust explosions 337
DVI (direct-to-vat inoculation) 174
DVI culture 272
DVSs (direct-to-vat starters) 255
DWS (diffuse wave spectroscopy) 211, 236

electrical
   conductivity probe 193, 196
   conductivity sensor 197
   heating 131
electrodialysis 311, 312
electronic nose 223
emulsifiers 137
emulsion breakdown 137
enclosed cheese vats 208, 209
energy demand 95
Entercoccus faecalis 154
Enterococcus faecium 154
EPSs (exopolysaccharides) 191
equipment replacement value (ERV) 320
ERV (equipment replacement value) 320
ESL (extended shelf-life) milk 306
evaporation 101, 312
evaporator tubes, fouling of 105

INDEX

evaporators 319
exopolysaccharides (EPSs) 191
expert systems 321
extended shelf-life (ESL) milk 306

falling-film evaporators 95, 96, 105, 112, 320
FAME (fatty acid methyl ester) analysis 148
fast cooling 279
fat
    content 125, 164
    crystallisation of 136
    globule membrane 134
    globules 119, 124, 132, 134, 135, 306
    in dry matter (FDM) 228, 251, 288
    in whey 213
    replacers 137
    standardisation 157, 165, 168
    standardised milk base 173, 191
fat-casein ratio 228
fatty acid methyl ester (FAME) analysis 148
FDM (fat in dry matter) 228, 251, 288
FDV (flow diversion valve) 130
feeding augers 282
fermented milk products 155
fermented milks 152, 156, 163, 165-67, 173, 181, 193, 197, 200, 201
Feta cheese 143
Feta produced by UF 310
FGD (flue gas desulfurisation) 338
fibre optics 211
fibre-optic probes 236
filatrice 282
filling lines 184
filling machines 90, 177, 188
fines (curd particles) 212, 235
flocculation 334
flow diversion value (FDV) 130
flow transmitter 126
flue gas desulfurisation (FGD) 338
fluid-bed dryers 98, 99, 192
foil-ripened cheese 229, 241
form–fill–seal packing machines 285
formic acid 160
fortification 158, 166, 168
    of milk 164
fouling layer 103
fouling of the evaporator tubes 105
fouling rate 117, 192
fractionated milk fats 135
fractionation 142
freeze-dried starter 204
frequency inverters 134

fresh cheeses 250, 252, 253, 310
fromage frais 184, 252
frozen yoghurt 193
fruit flavour yoghurt 184
fully-automated fermented milk-factory 165

gaseous-borne waste 332, 333
GDL (D-glucono-δ-lactone) 255
gel
    firmness 212
    strength 210, 234, 235
gelation time 211
Gelograph 211, 236
Geotrichum candidum 156, 252
Ghee 135, 147
Gouda-type cheeses 242

HACCP (hazard analysis critical control point) 292
hardening vats 284
hazard analysis critical control point (HACCP) 292
HBW (high-bay warehouse) 178, 181
heat
    classification 115
    coagulation time 115
    exchange performance 129
    recovery steam generator (HRSG) 329
    transfer coefficient 106
    transfer equations 106
helical screw compressors 327
high milk fat ingredients 119
high temperature for short time (HTST) 129
high-bay warehouse (HBW) 177, 181
higher voltage power supplies 324
high-pressure nozzle atomisation 109
high-protein powders 99, 101
homogenisation 99, 132, 134, 157, 159, 165, 167, 168, 172
    efficiency 135
homogenised milk 93
homogeniser 88
    efficiency 93
horizontal
    gang presses 220
    stretchers 292
    vat design 209
hot air 328
hot water 329
    boilers 327
hot-wire sensors 197-99
HRSG (heat recovery steam generator) 329

HTST (high temperature for short time) 129
hydrogenation 142

in-container sterilisation 132
in-line
    launch housing 185
    standardisation 170
insolubility index 116
instantaneous syneresis 263
Instron Universal Testing Machine 288
interesterification 142

Kefir 155
Kluyveromyces spp. 251
Koumiss 155

LAB (lactic acid bacteria) 197, 199, 225, 251, 276, 310
α-lactalbumin 314
lactic acid 160-62, 199, 206, 252
lactic acid bacteria (LAB) 197, 199, 225, 251, 276, 310
lactic
    butter 137, 145
    fermentation 256
Lactobacillus acidophilus 153, 160-162
Lactobacillus delbrueckii subsp.
    bulgaricus 154, 160, 161, 273, 276
Lactobacillus delbrueckii subsp. *lactis* 276
Lactobacillus lactis subsp. *cremoris* 153
Lactobacillus helveticus 276
Lactobacillus paracasei biovar. *shirota* 153
Lactobacillus paracasei subsp. *paracasei* 153
Lactobacillus reuteri 153
Lactobacillus rhamnosus 153
Lactococcus lactis subsp. *cremoris* 160
Lactococcus lactis subsp. *lactis* 153, 160, 225
Lactococcus lactis subsp. *lactis* biovar
    diacetylactis 153, 160, 226
β-lactoglobulin 159, 173, 314
lactose 100, 101, 109, 161, 199, 235, 251, 310
    crystallisation 99, 101
laser light scattering 211, 236
late blowing 228, 306
Leuconostoc lactis 225
Leuconostoc mesenteroides 156
Leuconostoc mesenteroides subsp.
    cremoris 153, 160, 225
lipase powders 278
lipolytic activity 277
lipolytic enzymes 279

liquid waste treatment 332
liquid-borne waste 333
liquid-vapour interface 106
long-life stirred yoghurt 188
low-concentration UF 255
lower-fat emulsions 139
low-fat cheese 310
low-moisture Mozzarella 288
lysozyme 229

maintenance operations 320
margarine 137
    manufacture 143
    spreadability 136
maturators 281
mean time between failure (MTBF) 320
mechanical
    stretching machines 267
    vapour recompression (MVR) 96
mechanisation of the fat recovery 123
mechanised
    cheese vats 213, 223
    enclosed vats 215
    salting vat 285
mellowing belt 221
meltability 288
melting properties 276
membrane
    filtration 158, 170, 244, 296, 303, 314
    fouling 192, 302, 305
    modules 299, 300
    processes 313
    processing 157, 168
MF (microfiltration) 231, 300, 305-307, 313, 316
microbial
    coagulant 210
    fouling 102
microbiological fouling 104
microfiltration (MF) 229, 231, 300, 305-307, 313, 316
    bactocatch system 305
microparticulate whey proteins 137
microprocessor software 201
microstructure of curd 292
milk
    collection 165
    composition 206, 255
    permeate 304
    powder 99, 100, 108, 304, 332
    protein concentrate (MPC) 305, 313
    reception 165

separation 124, 134
separators 123, 127
standarisation 228, 304
milk fat 119, 120, 122
    crystallisation 148
    fractionation 148
milling acidity 205
mineral fouling 104, 111
mixed emulsion 137
model-based control 111
models 113
moisture sensors 218
Mollier HT-diagram 248
motor-driven churning 140
moulage à la louche 259
mould transport system 242
moulding 260
    boxes 261
    carousel 283
    machines 283
mould-ripened soft cheese 250
moulds 260
Mozzarella
    balls 285
    low-moisture 288
    soft 270, 271
MPC (milk protein concentrate) 305, 313
MTBF (mean time between failure) 320
multilayer plastic bags 279
multiple blockformer tower 221
multiple-reflectance NIR spectra 218
multistage dryer 98
MVR (mechanical vapour recompression) 96
MVR evaporators 97
Mycobacterium paratuberculosis 231

nanofiltration (NF) 300, 304, 312
natamycin 245
near infrared (NIR) spectroscopy 212, 263
neural network models 113, 197, 201
NF (nanofiltration) 300, 304, 312
NIR (near infrared) spectoscopy 212, 263
NIR moisture data 219
nonproduct operation 318

ohmic heating 131
oil-in-water emulsion 122, 126, 137
oil-soluble flavours 139
open rectangular vats 258
open-type cheese vat 231
OST vats 213, 214, 215, 232, 233, 235
O-vat 265

packaged milk 91
    products 91
packaging 89, 91, 100, 175, 186
packaging line monitoring system
    (PLMS) 90, 91
palletiser 177
particulate emission limits 337
Pasta Filata cheeses 266, 269, 276, 279, 280,
    283, 285, 288, 289
pasteurisation 88, 89, 128, 129, 172, 231,
    255, 316
Pearson's square 124
Penicillium candidum (camemberti) 252
permeate flow 301
permeation resistance 302
PFPPs (preformed plastic pots) 177
pH probe 193, 194, 196, 197
phase inversion 137, 145
phase inverted cream 145
PHE (plate heat exchanger) 88, 171-74, 188,
    192, 283, 285
physioco-chemical predictive models 111
PID (proportional-integral-derivative) 113
    controllers 112
pigging system 184, 185
Pizza cheese 270, 272-74, 276, 283, 285, 288
plant commissioning 318
plastic fat 136
plastic vats 261
plate heat exchanger (PHE) 88, 171-74, 188,
    192, 283, 285
PLMS (packaging line monitoring
    system) 90, 91
polysulfone membranes 313
polyunsaturated fatty acid (PUFA)
    margarines 143
potable water 324
powder production 97
predictive maintenance 321
predictive models 116
pre-emulsion 137
preformed plastic pots (PFPPs) 177
pre-pressing vats 240
pressing 205, 242
Pressvatic® 240
probes 199
programmable logic controllers (PLC) 184,
    185, 206, 210, 223, 280
propellant valve 185
proportional-integral-derivative (PID)
    controllers 112, 113

INDEX

protein
  denaturation 115
  standardisation 304
protein:fat ratio 270
Provolone 283, 285
psychrotrophic bacteria 228, 259
PUFA (polyunsaturated fatty acid)
  margarines 143

Quarg 188, 263, 310

raw milk 124, 157, 228, 277, 307
recombination 171
recombination plant 88
recombined butter 142
recycling of cleaning solutions 316
redox potential 229
regeneration efficiency 129
rehydration 171
relative humidity 248, 278, 279
rennet 209, 257, 265, 271
  dosage 206
  dosing system 233
  gel 206, 212
  gel strength 205
  paste 276-78
re-using water 332
reverse osmosis (RO) 158, 170, 300, 303
Reynolds number 106
ripening 260
  room 261
  time 251
RO (reverse osmosis) 170, 300, 303
robot mouleurs 259
rotating atomiser 98

salt
  delivery 220
  nozzles 219, 220
salting 205, 282
  belt 219
  systems 243
salt-in-moisture 220
  values 219
salt-resistant lactobacilli 316
salt-tolerant microorganisms 243
Sanipress 242
saturated fatty acids 121
SCADA (supervisory control and data acquisition) 206, 211, 219
scald programme 212
scalding 204, 212, 215, 226, 232, 259

scanning electron microscope 292
scheduled maintenance 320, 321
scorched particles 100
scraped-surface
  cooling 137, 141, 142
  heat exchangers (SSHEs) 139
  technology 142
second law of thermodynamics 328
sedimentation 334, 335
self-de-sludging system 229
semi-hard cheese 225, 229, 231, 243, 306
sensing forks 219
sensors 167, 193, 198, 201, 206, 259, 282
set yoghurt 194
SFC (solid fat content) analysis 148
shaping machine 275
shattering 212, 214, 234
shear damage 134
short-path distillation 150
shredding mill 215
silo 165
single-effect vacuum evaporation 170, 171
site steam pressure 326
skimmed milk 125, 127, 128, 134, 143, 166, 168, 228, 256, 258, 306
  powder (SMP) 88, 99, 100, 172
smart sensors 201
smear
  bacteria 252
  cheese 251
smouldering milk powder 114
SMP (skimmed milk powder) 88, 172
sodium nitrate 229
soft fresh cheeses 256, 263
soft Mozzarella 270, 271
soft ripened cheeses 250, 251, 254, 256, 259, 263
solid fat content (SFC) 121
  analysis 148
specific surface area (SSA) 91
spiral-wound systems 313
spore-removing efficiency 229, 230
spray dryer 95, 98, 100, 108-10, 113, 114, 319-21
spreadability
  of butter 136
  of margarine 136
  of yellow fat spreads 136
spreadable butter 141
SSA (specific surface area) 91
SSHEs (scraped-surface heat exchangers) 139
stabilised yoghurt 188, 193

INDEX

standardisation 164, 166, 204, 255, 256, 307
  of fat content 167
  unit 125
starter cultures 157, 159, 167, 174, 197, 200, 201, 204, 205, 209, 219, 226, 251, 253, 260, 268, 271, 272, 274
  thermophilic 272
  whey 277
starters, freeze-dried 204
statistical process control 91, 201
steam
  barrier 131
  distillation 135
  infusion 96, 131
  injection 96
steam-making boilers 326
sterilisation 89, 99, 128, 130, 229, 231, 319
stirred yoghurt 194
stirring 206, 209, 213, 223, 232, 235, 254, 256, 259, 272, 280
stock rotation 181
strained yoghurt 188, 191
Streptococcus thermophilus 102, 154, 160, 161, 272, 276
stretched curd 277, 279, 283
stretcher machine 280
stretchers
  horizontal 292
  vertical 292
stretching 289
  water 271
supercritical carbon dioxide (CO_2) 148
supercritical CO_2 extraction 150
superheated water 131
supervisory control and data acquisition (SCADA) 206, 211, 219
surface
  microflora 251
  mould cheese 251
  ripening culture 252
sweet cream butter 145
sweetened condensed milk 99, 100
syneresis 194, 211, 218, 235, 251, 253-56, 261
  potential 255

TCN (triglyceride carbon number) analysis 148
temperature
  probe 193
  sensor 199
texturisation 208
  of fatty emulsions 142
THE (tubular heat exchanger) 188
thermal
  economy 172
  efficiency 95
  flux 113
  vapour recompression (TVR) 96
thermisation 128, 228, 257
thermophilic
  bacteria 103
  cultures 154
  starter cultures 272
Thermo-Quark process 256, 257
Thermus thermophilus 102
Tirtiaux Florentine™ continuous-bed vacuum filter 148
total productive maintenance (TPM) 320
TPM (total productive maintenance) 320
triglyceride carbon number (TCN) analysis 148
tubular heat exchanger (THE) 188
TVR (thermal vapour recompression) 96

UASB (upflow anaerobic sludge blanket) digesters 336
UF (ultrafiltration) 170, 191, 192, 255, 267, 286, 300, 304, 305, 308, 313, 333
UHT (ultra high temperature) milk 100, 128, 130, 159, 172
ultra high temperature (UHT) milk 100, 128, 130, 159, 172, 313
ultrafiltration (UF) 158, 170, 191, 192, 255, 267, 286, 300, 304, 305, 308-10, 333
uniform transmembrane pressure (UTMP) system 303, 306, 307
UTMP (uniform transmembrane pressure) system 303, 306, 307

vacreation 135
vacuum
  evaporation 95, 168
  pasteurisation 135
  sealing 222
variable flux operation 301
variable speed drives (VSDs) 325
vegetable oils 137
vertical
  blockformers 220
  stretchers 292
volumetric flow 127
votator 142
VSDs (variable speed drives) 325

warehouse management system (WMS) 177, 178
wash water dosage 234
waste
    handling 331, 332
    heat 331
    management systems 331
water activity (aw) 248, 252
water-in-oil emulsion 137, 141
water-soluble flavours 139
wetting rate 106, 107
whey 191, 204, 212, 215, 218, 235, 237, 240, 252, 254, 257-59, 265, 281, 312
    butter 145
    cream 228
    drainage 208, 215, 236, 237
    draining screens 265
    drawing 219
    powder 99, 101, 116
    protein concentrates 158, 313
    proteins 115, 137, 159, 310, 314
    separation 205, 206, 232
    starter culture 277
    strainer 233
whipping cream 134
white-box models 113
WMS (warehouse management system) 177, 178

yeast 251, 252, 316
yellow fat products 142
yellow fat spreads, spreadability 136
yoghurt 152, 154, 156, 158, 160, 161, 163, 166, 168, 170, 173-75, 177, 181, 184, 186, 188, 191, 192, 194, 197, 313
    measuring apparatus 193
yoghurt-fruit blend 174